Mobile Evolution

*Insights on
Connectivity and Service*

Mobile Evolution

Insights on
Connectivity and Service

Sebastian Thalanany

CRC Press
Taylor & Francis Group
Boca Raton   London   New York

CRC Press is an imprint of the
Taylor & Francis Group, an **informa** business

CRC Press
Taylor & Francis Group
6000 Broken Sound Parkway NW, Suite 300
Boca Raton, FL 33487-2742

© 2015 by Taylor & Francis Group, LLC
CRC Press is an imprint of Taylor & Francis Group, an Informa business

No claim to original U.S. Government works

Printed on acid-free paper
Version Date: 20150106

International Standard Book Number-13: 978-1-4822-2480-1 (Hardback)

Visit the Taylor & Francis Web site at
http://www.taylorandfrancis.com

and the CRC Press Web site at
http://www.crcpress.com

Mobility Inspired Knowledge Knowing and Insight

Along interdependent vistas of experience,

Connectivity and service blend in sentience.

Contents

Foreword

When Sebastian approached me with the invitation to write the foreword of this book, I felt very honored and indeed excited! Coming from the standards world as I do, it was a very nice present having standards introduced as one of the key levers toward the evolution of a whole ecosystem rather than this very arid and unfriendly set of stringent rules used in the development of technology.

This book is a living testimony—if we can consider a book a living entity by nature of the relevance and *liveliness* of its contents—that standards go well beyond technology and rules. And their lives impact and change human culture and behavior far quicker than nature can do.

In the last 25 years, we have all witnessed one of the largest yet quickest changes in human behavior. We have been a fundamental part of this change and are still living it half enthused, half amazed. Let me show this with a very basic, day-to-day example. Twenty-five years ago, if we wanted to make an appointment with a group of friends, we used to agree beforehand the place and time and that was it. If somebody wouldn't arrive in time, we would allow for some 5–10 minutes courtesy and then we parted—understanding that the one coming late could have had an unforeseen problem of the last minute. No one carried a mobile phone, as this was only meant for rich businessmen—a scepter of status.

Today, nobody can imagine that if someone is late—or cannot make it—to an appointment, he is not able to inform the others. In fact, the rules for making appointments in the day-to-day life have changed. The one without a mobile phone simply is out of the society, as the society has interwoven mobile communications into its rules.

The democratization of mobile communications is the result of a success story in the standardization arena called GSM, which started in Europe and has since extended and evolved worldwide through different technology acronyms—GSM, HSCSD, EDGE, GPRS, UMTS/3G, HSPA, LTE, LTE-Advanced, etc. —so, what next, if anything?

At the same time, it has been very interesting to analyze what technology trends have enabled, how the fact of democratizing mobile telecommunications, that is, making mobile phones available (or should I say accessible?) to everyone, has transformed what was a nice-to-have accessory into a must-have body appendage. People can walk out of their homes without their wallets, never without their mobile (smart) phones, whose evolution, by the way, also sheds some interesting reflections.

In its origins, a mobile phone was the device that allowed its owner to make a call to anyone from anywhere at any time (*selling proposition*). This nice device—and the service behind it—evolved into a masterpiece among everybody's accessories (wallet, credit cards, watch, etc.) that enables anyone to contact the owner at anytime, anywhere they are. Now the mobile device (no longer a restricted to just being a phone) has become a personalized and lifestyle enhancing accessory. Someone not having this personalized window to information is deprived of the convenience of mobility empowered lifestyle advancements. The mobile device at its fundamental level is a building-block in an evolving fabric of IoE (Internet of Everything).

So it seems obvious that the democratization of a given technology has produced, as a result, changes in human behavior that request additional features and enablers to the technology (coming from social relation needs), which are thereafter developed by leveraging on standards (making technology quickly available at reasonable costs), which then provoke some more behavioral changes. Indeed, the evolution and democratization of the technology have come hand in hand with a not less impressive change in cultures and attitudes—our children already write more than talk to their friends, using their thumbs instead of their tongues?! A one-day trip by public transport in any big city is a very clear example of what these changes mean.

This incredible journey has just started—what's just 25 years in history, a mere quarter of a century?—and we are already experiencing quite profound changes in human behavior. The evolution—shouldn't we actually say revolution?—keeps on going. Throughout this book, we will follow the author's experience and insight into the past, present, and future of mobile communication technologies, closely knit with the evolution and requirements of those whom the aforementioned technologies serve.

Mr. Luis Jorge Romero
ETSI (European Technical Standards Institute) Director General
ETSI Headquarters
Sophia Antipolis, France

Preface

An idea is a point of departure and no more. As soon as you elaborate it, it becomes transformed by thought.

—**Pablo Picasso**

This book explores the rich nature of the mobile information landscape, influenced by culture, vision, and innovation expressed through a fabric of insights and concepts. The convergence of technology mediation and human factors offers a virtually unbounded potential for advancements in the quality of experience of mobile multimedia products and services. They are reflective of a systems approach to innovation that animates imagination, imagery, scenes, and stories that entertain and educate. Scenes and stories that have spurred a contagious fascination through the ages reveal a compelling potential beyond archetypal boundaries. Just like music.

Technology changes rapidly; cultures endure, change slowly. The considerations that bridge the gap between technology and cultures are critical for an organic market adoption of the fruits of technological innovation. This book is a narrative on the exploration of this gap, through the fabric of concepts and insights: a fabric in the context of mobile broadband—evolution of LTE (long-term evolution), IP multimedia services, human factors, and interdependence. It explores a synthesis of the converging and contrasting patterns, across the landscape of technologies and human aspects that enable and sustain the evolution of experience in a mobile Internet.

Concepts—an intangible medium, invisible, like air. Yet indispensable. Amorphous. They weave experiences to manifestation in a changing world. They shed light on the nature and evolution of technologies. The fabric of concepts unveils the nuances of technology standards, innovation, and challenges. The context and concepts are woven in terms of standardization and innovation to promote connectivity and services in mobile communications.

Insights are a revelation of the nature of technology evolution derived through concepts and interdependence. They provide an expose of perspectives and interpretations across linkages that span technology mediation and human factors. Akin to life's journey, linkages are worth exploring to both enable and enjoy the fruits of mobile broadband communications.

In a nutshell, we explore the theme of interdependence—its impacts, implications, challenges, and opportunities. Interdependence motivates and inspires innovation in an intrinsically *open* communication landscape. Technology enables, while market realizes. Forward-looking strategies for innovation and sustainability, driven by the considerations of interdependence, are likely to thrive, while rigid, monolithic approaches, such as first-to-market, are prone to value erosion and commoditization. Abstractions of emerging trends and technologies provide an essential framework toward generalizations amenable to widespread applicability, in the intersection of technology and human factors. This interdependence demands not only a shift away from the silo-oriented thought and technologies but also one that embraces lateral thinking across a landscape of intersections. A model to harvest thought in the next-generation mobile ecosystem. A quest that enables enhanced lifestyles on a global scale.

Cultural Gardens

The Industrial Revolution marked the beginnings of a new era of conveniences, spurred by thought and technological innovation. The city of Cleveland, one of the prominent centers of this new chapter in history, attracted a colorful and rich microcosm of racial and ethnic heritage. It mirrored a pluralistic landscape that embraced the diversity while recognizing the profound power and meaning of the pervasive connectedness of all that is human.

The Cleveland Cultural Gardens depicts a tapestry woven out of the individual and collective identities of people in the making of the United States. Historical and imaginative perspectives are embodied in the depiction through symbolic and narrative representations. The images, shapes, and textures provide unique glimpses that shaped and molded the evolution of the twentieth-century United States—a walk through the passages of time, punctuated by the crossroads of the Industrial Revolution, World War I, the Great Depression, World War II, and the Cold War.

Beyond the boundaries of the tangible displays that portray these events in space and time, the observer is guided toward the perceptions of the significance of these events in the context of human existence, such as the social and cultural aspects of emerging communities across the United States and communities with hopes and dreams in new pastures that also brought with them the memories, perspectives, and traditions from the lands they left behind.

Situated between the University Circle, Cleveland's cultural center to the south, and Lake Erie to the north is the tranquil and serene environment of the Cleveland Cultural Gardens. In this ageless spirit, the vision of the Cleveland Cultural Gardens was born. The architect, Leo Weidenthal, also editor of the *Cleveland Jewish News*, believed that "True cultures impose no barriers of race or creed."

The behaviors and lifestyles were reflections of the plurality colored by the respective cultural ingredients, inherited from their native lands. A rich tapestry of behaviors, woven from these ingredients in the incessant tides of change, is both a snapshot in time and a unique representation of the individual and the collective.

The metaphor of the cultural gardens depicts a collage of change and evolution in the realm of human existence, while allowing the observer to experience the tangible display of visual symbols, as well as an emotional and perceptual connection to the people, places, and events, through the stories frozen in time.

The observer, the actors, and the stage become one.

Experiences, intangible yet real in the imagination, informative, compelling, and memorable, endear and motivate human existence, in all dimensions.

In a nutshell, an interdependent experience is one in which the realm of mobile communications continues to emerge and evolve, with interdependence as a central theme.

Boundaries and Beyond

The industrial revolution manifested the realization of a variety of lifestyle-enhancing products. It was fruition of technological ideas and innovation from the preceding centuries. Among these products, the quest for transcending the barriers of time and space appeared in the form of the automobile—a journey machine that allowed new experiences, ideas, and an expansion of perspectives; a personalized convenience; and an instrument that enabled human beings to meet, socialize, work, and play at different venues. It is an intersection that permitted both an understanding and a preservation of the personal, while promoting the innate desire to socialize for work and play. Nomadicity—in its uniqueness.

While technological innovations are ushered in an era of unprecedented conveniences, the elements of progress, in understanding and ideas, were being shaped by the bounded and rigid aspects as well. The inherently mechanical, silo-oriented, clinical, narrowly focused nature of the processes, while being essential for mass production, also shaped the thought landscape for decades. The tremendous success of established methodologies lent credence to their applicability for a wide variety of solutions, as products continued to evolve, in terms of value and complexity. It was as if the building blocks of these methodologies were unchangeable tenets. Tenets adhered to, in the face of endless change as the decades unfolded, are limiting and constraining in nature.

Innovation and product commoditization fed each other to create new products. In turn, value continued to shift toward more complex products. Embellished capabilities are ushered in an era of conveniences, in a variety of arenas of human existence. Product varieties proliferated.

The tenets that fueled the methodologies continued to work quite well through the later part of the twentieth century. New products provided enhancements over previous renditions. Enhancements continued. The manner in which ideas were conjured and applied largely remained unchanged. Expectations were mostly shaped mutually by both product designers and consumers. The elemental nature of the approaches for the creation of new products was constrained by the implicit belief systems that dominated the industry. Business models trapped in the economics of current products were primarily driven by current profitability rather than future potential. It was a model that stifled change and diminished motivation for any change, or at best yielded to incremental change. What was worse was that even when incremental changes in products and services occurred, the underlying philosophies remained stagnant.

Oneness

The existence of unifying principles in the world around us has been examined by sages and philosophers over the ages. A seventeenth-century study of mechanics postulated that the universe of forces could be reduced to the attractive forces between elemental particles. Newton's *Principia* suggested the unified nature of seemingly distinct and different forces. The universal law of gravitation was a foundational framework that shed light on the common thread across multiple phenomena, such as Kepler's laws of planetary motion, the ebb and flow of oceanic tides, and Galileo's work on terrestrial gravity. Max Planck[*] attributed the existence of all matter to a universal force that holds the atoms together.[†]

---

[*] Max Karl Ernst Ludwig Planck, Theoretical physicist who originated quantum theory—Nobel laureate in Physics, 1918.

[†] Das Wesen der Materie [The Nature of Matter], speech at Florence, Italy (1944). (From Archiv zur Geschichte der Max-Planck-Gesellschaft, Abt. Va, Rep. 11 Planck, Nr. 1797.)

So it is in the affairs of human existence—different styles, different behaviors, different choices, different cultures, and different communities. Yet the powerful unifying threads—emotions, perceptions, social interaction, memories, and experiences—defy all objectively observable boundaries.

Social interaction and communication are pivotal unifying threads that provide pathways to learning, discovering, sharing, and expressing. These provide a framework to both customize and personalize a rich variety of observable boundaries. The unifying and universal threads are few. The observable rendering is rich and diverse.

The abundance and beauty of the natural world around us provide powerful insights through revelations cast in simple, unchanging, and unifying threads. The unchanging act as a catalyst for the changing. Communications, with its incredible choices, styles, and modalities, provides a framework for human social interactions—an integral part of the world and nature around us. The leveraging of the unchanging, unifying threads is therefore central in the realm of personal communications. The evolving world of mobile communications is at its core—personal and experiential. It is an intersection of a multitude of human factors and technology innovation.

Communities

Communications enable communities to be expressive and to stimulate the diversity of thought that is pivotal for the growth of social awareness—a catalyst that spurs innovation and lifestyle enhancements.

Cooley, in *Social Organizations: A Study of the Larger Mind*, stated the following about communications with respect to human nature and society at large: "…the mechanism through which human relations exist and develop—all the symbols of the mind, together with the means of conveying them through space and preserving them in time." Connectedness is foundational in communications. People, relationships, interests, and experiences form the actors and symbols, as Cooley alluded, in the landscape of connectedness. Social networks are an exemplification of linkages across communities in the realm of connectedness. It is in this realm that the notions of context play a significant role. The symbols of relationships, interests, and experiences weave an interdependent web. It is a web of potential and possibilities that are realized through the virtual dimension of mobile communications. The framework of mobile communications provides content, context, and linkages that enrich the lives of people and in turn their communities.

Multimedia, as the name implies, synthesizes words, audio, and visual effects to both capture and render an unbounded assortment of experiences. In turn, these experiences are richly colored by the cultures and traditions native to diverse communities.

Transmutation

The landscape—of thought—shapes concepts through imagination. How do concepts shed light on interdependence?

Experience is a fabric woven through diverse and accumulated changes. The progression of ideas continues to be shaped by change, fueled through thought. Concepts appear in the thought realm. The crystallization of concepts demands an examination of intrinsic details, as separated from extrinsic details. The former identifies the essential characteristics of a subject domain space, while the latter describes the features and capabilities of the subject domain space. Product and service design require an understanding of both the intrinsic and extrinsic details, and typically the emphasis is primarily on meeting a defined set of market requirements and considerations. It is necessary and useful but has a narrow vision.

On the other hand, an abstraction of the intrinsic details, beyond any immediately known market demand, for a generalized perspective, is an activity rooted in the realm of thought. Generalizations lend themselves to potential applicability beyond the boundaries of its origins. Unbounded in nature, generalizations fuel the thought landscape, toward new frontiers. They usher new perspectives that kindle innovation.

Although mobile connectedness has morphed in form and function over the years, it has essentially been technocentric, to which humankind was prompted to adapt in a rather jarring fashion. Every incarnation demanded getting familiar with new user interfaces, often with complex feature linkages and command sequences. Inherently nonintuitive, crutches were an accompaniment in the form of one-size-fits-all-oriented external help resources: user manuals, online help, customer service, etc. With these inherent barriers, usage naturally gravitated toward basic features such as messaging and voice. Device capabilities advanced. Adoption and usage lagged. Expectations thwarted. Growth constrained.

The iPhone*

An idea that is expressive of a user-centric mobile communication landscape. Experience, strategy, and rich perspectives, unencumbered by existing, widely held views, are pivotal for the incubation of a sustainable vision. This creates a transformational potential, beyond the realization of any specific technology refinement. It reflects a symbiotic combination of the interdependent human and technology considerations. The iPhone ushered a sea change of thought, beyond a very attractive and user-friendly mobile device. It was about perceptions and user experience. These aspects are a crux for the creation of new market opportunities, driven by unfettered imagination. Lifestyle-changing applications and services are allowed to thrive, where the actors—end users, connectivity providers, and service providers—in the ecosystem are enabled to create and consume with relative ease.

Thought leadership is transformational, while the specifics—hardware and software—are subject to commoditization. A narrower version of this paradigm is limited to specific product—hardware/software—enhancements/optimizations realized through thought-driven innovation. It is the breadth of thought—a holistic persuasion—that spurs visionary innovation.

Mobile Information Age: A Shifting Paradigm

With the advent, and the continuing evolution, of the mobile information age, the leveraging of value in openness is a natural shift away from the status quo ante models for connectivity and services. The latter models were rooted in the premise that value is both created and sustained through the modalities of *getting to market first* and *product/service acquisition barriers*.

Predominantly, protection-oriented models lend themselves to commoditization in a relatively rapid fashion, particularly in an era where incremental technology-driven innovation is a catalyst toward lower costs and widespread affordability for consumers. While this direction hinges on a relatively focused set of capabilities for products and services, it is constrained in terms of

---

* A category of smartphones designed and marketed by Apple Inc., which debuted on June 29, 2007.

interoperability and extensibility of the underlying enabling fabric. It is a limited direction in breadth and scope, not a transformational paradigm.

The Internet

Virtual in nature, it is implicitly an expansive paradigm. The mobile information landscape is an integral part of the Internet—information on the go, personalized, powering a virtual world with a dizzying array of capabilities. In this model, the realm of ideas and imagination is a fundamental note. It separates commoditization from the boundless value resident in the thought realm. It is an incubator for open models fostering value creation and sustainability.

The distributed nature of information creation and consumption is pervasive in the Internet, specifically through mobile data usage—Internet of Things. The glue that binds and strengthens the proliferation of information in the Internet fabric consists of enabling technologies and human behaviors. These twin components foster unlimited potential in the realm of social interactions and lifestyle-enhancing applications. A plethora of mobile communication modalities—synchronous, asynchronous, and isochronous—through realizations such as e-mail, instant messaging, Skype, LinkedIn, Twitter, and Facebook, to name a few, are rampant as the mobile information age moves inexorably toward new horizons.

Internet of Things

It represents a massive distribution via a cloud model of connectivity and service, facilitating a malleable personalization. In this model, a cloud of networked resources for connectivity, computing, and content enables a distributed execution of applications. Estimates of the number of interconnected devices—with an unbounded proliferation of device variety—are anticipated to be in the vicinity of 30 billion or more by 2020*. New value generation will appear in the form of enormous novel service creation potential, in an Internet of Everything. In these horizons, devices serve as commoditized bridges of value-added services. Connectivity enabled to suit access to personal and public services—a localized, global experience.

Merging of technology and human factors—in the innovation and standardization endeavors—promotes customization in a Long Tail†–oriented market expansion. The implications of this shift are enormous, spanning strategies and business models, in the next-generation mobile ecosystem. This shift is likely to satisfy the elemental aspects of the human condition—personal, contextual, and self-expressive—in the shaping of a vision for standardization and innovation, in the evolution of the mobile Internet.

Thoughtscape: Ensuing Chapters

The flow of thoughts and ideas in the making of this book has been shaped through an unfolding of observations and experiences. The focus is on the intrinsic details, through the screens of time, as they have revealed themselves, in the author's journey. The subject matter is portrayed, with an emphasis on matter with no subject, to reveal the unchanging attributes, through the dawn of new

---

* ABI Research Report. *Wall Street Journal*, "Internet of Things Poses Big Questions," July 2013.
† A probability distribution has long tail, if a larger share of population is within its tail than that under a normal distribution.

horizons. The perspectives are intended to be a depiction of insights into the technology and social evolution rapids, as they carve new tributaries, streams, and rivers, into the endless and emerging seas of change.

From search engines to service capabilities, personalized mobile communications have realized the virtualization of information through the artifact of data. The virtual nature of data, enabled through technological advancement, resembles thought, in the language of observation, perception, and expression. It is an agile template that scripts the formulation and articulation of creative thought.

As in life, data transactions are rich in texture, diversity, and representations. Simultaneously interlinked and isolated, it provides a malleable framework where both the artist and the subject are allowed to interact and communicate. This is the stage where heuristics reign to examine and synthesize the ideas behind products and services. Generalized approaches foster widespread applicability and evolution. This contrasts with relatively rigid algorithmic approaches for the crafting and delivery of products and services.

Predictions are that the traffic over the Internet is estimated to exceed 130 Exabytes (1.3 billion Gigabytes) by 2017, with an annual mobile user-driven data growth rate of 79%,* which is projected to be three times faster than the traffic over the fixed Internet.

These projections usher an unbounded virtuality and unprecedented connectedness across humankind.

The opening two chapters depict a brief history and the critical role of standardization. The subsequent two chapters delve into the connectivity and service aspects through the lens of concepts, standardization, strategy, and innovation perspectives. The narrative concludes on the central theme of interdependence—a renaissance in the mobile information age.

Boundaries fade to enable value and uniqueness, in the shifting era of the Long Tail. The particulars must be sufficiently abstracted for effective generalizations, necessary to effectively mine the gaps of an emerging interdependence.

The author's journey through the narrative has been a delightful one. Perhaps the thoughts and ideas, examined through an experiential lens, will reveal unique and shared insights for a practitioner or an enthusiast, shaped by their own journey.

---

* Cisco Visual Networking Index. Forecast and Methodology, 2012–2017, Published May 2013.

Acknowledgments

As in the reflection "No man is an island, entire of itself,"* this book has been realized through conversations with colleagues and exploration of ideas, in the author's experience across the industry.

- Michael Brenner, Steve Crowley, Mikki Jang, Eileen McGrath, and Musa Unmehopa for the inspiration
- Michael Irizarry and Narothum Saxena for the enthusiasm
- Claudio Taglienti for the persuasion
- Kevin Lowell for the encouragement
- Mark Hammond for initiating the journey
- Sarah Tilley for appreciating the Preface that set a direction for the narrative
- Rich O'Hanley for the constructive criticism, and the endorsement for publishing a journey of imagination
- Family and friends for the interest
- Moray Rumney for review and perspective
- Paul Abraham, Vinithan Sedumadhavan, and Jennifer Stair for reviewing and editing the manuscript

And many more that have shaped and influenced the author's journey in the evolution of mobile communications.

Cover design:

Mikki Jang and Elizabeth Thalanany

---

* John Donne.

Author

Sebastian Thalanany is engaged in the various facets of the evolution of mobile communication—technology, leadership, standardization, research, strategy, and human factors.

Chapter 1

Genesis of Personal Communications

Communication leads to community through understanding, relationships and mutual value.

—Rollo May

1.1 Nature of Communications

Human beings have an innate need to interact with one another to support, nurture, and enrich lives. Various forms of communication through sight, sound, and perceptions reveal the world around us. These revelations provide a framework that shapes our understanding of one another through the acquisition, processing, and distribution of information. At the same time, the various modalities of communication are being shaped by our behaviors. This is an interdependent cycle that is in a state of continuous evolution toward new experiences and a meaningful existence. The nurturing impacts on well-being through connectedness in the social context* are a pivotal motivator for the creation and consumption of innovative communication capabilities.

Through the ages, through the present, the saga of human communications has been motivated by the instinctive desire for self-expression, while being immersed in the diverse aspects of a collective existence, which motivates one's empathy for all.

Over the ages, voice communications has been established as the most widely accepted form of human communications. The nature of voice communications has evolved, from the simple gathering of individuals at geographically coincident locations to spatially disparate locations enabled by the digital era. The organic nature of the evolution and popularity of voice communications stems from the implicit significance of voice communications in human existence. It does not require advertisement. Its value is pervasive. The advent of mobile communications has augmented

---

* Lewis, T., Amini, F., and Landon, R., *A General Theory of Love*. New York: Random House, 2000.

this value through a high level of personalization, which is compatible with a multitude of existing and potentially new human behaviors.

Human behaviors may be identified in terms of communities that are organized around cultural traditions, beliefs, and interests. Communities exist in a variety of forms, which are based on diverse motivators, for example, interest groups, neighborhoods, and industry, among these different types of community or Gemeinschaft as envisioned by Tönnies.* A Gemeinschaft of the mind is one where the structure of the community is built around people with a common interest. This common interest is manifested in terms of an intellectual, emotional, or spiritual resonance. Data-oriented mobile communications is a powerful enabler for enhancing the level of sophistication of information exchange pertaining to the robustness of virtual communities or a "Gemeinschaft of the mind." Such communities are not tethered to geographic boundaries or neighborhoods.

The dawn of the information age, propelled by the ubiquitous Internet, followed by the emergence of data-oriented mobile communications, lends credence to the notion of a departure from the trodden paths toward new horizons inspired through technology innovation.

Do not go where the path may lead; go instead where there is no path and leave a trail.

— Ralph Waldo Emerson

This metaphor resonates within the realm of human affairs, as well as in the business models and technology enablers, in the mobile communication landscape.

Lifestyle patterns are modulated by a variety of events with distinct temporal constraints, for example, real-time, non-real-time, periodic, and disruptive. The personalized nature of mobile communications accommodates the complex demands generated by the lifestyle patterns. This accommodation in turn provides an environment where advances in lifestyles and the evolution of mobile communications are mutually supportive. The existing and emerging lifestyle patterns serve as a research arena for the identification of potential innovation and standardization aspects in mobile communications. It provides a framework of services to enable our minds on the exchange and access information and to form relationships that nurture our health and well-being.

The element of openness in business models and technology enablers fosters a realization of the various modalities of human communications in an unencumbered fashion. The power of openness together with a rich diversity of service choices allows a natural expansion of markets, since it empowers user-centric behaviors and preferences. In an interconnected world, this is potentially a significant ingredient for long-term business sustainability and economic stability in a dynamic market environment.

1.2 Sociological Perspectives

Communication modalities, whether it be in the form of sight and sound as a recipient or in the form of interactions, provide pathways that resonate with the uniqueness of each human being. At the same time, universally appealing notions serve as the glue, without an impairment of

---

* Tonnies, F., *Community and Society*, translated and edited by C.P. Loomis. East Lansing, MI: Michigan State Press, 1957.

uniqueness. This is an intangible, which is beyond the realm of the rational mind, which is mired in a structured sequence of actions and events. Untethered communications allows a weaving of the elements of the human spirit with neuroscience.

The complexities of human connectedness are akin to the shapes and forms that abound in nature. The notion is similar to the use of fractals to describe these shapes and forms such as coastlines, mountain ridges, plant leaves, and flower petals. Fractals postulated by Mandelbrot provided elegant models that reflected the brilliance and creativity of nature, as compared to the artificial and highly structured nature of Euclidean models, which provide simplified models that lack the rich texture and essence of the entities they represent. In his book *The Fractal Geometry of Nature*, Mandelbrot states, "Clouds are not spheres, mountains are not cones, coastlines are not circles, and bark is not smooth, nor does lightning travel in a straight line."

The feelings and emotions that are evoked through sight and sound have a complex and profound influence on well-being and behaviors. An understanding of these experiences has been facilitated through a research of scientific and practical knowledge. Studies have shown that individual and interconnected experiences are enabled through a resonance with the limbic system, which is an integral part of human physiology.

It is the fabric of untethered communications that requires an examination through a lens that unites the traditionally segmented arenas of the human dimension and the predictable, rigid aspects of technology evolution.

Untethered communications, in the form of mobile communications, has the *je ne sais quois* to weave capabilities and services that are of paramount value in the realm of personal and business relationships. It is these relationships that form the substrate of a community formed through a commonality of vision and purpose. The pursuit of shared objectives is realized through individual and collective participation of the members of a community. Participation in the form of a variety of rituals, both social and business-oriented that are a part of the nature of the community, promotes understanding, knowledge, and awareness through information exchange and creation of new information. Communities are shaped by the existence of social groups and cultural forms* and consist of a range of actors from the more dominant to the less dominant voices. Technology-enabled capabilities provide a playground for dominant and subdominant actors in the community to contribute effectively. The strength and sustainability of the community are bolstered through the technology-powered mobile communications.

One perspective of "What constitutes a community?" is based on an observation that it represents a collection of groups of people in a distinct geographic area. In this representation, communities are characterized by well-defined spatial boundaries,† where people gathered to live and work. This view is aligned with the transition of communities that were agrarian to industrialized, where the spatial boundaries could be reoriented through technological innovations in rapid transit across boundaries. Cities were removed from rural areas together with changes in behaviors and lifestyles through the conveniences of technological advancement. The technological advancements that were pivotal in these changes are *transportation and communications.*

The merging of transportation and communications, in the twentieth century, ushered in an era of connectedness that transcended the notion of spatial boundaries across communities.

---

* Freie, J.F., *Counterfeit Community: The Exploitation of Our Longings for Connectedness*. Lanham, MD: Rowman & Littlefield, 1998.
† Goldberg, S. and Haines, V.A., "Social networks and institutional completeness: From territories to ties," *Canadian Journal of Sociology*, vol. 17, no. 3, pp. 301–313, 1992.

Communities did not require a spatial attribute for characterization, because technological capabilities enabled like-minded people situated across disparate geographic regions to satisfy the demands of work and play. The essential ingredients that preserve the fabric of a community are allowed through a variety of communication modalities, across time and space, for an elastic flavor of social ties and information exchange.

This shift is similar to remotely watching broadcast television for a remote event such as a live concert or a game versus being physically present at the event. The experiences are distinct, while both are enriching and informative. One is not a replacement for the other. Both are attractive options and provide choices, which is the cornerstone of human existence. Technology is a catalyst for choice.

A pervasive thread that winds through the fabric of community is social capital that fosters a collective awareness of the interdependence, trust, and effective information exchange. Social capital engenders both individual and collective benefits. The notion of social capital is amenable to communities defined in terms of spatial boundaries as well as to virtual ones enabled through the mediation of mobile communications. The degrees of social capital across spatial and virtual communities are a function of the level of individual and collective interest and participation.

In virtual communities, a Gemeinschaft of the mind is a dominant thread that fosters the promotion of social capital. The accumulation of social capital in a virtual community is dependent on the quality and frequency of interactions. The quality of mobile-mediated communications is a function of the appeal of user-oriented services, in terms of the ease of use and the related experience.

A significant attribute of communities is the notion of mutual support. Among the various types of content within information exchange is support-oriented content, such as education, consulting, and commerce. These types of exchanges promote the social capital through a continual advancement of relationships, knowledge, and well-being. This is in contrast to the traditional type of social support in spatial communities, where exchanges are physically tangible, such as face-to-face meetings to discuss topics of mutual interest and purchase of items. Interactions among individuals[*] through reciprocal actions strengthen the growth of social capital within a community, whether virtual or spatial. Why is this significant? From both an individual and collective perspective, social capital enhances the overall well-being in a community.

In the shaping of a community, a sense of belonging is vital in the promotion of connectedness within the community. The power of connectedness is pivotal in inspiring an awareness of self-worth and stewardship.

The flexibility of virtual communities allows a variety of choices that enable interest and purpose-driven formation of communities. The experiences of interactions, virtual or spatial, profoundly influence our behaviors within the corresponding communities.[†] The promise of virtual communities is influenced primarily by primordial and deeply rooted experiences within human beings.

---

[*] Putnam, R., *Bowling Alone: The Collapse and Revival of American Community*. New York: Simon & Schuster, 2000.

[†] Freie, J.F., *Counterfeit Community: The Exploitation of Our Longings for Connectedness*. Lanham, MD: Rowman & Littlefield, 1998.

1.3 Advent of Mobile Communications, Standardization, and Innovation

> Life is a distributed object system. However, communication among humans is a distributed hypermedia system, where the mind's intellect, voice + gestures, eyes + ears, and imagination are all components.
>
> **— Roy T. Fielding, 1998.**

The concept of cellular communications, proposed by D.H. Ring (AT&T Bell Labs, 1947), was pivotal in the dawn of the mobile communications era. The notion of cell-oriented wireless communications, where each cell is served by a transceiver, allowed untethered communications and handovers across cell regions. This novel concept remained dormant, within the confines of research labs, for several decades.

In the United States, the Advanced Mobile Phone System (AMPS) evaluation and trials began in 1978, in Chicago and Newark. Licenses for the commercialization of AMPS were granted by the FCC.

Commercialization of the cellular concept, for mobile communications, occurred in Japan in 1979. The service was launched, in its rudimentary form, by Nippon Telegraph and Telephone (NTT).

The first generation (1G) mobile communication system, known as the Nordic Mobile Telephone (NMT), was launched in 1981, across Denmark, Finland, Norway, and Sweden. This marked the advent of voice-oriented mobile communications. Subsequently, in the United States, Illinois Bell introduced AMPS in 1983. The first mobile device in the United States was the Motorola DynaTAC, which appeared in 1983. Advances in the 1G system occurred in 1984 with the introduction of the concepts of centralized control of cellular transceivers and the reuse of communication channels across the transceivers.

The arrival of 1G systems marked a turning point on the history of human communications. The profound impacts of the role of mobile communications and its potential in the human ethos were beginning to be discovered.

Almost a decade after the arrival of the 1G, the second generation (2G) mobile communication system was introduced in 1991 by Radiolinja (Elisa Group), a mobile service provider in Finland. The 2G system was based on standardized specifications. These specifications were developed by the Global System for Mobile Communications (GSM), which was established in 1982 by the European Conference of Postal and Telecommunications Administrations. The specification development initiative for GSM was subsequently transferred to the European Telecommunications Standards Institute. The 2G systems utilized digital methods for both signaling and voice information. In the United States and Japan, 2G systems were standardized under the auspices of Telecommunications Industry Association (TIA). The GSM 2G systems utilized time division multiple access (TDMA) methods, while the TIA 2G systems utilized code division multiple access (CDMA) methods. In common parlance, these two systems are widely referred to distinctively as GSM and CDMA systems.

Because the 2G systems were built on standardized specifications, mobile phone users experienced the benefits of the inherent interoperability and roaming capabilities. The value proposition of standardization was beginning to be realized for both mobile phone users and service providers. Standardization facilitated new opportunities, in terms of choices and options, while preserving flexibility, for the nascent mobile communication ecosystem. Mobile phone users were empowered to select mobile devices to suit their preferences in terms of both form and function. Mobile service providers were afforded choices through a standards-compliant, multivendor environment to suit business directions.

With the transition from 1G to 2G systems, mobile communications continued to evolve to augment voice communications with data-oriented communications. These additional capabilities manifested in terms of the adoption of text messaging as either a complement or an alternative to voice communications. The initial text messaging capabilities appeared in Finland, in 1993. Other examples of an early adoption of data-oriented communications occurred in the form of mobile commerce and content, such as the payment for products in a vending machine and ring tones around 1998, in Finland. Mobile commerce to support payment systems, such as banks and credit cards, was introduced by the service providers Globe and Smart in the Philippines in 1999. A significant step, in the convergence of mobile telephony and the Internet worlds, occurred in 1999, with the i-mode Internet access service, introduced by NTT DoCoMo, in Japan.

Toward the end of the 1990s, the excitement, demand, and the potential of mobile communications underscored the significance of data communications and the related need for technology enhancements. The requirements and objectives in the transition beyond 2G toward 3G were captured in the International Telecommunications Union (ITU) envisioned International Mobile Telecommunication-2000 (IMT-2000) initiative. This initiative formed the basis of the 3G specification development for both GSM and CDMA mobile communication systems. The ITU-3G initiative was originally part of the Future Public Land Mobile Telephone System (FPLMTS) work that started in 1985.

The challenges in the transition toward 3G systems were largely related to the complexities associated with the nature of data communications and the experience of multimedia services that were proliferating over the wired Internet. In case of voice services, the parameters that define the mobile phone user expectations are well-defined and universal. In contrast, the mobile phone user expectations, with respect to data services, are variable and limited only by imagination. The variability of experience is a result of the peculiarities of data services, such as quality of service (QoS) guarantees that are impacted by the nature of the media content (real-time and non-real-time) as a result of network conditions, and mobility. Since data services are naturally amenable to high level of customization, diverse use-case scenarios, as a result of personal preferences and usage metaphors, the experience and expectations are prone to change. These changes in turn are triggering new requirements, standardization, and innovation for continuing enhancements toward the next-generation network (NGN) vision, as the 3G mobile communications landscape continued to mature. This vision embodies the convergence of the mobile and Internet worlds, with the Internet protocol (IP) as the ubiquitous and unifying protocol across wired and wireless communication technologies.

The promise of an enhanced data service experience is an ongoing pursuit, with encouraging growth of Internet-oriented multimedia services over 3G mobile communications. In the service arena, data services mark the dawn of an era of rich communication modalities that conform to the firmly rooted aspects of human communications and connectedness.

The texture of the mobile communication phenomenon has been significantly influenced by the demand and interest created by the advent of the iPhone product, from Apple in 2006. The focus of the iPhone on the user experience and data services is undeniable. It has created a new awareness around the complex aspects of the nature of human communications and behaviors, in communities across the globe. It has been a catalyst in the forging of new pathways toward nomadic networked communities.

The iPhone has created the impetus for a proliferation of data-oriented mobile devices, such as the smartphone. It is anticipated that this trend will lead to a convergence of mobile and portable computing devices as technology enhancements and usage paradigms continue to evolve in an interdependent fashion.

With the ongoing maturing and enhancement of 3G mobile communications and the emerging market trends toward a *data-centric*, *user-centric* world, ITU launched the IMT-Advanced initiative in 2005. This initiative is widely referred to as the fourth generation (4G). The vision is to meet the challenges of a service-oriented usage model, with the underlying technology standardization and innovation that are necessary to enhance and sustain the user experience of sophisticated IP multimedia services. The standardization initiatives to address the IMT-Advanced vision are being explored under the Long-Term Evolution (LTE)-Advanced activities. A precursor to LTE-Advanced is simply LTE, which is an advanced 3G technology, which has spurred considerable global interest and participation in the development of the related standards.

Technological innovation firmly planted in the human context is a compelling ingredient in the quest for communication modalities. In the social context, the search for a balance between personal vocations and the constraints of externalities demands diverse relationships. For example, the geospatial separation of places of residence, work, relaxation, and entertainment imposes constraints that must be managed through technology advancements for a fulfilling experience.

Vision is influenced through the collective learning and observation of communities, driven by diverse interests and skills. With the highly personalized nature of the mobile communications, this is a pivotal ingredient, in the shaping of a vision, for the evolution of the next-generation mobile ecosystem. This understanding is an evolving thread through each new vintage of the mobile ecosystem, for example, 4G, 5G, and beyond.

Why are open standards significant? The significance is embedded in collaboration that lends itself to a broad consensus in the industry. The benefits include

- Interoperability
- Wider distribution of products and services
- Promote a diversity of vendors and service providers
- Open access
- Accommodation of market trends
- Reduced costs for adoption
- Knowledge transfer
- Choices to support diverse business models
- Enable customizable services
- Seamless mobility

1.4 Amorphous Convergence

The interdependence between the creation and consumption of personal communication services inspires an amorphous convergence across a variety of aspects that are related to human concerns, evolution of standards, innovation, and advances in technology.

Mobile communications, embellished through standardization and innovation, continue to evolve and impact human social behavior and cultural change. Multimedia services are at the core of this evolution, where communities interact individually and collectively to discover and shape lifestyles, modes of interaction, and creation and consumption of content.

What are the choices? What are the elements of personalization? What is the landscape of enablers?

The subsequent chapters examine these aspects and explore the inherent trade-offs. It is anticipated that a revelation of these aspects will provide insights, with respect to the rapidly changing and evolving landscape of mobile communications and its significance for communities across the globe.

Chapter 2

Elements of Standardization

Communication leads to community through understanding, relationships and mutual value.

—Rollo May

2.1 Universality

The exploration and examination of ideas that promote interoperable implementations are the seeds that set the stage for identifying and specifying the necessary enabling actors and stories. The endeavor entails discussion and debate across competing or complementary ideas toward a consensus, which typically represents a convergent approach that embodies a representation of the related trade-off considerations. These trade-offs may be potentially diverse across different metrics, such as processing, latency, flexibility, scalability, and extensibility aspects among others. Approaches that allow a variety of implementation or deployment choices in the marketplace, while preserving interoperability, are inherently valuable in the process of exploration. The measure of success of a convergent enabling approach is its endorsement in the industry in a variety of implementations. The implementations may either create new markets or enhance an existing market.

The ingredients of a convergent approach are oriented toward both technological advancement and economic benefits for both the consumers and the producers of implementations that are aligned with a convergent approach, or in other words standards compliant. The economic benefits stem from the choices that accrue from a variety of implementations for a given function or a set of functions that are standards conformant. The implementation choices allow adaptability to different business models. This flexibility provides opportunities to optimize both capital and operational costs associated with implementations. The technological advancements appear as innovation in implementations that spur a competitive market environment for growth and expansion, in a foundation of interoperability.

A level of predictability that is afforded though the use of a consensus-driven convergent approach in the crafting of standardized specifications allows the consumers of the standardized implementations to focus on competitive implementations for a realization of specific business model objectives. This avoids both fragmentation and monopolies that result from proprietary implementations, where the consumer is adversely impacted in terms of the quality of experience, as well as in terms of the incurred costs.

The unique and multifaceted nature of a standardization initiative is an intersection of strategic technology and market considerations. The output consists of published documentation that captures the enabling building blocks of features and capabilities that serve as a framework for the development of implementation designs. The implementations include products such as mobile devices and mobile network entities, services, and deployments. The process of standardization is reflective of the complex and evolutionary nature of information, which is mediated by technology while it has widespread implications in a human context, especially from the perspective of a global mobile society, across a variety of markets and demographics. According to the International Telecommunications Union (ITU) view, the standardization initiative applicable in the context of the mobile information age has the following intrinsic flavor:

> Standardization is one of the essential building blocks of the Information Society. There should be particular emphasis on the development and adoption of international standards. The development and use of open, interoperable, non-discriminatory and demand-driven standards that take into account needs of users and consumers is a basic element for the development and greater diffusion of ICT (Information Communication Technologies) and more affordable access to them, particularly in developing countries. International standards aim to create an environment where consumers can access services worldwide regardless of underlying technology.*

The dominant theme in the establishment of a convergent approach in the evolution of a mobile broadband ecosystem is the development of compatibility in diversity across a multitude of network infrastructures, mobile devices, and the openness of interfaces across the connectivity and service layers of communication. The highly specialized nature of the constituent technology components requires significant expertise, experience, and awareness with respect to the technology components together with the social, cultural, behavioral, and business components that are vital for establishing a convergent approach in the crafting of the related specifications. These aspects serve as technology blueprints toward an interoperable design realization, with cross-functional implications that span policy, regulation, economics, and future market potential.

With information globalization, the fruits of standardization, while preserving unique and valuable regional differences, pave the directions toward uniformity in the landscape of economic opportunities through the beacon of technology innovation. Attractive reductions in the operational costs through interoperability and consistency in competitive implementations are among the pillars of benefits that inspire global participation and collaboration in the standards initiative.

The process of technology standardization is collaboration through diplomacy in the commercial realm. The goal is the propagation of evolutionary thought, framed within the commercial realm, where the participants play out this direction, while protecting and influencing the represented business models. While the business models mold the existing market profiles, the standardization process is cardinal in anticipating and shaping future market trends. This aspect maximizes the potential for related studies that reveal and guide specification development well in advance of implementations to match the anticipated market trends, as they continue to evolve and expand through connectivity and service scenarios. The effectiveness of the standardization

---

* http://www.itu.int/wsis/docs/geneva/official/dop.html.

process hinges on a deep exploration and understanding of the business models and their operation within the context of the evolving market trends. Examination of the multifaceted aspects—technology, culture, and usage behaviors—across the market landscape demands leadership skills that effectively blend these aspects to curate and shape the evolution of connectivity and service specifications. The intellectual property rights regime, for an equipment vendor or a service provider, benefits in terms of an enhanced innovation potential for implementation paradigms, through the dissemination of information motivated by emerging specifications. The development of specifications is complex, as a result of such multifaceted consideration, together with the need to formulate appropriate strategies and the establishment of trust through collaborative partnerships and engaging participation.

The leadership element, within each participant is shaped by style and personal characteristics. The global nature of standards organizations composed of regional nuances, cultures, and behaviors demands a leadership style that is adaptive, accommodative, and interpersonal. These attributes are reflective of the rapid evolution of mobile communications—connectivity and service—that warrants astute and strategic thinking, evangelization, and consensus building.

Mobile multimedia architectural models, procedures, and protocols are among the prominent categories of ideas that sculpt the development of standardized specifications, which serve as a common fabric in innovative implementations. Since the adoption of standards hinge on the innate offerings of interoperability, it is paramount in innovative pursuits for persuasive product and service offerings. The formulation of convergent approaches that characterize standardized specifications is an art form of strategic abstractions, interpretations, and expression. The interoperability offered by standardized specifications promotes widespread exchange of information across societies on a global scale. Among the social benefits and user experience enhancements that accrue from the interoperable nature of standardized specifications, there is the enormous potential for economic growth and opportunity.

Openness in the standardization process promotes competition and free trade, while proprietary or narrowly specified closed standards promote fragmentation together with higher entry barriers that stifle market expansion potential. Interoperability coupled with openness fosters emerging communication and service modalities, such as digital cultural life, digital education, digital medical diagnostics information, digital health information, digital journalism, and digital commerce. Access to health and medical information repositories in a consistent and reliable manner is a vital benefit for wellness consumers, pharmacists, patients, and practitioners.

The common platforms of implementation that are a result of standardized specifications enable new opportunities across both economically developed and developing markets. This is a critical ingredient in the promotion of innovation that offers universal lifestyle enhancements and growth potential.

Participation in the development of standards consists of leadership in a combination of technology advancement aspects, systematic processes, and consensus building through an expose of various trade-offs related to market requirements—in terms of the features or capabilities being examined. Influencing the evolution directions of standardized specifications to meet business objectives in a globally harmonized fashion is strategically significant for establishing the potential for adoption and realization in the industry. Figure 2.1 depicts multifaceted aspects associated with the standards process.

The multifaceted aspects open access over a landscape of wireless and wired access network are essential among the considerations for mobile connectivity. Together with interoperability, these considerations provide a framework that inspires both collaboration and competition, while offering reliable, affordable, and an experientially attractive mobile connectivity and service.

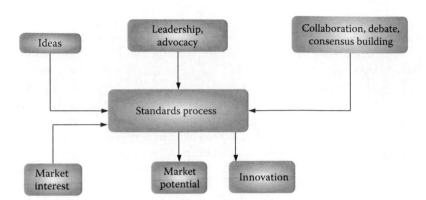

Figure 2.1 Contextual landscape of standardization.

In concert with standardization, a compatible regulatory regime and policies are vital for a realization of viable and relevant specifications. A nondiscriminatory policy for connectivity over radio-access networks is critical for an exploitation of open interfaces that allow plug-and-play paradigms to augment the user experience. Aesthetics and convenience are the bedrock in innovative implementations that leverage interoperable specifications. These twin attributes operate synergistically to harness lifestyles in a digital world, where ubiquity and consistency of connectivity prevail across different mobile networks. The notion of open access networks embraces a few concepts:

Promoting the possibilities for third parties to operate a network infrastructure.

Access availability for users is independent of the type of wireless access or the access provider.

Competition for operating access network infrastructures must not be constrained by monopolistic policies.

Coordination across access networks administered by different domains, whether these domains are owned by network operators or private or public entities.

Fair and nondiscriminatory terms for access to services, over different wireless access networks, enabled through emerging mobile broadband technologies, such a long term evolution (LTE).

Open access strategies, such as wholesale wireless access networks that inspire new business models and collaboration across the access and service value chain.

User-centric policies and regulations that are least intrusive in a market-oriented model that promotes universal access to mobile broadband to effectively leverage the benefits of the digital dividend, in next-generation systems.

Neutral schemes for packet-oriented information traffic management, across different types of wireless broadband networks, where the quality of service (QoS) treatment for the different types of information streams are objective, in terms of content, application, service, and device.

Creation of appropriate measures to manage the risks to privacy, confidentiality, and consumer rights that is in step with the evolution mobile technology capabilities.

Incorporation of security and fault tolerance through the adoption of distributed, decentralized, and autonomous architectural models in a complex ocean of Internet Protocol (IP)-oriented information creation, consumption, and proliferation.

Harmonization of wireless spectrum across all the ITU regions around the globe for mobile access ubiquity and a consistency of multimedia service experience to realize the potential of openness.

The openness paradigm spans the elemental attributes of the standardization initiative:

Examination of potential market drivers, from a technology perspective
Identification of nondesign-, nonproduct-, nondeployment-oriented building blocks
Examination of the nature of the building blocks, within the context of system requirements
Identification of the ingredients (e.g., protocols, procedures) necessary to meet the requirements of an envisioned system, together with relevant extensions/adaptations
Establishment of common directions, across potentially diverse, often conflicting requirements, to foster interoperability, a multivendor ecosystem, and market expansion
Crafting, management, and evolution of working-group procedures, processes, and cultures to shepherd, monitor, examine, verify, and validate the progress of features toward stable specifications for publication

The peculiarities of the standardization process are that it demands not only a deep understanding of the underlying technologies but also an awareness of the interrelationships with business, market, cultural, policy, and strategy. Insights into the technological design and usage scenarios are foundational in the creation of implementation-independent abstractions that shape the textual content and fabric of interoperable specifications. The value and the potential for the effective applicability of a specification are measured in terms of its function as a catalyst in the promotion of innovative product and service designs, with various trade-offs and choices, while preserving widespread interoperability. The specification framework directions also serve as guidance for researching and creating forward-looking business models, strategies, and roadmaps. Figure 2.2 shows a generalized model of constituent functions and interactions in a standardization process, which is reflective of global standards bodies, in the realm of mobile communications.

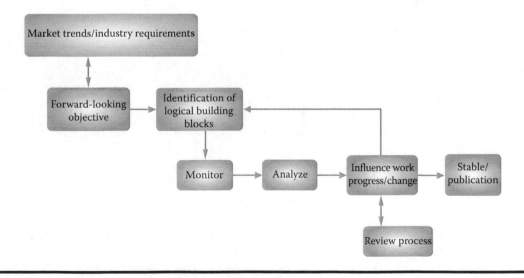

Figure 2.2   Landscape of the standardization process.

2.2 Architectural Role

The inclusion of a variety of perspectives and considerations is pivotal in the shaping of a framework of specifications that has significance in terms of applicability. The maximization of benefits in the industry occurs through a consensus process that espouses open, fair, and equitable considerations. This is an elemental ingredient in the realization of robust and well-defined specifications, which bears the blueprints for product and service implementations across diverse market segments. Consolidation of specifications augments the quality of implementations through a shift in the allocation of resources from an examination and application of panoply of specifications to technology innovation and research in the implementation space. With this understanding, implementation costs are contained and the innovation is unleashed.

These aspects within the standards process demand specific roles and behaviors with respect to the participants. It is a notion concurrent with abilities to mitigate risk factors, postulate trade-offs, and provide recommendations, where something is unknown, unproven, or untested—such as service-level requirements, trade-offs, scalability, interoperability, reliability, availability, extensibility, maintainability, manageability, and security. Best practices in this role include direction and guidance in the interpretation or crafting of specifications for designers, developers, integrators, field engineers, marketers, sales agents, and customer-care agents for compliance with the system requirements, architecture models, etc. The roles and behaviors embody a variety of attributes that are closely aligned with architectural perspectives that encompass strategy, forward-looking thought, interpretation, analysis, negotiation, trade-off, innovation, advocacy, and leadership.

The intrinsic characteristics of the roles and behaviors are reflected in an observation by Vitruvius, circa 25 BC: "The ideal architect should be a person of letters, a mathematician, familiar with historical studies, a diligent student of philosophy, acquainted with music, not ignorant of medicine, learned in the responses of jurisconsults, familiar with astronomy and astronomical calculations."

A designer is concerned with what happens, for example, when a single action or event occurs, while an architect is concerned with what happens when multiple actions or events occur. The latter role is multifaceted and embodies the cross-disciplinary nature of the standardization process. The guiding principles that capture these characteristics are

- Advocate motivators for new considerations
- Evangelize and create awareness for ongoing work
- Decouple strategies/logical models from design and implementation
- Develop specifications, grounded in industry-wide consensus

The intersection of regulatory, business, and forward-looking considerations pose challenges that are often conflicting. These challenges demand the application of cross-disciplinary skills and experience that transcend the traditional boundaries of expertise that includes a broader understanding. Broad perspectives reveal insights and applicability in the realms of technology, strategy, and consensus building and increasing levels of significance through the dawn of evolution in the information age, where the awareness of interdependence is vital.

In these avenues of thought, the prisoner's dilemma* expounded by Robert Axelrod is particularly revealing in the standardization process. De jure standardization initiatives offer benefits on a broad scale through the artifacts of cooperative behaviors that lead to consensus on technological

---

* http://www-personal.umich.edu/~axe/research/Axelrod%20and%20Hamilton%20EC%201981.pdf.

ideas and proposals that are explored during the specification development process. Such initiatives transcend the "prisoner's dilemma" conundrum. In the emerging mobile broadband arena, prominent examples of de jure standardization include the Third-Generation Partnership Project (3GPP), Internet Engineering Task Force (IETF), IEEE, Open Mobile Alliance (OMA), World Wide Web Consortium (W3C), and 3GPP2.

The prisoner's dilemma in game theory illustrates the benefits of cooperation through a non-zero sum game. In this game, two players may opt to either "cooperate" or "defect." When both the players cooperate, then there is collective benefit. When neither cooperates, the results are mediocre. If one of them cooperates, while the other does not, then although one may gain temporarily, the negative impacts are guaranteed in the longer term. Figure 2.3 expands the logic behind the prisoner's dilemma model of strategic moves and the corresponding outcomes for the players engaged in the game.

The cross-disciplinary aspects of standardization invite the strategies and inferences from the logic in Figure 2.3, for considerations in leadership and problem solving that reveals the often intangible and hidden directions toward consensus through cooperation. Different modalities of communication foster the progress of specification development through the exchange ideas and trade-offs that serve as a bridge toward consensus. Among the modes of communication, the most significant is the face-to-face interactions and discussions, since the nature of changing information in the evolution of ideas and perspectives demands the richness of expression that is created through the physical presence of the participants. This is especially crucial on a global stage, where cultural, behavioral, linguistic, strategic, political, market, and business views modulate the information exchange in the crafting of the enabling and evolving mobile technology specifications.

The necessary—formal and informal—information exchanges hinge on democratic principles. Agreements on specific architectural or procedural capabilities are established through a combination of negotiations among the participants, as well as through moderated negotiations. Universal inclusions among the participants, together with equal opportunities to weigh in on the various options and trade-offs, are among the essential aspects of the standards process. This process is moderated and adjudicated by the leadership in the corresponding technical specification working groups.

Accessibility to information is a critical ingredient in the social, cultural, and commercial arenas. Mobile multimedia communications continues to establish itself as a cornerstone in the affairs of human communications, in virtually all kinds of different usage scenarios. With this backdrop, the creative crafting and application of standardized specifications is an imperative from the perspectives of design, operation, cost, interoperability, and user experience. Public interest in

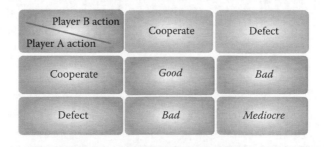

Figure 2.3 Illustrative model of the prisoner's dilemma.

technology-mediated communications is an integral component in the integrity and well-being of assorted communities. Interoperability, backward compatibility, and forward-looking capabilities are necessary considerations for a vibrant and relevant framework of specifications. While interoperability is implicit, forward-looking ideas are fundamental to the standards endeavor.

The backward compatibility aspect is preserved during the introduction and life cycles of new ideas that translate to new market potential. Where forward-looking ideas reveal an archetypal transformation, backward compatibility is traded off in the interest of technology evolution, which is the engine of new opportunities in the human journey. These changes lead to market expansion and shape the future potential of new capabilities. The introduction of a packet-oriented, distributed all-IP architecture is an example of a transformative paradigm in the third-generation vintage, in mobile communications that owes its heritage to a circuit-switched, centralized model.

This transformation rooted in the nature of IP demanded breaking away from the legacy hierarchical models that ushered the notion of mobility in a predominantly voice communication world inherited from circuit-switched wireline telephony. In turn, this shift invited the adoption and the continuing evolution of compatible architectural models into the standardization process for mobile communications. New interfaces and adaptation to the nature of wireless mobility were envisioned to capture and leverage the benefits of the distributive nature of IP. Flatter architectural models provide the benefits of resource sharing coupled with reduced signaling overhead leading to reduce latencies, especially for real-time multimedia services, such as voice and video information transport.

Organizations and groups of any kind have within them an identity, a culture, and unique behavioral patterns that are a human characteristic. So it is with standards organizations and the subgroups and individuals that compose the organization. Adaptability of personal presence and styles is of paramount significance to influence collective styles, relationships, and the flow of discussions toward the adoption of globally valuable ideas. Although these endeavors often take time and are often laborious, the potential benefits far outweigh the challenges and are a motivating factor. In this light, an architect's role is to enlighten and elucidate. The nature of the topics being examined range from relatively simple to complex, where an understanding of diverse subject matter—cross-protocol layer, algorithmic ingredients, radio-frequency aspects, and regulatory, business, market, and strategic considerations—is vital to debate and adjudicate trade-offs for effective recommendations.

Aside from the interplay of technology, business, marketing, and strategy, the procedural framework in the specification development process is important in terms of the logistics of specification creation and in the life cycle in terms of revisions and updates. The review process of the specifications demands the collection of comments and their subsequent resolution toward publication, through the specification validation and verification phase. The review process is by no means trivial, and it entails a level of rigor in the crafting and the language of the specifications as much as the content of the subject matter described in the specifications. These incremental steps include multiple conversations across the contributors—formal and informal—and the effort is directly related to the nature of the various ingredients of the specifications, the level of complexity, and controversy that may be attributable to the specific ingredients that constitute the specifications.

Aside from the technology, business, marketing, and strategic element of the specification development process, the procedural framework is vital. It defines the logistics of the progress of the specifications from a draft status to a finished product through the publication event. The procedural framework entails conversations—formal and informal—across the contributors engaged in the crafting of the specifications. This involves the collection and review of comments associated with the ingredients that constitute the contents of the specifications. The language, style, and the organization of the content are critical measures of the readability and the clarity of the finished specifications. Linguistic dexterity is a significant attribute of an architectural role, reflective of multifaceted considerations, while crafting text associated with protocols or procedures that are described in a specification. During the review phase, as well as during the original crafting phase, careful attention to the robustness and nonambiguity of the content in conveying the intended ideas is an important metric, with respect to the quality of a specification. The validation and verification phase is the final scrub period for both language and technical correctness, before a specification is published. A published specification serves not only as a guide for the ideas, features, or capabilities but also as a description of interoperable definitions, enabling algorithms, interfaces, and architectures, as applicable. The specifications provide the requisite guidance for designers and conformance testers, without constraining innovative product or service implementation.

An extension of these multifaceted capabilities is relevant in the context of providing direction or moderating the flow of conversations or deliberations in the making of a specification, incubated in ideas and imagination. These skills translate to leadership prominence in terms of chairmanship within or across organizations that imply different levels of impact associated with a formal title. With the increasing levels of complexity and interdependence influenced by the evolution of technology, informal roles of leadership and influence across different domains of interest, within or across standards organizations, are significant as well. Such roles are driven by interest and relationships, which are catalytic in nature and serve effectively in bridging gaps across polarized parties within or across standards organizations.

The impact of participation and engagement in conversations, debates, and discussions is tangible in terms of the potential to establish collaborative and influential relationships. The modalities of presence-oriented communication may be articulated through a variety of tools including emails, instant messaging, proposals, recommendations, and face-to-face meetings. The nature of new ideas and the inherent complexities ensuing from multifaceted considerations—technology, business, cultural, social, interpersonal, markets, and strategies— underscore the imperative of face-to-face meetings. It is through the face-to-face conversations and debates that the unspoken ingredients of human communications are revealed. It establishes a common framework of understanding and negotiations, in a world for competitive business models and strategies. This is an essential tenet in the making of global interoperable, locally applicable specifications.

The presence capital tends to get eroded over time, if participants are absent from the various types of specification development discussions and conversations. The presence capital translates to the potential benefit of prominence and the scope of influence in future discussions. Activities for participation across a plethora of standards development organizations (SDOs) demand careful attention to selective participation, where the prominence capital is not eroded significantly, after the initial investments of participation and contributions, in a standards organization—unless warranted by the emergence of new organizations in a landscape of ceaseless change and innovation. Participation in ad hoc focus groups and special interest groups to move the ongoing work or to establish new endeavors reflect opportunities to build prominence capital and relationships toward collaborative partnerships. Noticeability, during face-to-face meetings, is the simplest step in the accumulation of prominence capital. It lends itself to better receptivity, as an artifact of human behaviors, and in turn translates to opportunities to provide direction toward a consensus or compromise, where conflicting views are presented.

G.K. Chesterton mused, "Compromise used to mean that half a loaf was better than no bread. Among modern statesmen it really seems to mean that half a loaf is better than a whole loaf." Breaking bread—informal or formal conversations—over social events or food at face-to-face meetings provides a common and appealing stage to clarify perspectives or positions toward a deeper understanding and effective results in the progression of specifications or ideas. Pleasantries and interpersonal interest in conversations serve as catalysts, as well as to an enjoyable experience—one that mirrors the specification development objectives through an enabling of products and services that embellish the mobile user experience. Effectiveness of specifications hinges on a union between technology and art, specifically in the emerging realms of mobile multimedia services, where personal adaptability, experience, and convenience reign. Listening, interpreting, and discerning are elemental in the building of forward-looking frameworks. To ask both "why" and "why not" is an essential signature in the standards and innovation horizons. The framing of information in different contexts is elemental in the elucidation and evolution of subject matter encapsulated in the information.

Collaborative partnerships, in the landscape of ideas and exploration, are the fuel for the creation of new capabilities and their incorporation into specifications. The ability to delve into the unspoken and the unwritten is significant in the discussions and debates. These attributes are inferred and understood through frequent interactions with the participants with experience, the knowledge of the history of the working groups, and an established level of prominence and leadership. At the same time, being open to radically new and different ideas promotes valuable counterpoints in the moderation and the flow of the discussions toward the enhancement of value in specification development. Traditions and cultural nuances, in the realization of standards process, while necessary, are subject to change in a dynamic technology and market evolution environment.

Initiatives to pursue ideas and issues that are relevant to the industry as a whole, beyond specific business preferences, is one that enhances reputability and respect in the standards development community. It embodies philanthropy of ideas through volunteering, encapsulated in a passion for problem solving. It portrays leadership qualities that inspire similar behaviors in the research, creation, and applicability of specifications. The architectural role invites a level of multifariousness through an active engagement in shedding light on technology trade-offs as well as an appreciation of the standards process from a practical vantage point. Diligence and commitment are the pillars of cohesion and collaboration in the working groups. The breadth of perspective painted by an architectural brush both influences and shapes the ideas that guide the specification of new horizons in the evolution of the mobile Internet.

Understanding and interpretation of the terminology and traditions are integral to the role of the architect in shedding light and lucidity across the ingredients of the underlying technologies and the trade-offs. This not only is pivotal from a research view but also inspires the introduction of new ideas into the discussion and debates that shape the texture and content of the related specification process. While an alignment with the processes provides a framework for tangibility, progress and consistency of the specification content, flexibility, and openness—even radical—are the essential seeds of creativity that drive innovative possibilities in the specification content. In this regard, strict congruence with process must coexist with "changeability" in the mechanics of "how" new ideas are debated, discussed, and adopted. This is adherent with a vision of innovation, while allowing flexibility in the details.

Definition of terminology is the essence of specification development. It symbolizes a data structure that is the scaffolding of the specification content. It sets the stage on which the actors within the specifications play out new features, capabilities, and stories—procedures and usage scenarios—and frequently return to revise and update an understanding of both the terminology and the specification content, through the publication phase. The styles and sequencing of the specifications are also influenced by the terminology, which must be carefully selected and explained. The specification content, while conveying the relevant content, must also be clear and readable ranging from high-level requirements and usage scenarios to architectural models, interfaces, algorithms, and message formats.

Familiarity with the terminology is a requisite for a proper interpretation of the specification content. Sometimes cryptic, sometimes rooted in historical precedence, the terminology tends to be obfuscated. Simplicity, with clarity of meaning, and applicability are the hallmarks of well-crafted terminology, which then becomes the vocabulary of the spoken and written matters associated with the landscape of specifications.

Impressionability in conversations and presentations—formal and informal—is essential in shaping an umbrella of influence. The architect's role encompasses that of a catalyst across a variety of topics and views, to interpret, diagnose, and propose ideas effectively. As a catalyst, there is an awareness of the personalities and behaviors in the standards community to leverage the individual interests and information. There is recognition of interdependence across seemingly disparate silos of activity toward synergy through collaboration, driven by the complexity of the emerging mobile broadband systems. A notable example in this emergence is bringing together of regulators and technologists in a fragmented spectrum landscape for effective and viable solutions for input into interoperable specifications. The catalyst component is rooted in a personal interest in the views of others, a desire to help, and a conviction of vision concerning an overall strategy.

During the inception of an idea or capability, extracting the essence together with a tolerance for ambiguity is elemental in fostering the progress of discussions that leads to the creation and adoption of content in a specification. The ability to listen and to extract the intent as well as the specifics of a discussion or presentation is paramount. Selective and insightful interjections into discussions—written or verbal—weave the fabric for the effective progress of specification development, as well as the architect's influence capital. Strategic compromises, through substantive negotiations, hold the keys to foster consensus among proposals that are conflicting, in terms of the technology or applicability underpinnings.

Decentralization and a loose coupling among collaborating partners, in the standards community, are central to the theme of creative thinking and application. In this environment, fluency in the art of verbal and written communications, tempered with interpersonal awareness, is cardinal. Social and cultural awareness lends itself to interpreting and expressing intent and views on behalf of participants that may be challenged in the fluency of expression, in a cross-cultural global landscape. Such initiatives exemplify collaborative and partnership directions that augment the influence capital that fosters cohesion, while inspiring creative thought and goodwill. Coauthoring proposals, editing specifications, and presenting proposals are illustrative of collaborative mechanics in the standards process.

2.3 Knowledge Processing

In the oceans of information, mobile communications are unique because it not only serves as a window into these oceans but also provides a personal experience—an intersection of the prolific and the particular. The creation and commercialization of knowledge in the form of mobile technology mediation—networks and devices—span an entire life cycle, of which the research and transformation of knowledge and understanding into standards is an early segment in this life cycle. The application of technology knowledge, through creative ideas, ranges from paradigm shifts to incremental enhancements. These are selectively captured in specifications without encumbering innovative designs, while enabling openness and interoperability. The life cycle of the standards process spans prestandardization investigation, planning, research, advocacy, creation, adoption, publication, education, and maintenance.

An idea is the seed of any standardization initiative, around which collaborative effort is woven. A rough consensus on the relevance of a candidate topic for standardization is established, as a precursor to the scope, definition, and progress of the content through proposals and a collective arbitration of perspectives. Global and regional partnerships are critical in the promotion of agreement on the specification content. The accommodation of different views coupled with the appropriate levels of rigor and moderation among the participants and the leadership within the working groups serve as a vital platform in the progression of consensus—sometimes through unequivocal agreement, sometimes through workable compromise. The standards process is grounded in guiding principles that include openness, due process, balance, collaboration, advocacy, a right to appeal, and consensus. The promotion and evangelization of a published specification occur through liaisons with the industry. Formal industry forums or conferences also serve to advertise the published specifications that serve related industry requirements and implementation. The collective wisdom of the working groups, engaged in the specification development work, is catalyst in the resolution conflicting positions as well as in the examination of trade-offs for instilling value in the specification content. Between conflict resolution options of a majority-vote and a consensus approach, consensus is recommended despite the potential challenges for bridging conflicting positions through a search for creative ideas. The latter sustains the vision of openness and collaboration on a global scale, for standardization without encroaching on design innovation in the realization of products and services.

The measure of effectiveness of a standards process or organization is directly proportional to the desirable attributes of openness in all aspects—participation and architectural models—procedural equitability, transparency, and the level of participatory engagement. These attributes significantly enhance the potential for a fair and effective incorporation of the interests associated with a larger number of stakeholders. This satisfies the intent of the standards process to serve and create industries, which represent its clients.

Among the considerations for openness is the selection of a technology architecture, feature, or capability for standardization. This selection determines whether a technology facet is in the realm of implementation design and innovation, or in other words open or proprietary. In some cases, the time to market or to create a new market a level of integration across the connectivity and service layers may demand an emphasis on "proprietary" with some open building block protocols. Apple products—iPhone, iTunes, etc.—among others represent examples of a combination of open and proprietary capabilities. Excitement and adoption of such products in the market spur the need to open the proprietary components for broader market participation, through standardizations and innovation cycles. Google Android products and services represent the shift toward more "openness" in the building blocks. The level and nature of shifts toward openness are a function of the appropriate shifts in the business models to mine potentially new revenue streams—revenue sharing, new collaborative partnerships, etc.—in a paradigm of change toward openness.

Standardization of architectures, interfaces, and procedures produces consistency, which in turn translates to improvements in the quality, reduction of product realization lead times, reduction of errors, and a minimization of the time to market. It enables a building-block approach toward a customization of implementations, where conformity to the standardized ensures predictability of behaviors at a system level. The development of products and services across borders on a global stage is facilitated, where the creation and consumption are separable without any impact to expected behaviors—an imperative for a vibrant and interdependent global economy.

The use of specification information repositories, containing standardized technologies and terminologies, and an active engagement in the standardization process offer enormous insights and enhance the sustainability of existing and long-term endeavors for businesses. These translate into a reduction in research and development costs through a leveraging of information sharing and collaborative ventures that are implicit through an active engagement in the standards process that include face-to-face and virtual discussions and debates.

The guiding tenets for standardization in the progress of mobile technologies embrace the following:

Layering of functionality within architectural models
Optimization of procedures for connectivity and service
Loose coupling of functional layers
Tight cohesion within a functional layer
Coordination across disparate system of connectivity and service

Consensus-oriented partnerships that span the globe, across public and private endeavors, are significant thread in a plethora of ingredients that shape and characterize the nature of standards organizations. It is this essence that fosters the potential not only to promote beneficial studies and specifications but also to inspire innovation and implementation. The streamlining of industry processes and human lifestyles in terms of cost and experience is among the compelling benefits of the standardization process. The cost reductions, in return for the investments in the standardization process, are compelling motivators that enhance the business revenue potential and reputability in the industry. The indirect benefits of an active engagement in the standardization process include the potential for new business ventures and collaboration opportunities, across the supply chain for a realization of products and services.

Ideas transcend and propel manifestation—from the zip to the slide to unlock idea on the iPhone. It is the engine that inspires innovation and invention in free markets, with appropriate arbitration, where the ideas must be valued and credited. It is the seed that leads to a harvest of creative designs—bedrock of evolution in the standardization process. The standardization process invites conformity in a sea of innovative ideas and perspectives. It differentiates and amalgamates business models, technologies, and strategies toward a sustainable market potential. Competition and collaboration coalesce to foster revenue potential and optimize both operational and capital expenditures through the availability choice in product and service implementations (Figure 2.4).

Mobile connectivity and services standards blend the factual and the experiential through an exploration ideas, debate, and consensus, which are mapped into a vision, guidelines, requirements, architectural models, protocols, and procedures. This is a behind-the-scene scaffolding—a framework that propels innovation, invention, and implementation realized through mobile product and service designs. Compatibility promotes lifestyle enhancements, through the notion of plug and play—universal sockets for electric bulbs, universal power adapters, and universal protocol such as IP. The universality and third-party contributions to the various aspects of mobile connectivity and service are a direct consequence of the logical and virtual nature of IP, which allows widespread compatibility across assorted physical interfaces and hardware/software renditions.

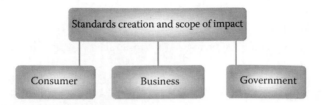

Figure 2.4 Participatory landscape of standardization.

Universality and consistency of performance and the resulting experience are among the significant aspects of the value proposition in standardized products and services.

Thinking along the lines of logical design constructs, without constraining implementations, is elemental in the landscape of standardization. Charles Burnette* refers to this notion as "a process of creative and critical thinking that allows information and ideas to be organized, decisions to be made, situations to be improved, and knowledge to be gained." Creative and design-oriented thinking, while being conduit for technology evolution—especially in the multifaceted nature of mobile communications evolution—is the essence that propels innovation, independent of the domain of application, form, or function. It is a fabric that transcends boundaries, labels, or silos of subject matter. It is the formless ingredient that resonates with the observation—"the Tao that can be told is not the eternal Tao†"—that is the source of products, service realizations, and human experience. The standardization process is reflective of this elemental theme and embodies a variety of related attributes—collaborative, consensual, interpersonal, inductive, deductive, abductive, interpretive, integrative, cooperative, experimental, exploratory, and evolutionary.

The landscape of fourth-generation mobile communications is one that is textured with the various modalities of connectivity—wide area, local area, and proximity—and services (assorted media, applications, and usage experience). It is one where mobility, with all the latter attributes, is subsumed into the Internet experiential plane. The resulting arena demands the exploration and use of multifaceted tools to address the associated demands for capacity and coverage in packet data–oriented fabric of communications, propelled by higher data rates, and dynamic resource allocation to negotiate and satisfy application demanded QoS profiles, on a communication session discriminated basis. The 3GPP standardization process establishes an interoperable basis, through the development of related specifications that shape the evolution of mobile systems, while being aligned with forward-looking perspectives, in an ever-changing and dynamic market environment. The system architecture evolution (SAE) endeavor—standardized framework underlying the LTE vision—began as an initial framework of ideas, within 3GPP, to augment the mobile experience, while adopting Internet principles, formulated the evolved packet system (EPS), and is reflective of directions toward all-IP mobile system architecture. This paradigm shift mirrored a sea change, from the legacy architectures that had prevailed for decades in an exclusively voice-centric world of mobile communications. This shift also embraces the notion of heterogeneous radio-access coverage footprints—ranging from wide-area coverage to small-area coverage—to meet the simultaneous demands for both capacity and coverage. The EPS model embraces a high level of distribution, autonomy, self-organizing capabilities, and various modality user entities—machines and humans.

---

* http://www.idesignthinking.com/01whyteach/01whyteach.html.
† Mitchell, S., *Tao Te Ching*. New York: Harper & Row, 1988.

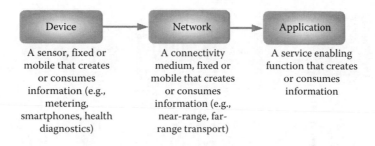

Figure 2.5 Convergence of M2M and M2H services—Internet of Things.

Figure 2.5, depicts the enabling directions of the evolving SAE framework, to meet the enormous scale and diversity inherent in the nature of the Internet of Things.*

The complex nature of technologies, user experience, and markets that comprise the evolving mobile communications horizons invites a widespread cooperation, collaboration, and partnerships. The LTE vision for the next-generation mobile ecosystem pivots on wide and diverse global partnership for a viable realization of connectivity and services. In this vision, the 3GPP is predominant global standards setting organization, with a variety of partnerships that leverage the rich and broad expertise imperative in a challenging and evolving mobile evolution.

Figure 2.6 depicts the list of global organizational partners (OPs) under the umbrella of 3GPP that serves to advocate regional demands through input contributions into the global fabric of evolving mobile standards

Figure 2.7, shows the industry partners that provide information on markets requirements for further consideration in 3GPP, from the perspective of a potential for standardization. The decision on whether a requirement is a candidate for standardization is carefully examined through debate and discussion, where the attributes of distinction between "standardization" and "design" are applied. If it revealed that there are no underlying interoperability impacts that affect protocols, procedural frameworks, architectures, or interfaces, then it is deemed that the requirement is in the realm of design innovation or invention for competitive implementations.

Figure 2.6 Global OPs—3GPP.

---

* http://www.internet-of-things-research.eu./pdf/IoT_Cluster_Strategic_Research_Agenda_2011.pdf.

Figure 2.7   Prominent global market representation partners—3GPP.

The 3GPP LTE initiative for a standardization of the connectivity capabilities is a global framework for convergence, for the delivery of mobile services, through a common and well-defined set of global specifications. The flexibility of design and information dissemination is open and available to participants of any denomination. It is the richness of ideas, discourse, conflict, and consensus that imparts enormous value in the standards endeavor. An engagement in the exploration of technology ideas and concepts, within the standards process, provides early research opportunities that capture the essence for a critical revelation of motivators and the mechanics, which are vital in the formulation of strategies, and business models—vital in the creation and realization of new or enhanced market opportunities. Further benefits include an establishment of credibility and influence in the related specification development initiatives.

The expert members of SDOs (European Telecommunications Standards Institute [ETSI], TIA, ARIB, and others), in the mobile evolution space, have a tangible impact on the creation and progress of specifications in an amalgamation of regional and global perspectives—directions that infuse quality of life enhancements in a user-centric world. The conformity and compliance of designs with the relevant product or service implementations are a function of the industry requirements. These requirements are identified by the consumers—mobile connectivity providers, mobile service providers, and value-added service providers in the mobile ecosystem—of products and services. The power of conformity and compliance is a realization of consistent service behaviors, service experience, flexibility of deployments, lower operational costs, and market expansion through innovation and invention in a fabric of interoperability. The ability of a product or service to interoperate with a variety of different and potentially competitive designs is the vortex of creative thought in crafting virtually unbounded technology, strategy, and business models to suit the demands of regional and global markets. Established third parties are typically utilized for a validation and verification of conformity and compliance, such as for mobile device certification.

Global entities such as the Global Certification Forum (GCF) provide an effective environment for standards compliance, and in turn conformance with regulatory constraints, such as spectrum allocations on a regional and global scale. Such global certification endeavors minimize postproduction costs, while enabling standards compliance across disparate regulatory regimes. Reduced certification costs in turn enable investments in research toward the evolution of enhanced mobile devices and services.

The 3GPP LTE/SAE standardization endeavor in the evolution of the mobile broadband ecosystem embodies the following vision:

Reduction in information transport cost, in terms of cost per bit.

Increase in peak data rates, with linear scaling as a function of available spectrum.

Flexibility in the usage of the existing and new frequency bands, driven by market evolution.

Reduction of service delivery costs.

Enhancement of user experience.

Distributed and flat architectural model, with access network interface simplification and openness.

Power consumption reduction for mobile devices, through optimizations in procedures for signaling and payload.

Latency and throughput optimizations for voice-over-IP and IP multimedia services with various application-specific combinations of real-time and non–real-time payload.

The latency targets for session state transitions from idle to active are less than 100 ms and less than 5 ms for the transfer of small packets.

The establishment of consensus is the sine qua non of the standardization process. It is reflective of individual and collective needs of various stakeholders—private or public—across interdependent communities on a regional and global basis. This is significant in a virtual world coupled with regional and global mobility, where the traditional framework of space-time barriers—cities, countries, and time zones—while being contextually relevant, vanishes in terms of viable ideas for a realization of mobile devices and networks.

The emergence of technology ideas and its dissemination through the standards process is vital ingredient in shaping of business models and strategies. Whether or not specific developments or features in the specification of architectural models, interfaces, and procedures are adopted for implementation, the inference of possibilities revealed through the standardization process that provides value to the evolution of business and technology. The exercise unveils trade-offs and risks pertaining to capabilities and features, within a common fabric of interoperability and economies of scale. An astute application of inferences is a pathway for new ideas that elevate business potential through an accumulation of intellectual property in the arena of design and implementation. At the same time, a reduction of operational and capital expenditures through interoperability, lowering of entry barriers, and a widespread participation in the creation of products and services are compelling benefits of the standardization endeavor.

With the virtually unbounded demand for choice-driven personal experiences conveyed through the mobile devices, smaller, attractive form factors with dense functionality are emerging conduits of information consumption and creation. The density of functionality realized through highly packed components in integrated circuits—silicon chips—continues to improve through innovative design, along the notions projected by Gordon Moore.* These aspects together with the attribute of portability and mobility demand enhanced energy consumption and charging efficiencies for practical and appealing experiences. A balance of tension between the increasing demand for small form factor mobile devices, wireless sensors, etc., and the costs of design innovation, including energy consumption efficiency, to meet the increasing demands will be a necessary sustaining factor in the continuing relevance of Moore's law. Jonathan Koomey observed that for

---

* Mack, C.A., "Fifty years of Moore's law," *IEEE Transactions on Semiconductor Manufacturing*, vol. 24, no. 2, pp. 202–207, May 2011.

a given demand of computing power, the required battery capacity to support this demand for untethered devices continues to decline.*

The rise of functional densities and enhanced energy consumption efficiencies are especially significant in the proliferation of smartphones, tablets, sensors, etc. These trends serve as a catalyst in machine-to-machine (M2M), human-to-machine, and human-to-human communications, under the umbrella of the Internet of Things. The continuing enhancements in computational efficiencies, signal processing, radio-link optimizations, protocol improvements, and system-level architectural evolution have been instrumental in the energy consumption reductions, while simultaneously augmenting experience.

In the storm of a rapidly changing landscape of ideas, innovation, products, and services, the marketplace tends toward chaos—resulting in high costs, limited choice, limited interoperability, and a decline in the consumer experience. The Internet exemplifies the enormity of the challenges and complexities, both from a technology and social perspective, which has been effectively managed and evolved through global standardization initiatives. The collaborative work across the W3C and the IETF together with their various affiliates is remarkably illustrative of the role and benefits to society at large in fostering technology evolution, innovation, and market expansion.

The life cycle of the specification development process has its origins in the exploration of new ideas derived from innovative technology endeavors and market trends. Its leading edge lies in discussion, debate, and consensus around new ideas, concepts, and perspectives toward potential candidates that are suitable for further study and specification. Measures of relevance and significance of the specification development process consist of either interim specifications that promotes either a change of direction or a convergence toward specifications that are adopted in the market. The engagement in the participatory endeavor of specification development spans corporations, industry associations, government entities, nonprofit organizations, and academic institutions. It is the openness of debate and discussion that embellished the quality of specifications through an infusion of interest and differing perspectives for a convergence to satisfy both common and different objectives. The objectives of open standards and open source are complementary. Open-source initiatives are typically associated with an implementation or a deployment. These initiatives are also applicable within the framework of an open standard specification, since the objectives of an open standards and an open-source initiative are inherently orthogonal.

Open standards are based on consensus, where the development of specifications is guided by a variety of diverse, often conflicting, requirements that are driven by regional and global market trends, through debate and discussion. The specifications that result from an open standards initiative are expected to be crafted such that no infringements on implementation-specific aspects occur, although the debate and discussion associated with the development process may cover the implications of an implementation, or a deployment. An open standard specification always contains normative text and in some cases contains informative text to exemplify applicability, through the use of case scenarios. The intent of open standards is to promote interoperability across logical functions, through the use of well-defined reference points, interfaces, configuration options, protocols, procedures, and APIs (application programming interfaces). In turn, these directions promote a ubiquitous framework for market expansion and innovation, while simultaneously reducing costs, independent of business models. Lifestyle enhancements, public safety, regional market and regulatory nuances, augmented implementation, and operational efficiencies are among the prominent drivers that demand diversity, while at the same time demanding a consistency of service experience. Global collaboration is a vehicle for exploring, studying, and satisfying the emerging

---

* Koomey, J., "Outperforming Moore's law," *IEEE Spectrum*, vol. 47, no. 3, p. 68, March 2010.

and challenging requirements. The standards process provides a consensus-oriented platform for global collaboration, where the specifications allow for well-defined interoperable consideration, in a landscape of diverse and sometimes conflicting requirements. This is pivotal in the promotion of a service experience that converges toward consistency in design implementations and system deployments, while lowering costs and expanding markets and revenue opportunities.

The standards process allows a graceful departure from disparate and targeted solutions that are inherently problematic in an increasingly interdependent world, while a realization of its distributed nature is preserved. The smart-grid endeavor embodies the notion of interconnectedness and interdependence of an end-to-end energy generation, delivery, and consumption system. A leveraging of this notion unleashes the synergy across the traditionally disparate administrative domains for flexibility, load balancing, enhanced efficiencies, and new revenue share opportunities through innovation in communications. Mobile communication–enhanced energy delivery systems, under the umbrella of the smart grid, cover a plethora of smart energy devices, in the form of smart meters, energy management, energy usage monitoring, vehicle battery charging, etc. Both near-range and distant-range fixed or mobile communications are relevant in improving the reliability, efficiency, cost minimization, capacity utilization, and resiliency of energy generation and distribution systems. The IEEE 2030 standard is an example of a smart-grid interoperability framework, together with NIST.* The contemporary and changing nature of the role of SDOs, from a strict focus on narrow domain of technology to one of an interdependent, collaborative awareness, is vital for the creation of significant and relevant specifications in the face of unprecedented change inherent in the information age. Liaison relationships with assorted industry organizations are paramount both from an ongoing progression of specification development and from an exploratory perspective, while avoiding a duplication of initiatives.

Standards creation is a process that finds its way in the various facets of technology evolution, in its various incarnations, where its relevance and applicability in mobile communications evolution is universal. This aspect is a reflection of the broad-scope mobile services in the human context, wrapped in virtually unbounded costumes of lifestyles. Akin to the example of the role of standards-driven smart-grid enablement, interoperability in the information distribution and management space is essential to manage and optimize the complexities related to any traditional services—health care, medical treatment, nutrition, education, transport, commerce, trade, banking, sales, and distribution. Wireless interoperability offers new capabilities to envision and enhance market expansion, through technology innovation and design. It is the interoperability and market expansion that incubate new opportunities and competition, while avoiding the constraining forces of commoditization, through the boundless power of ideas. Transformational business models, together with collaborative partnership, incubate ideas spawning the enormous potential for new product and service vistas. Beyond the rubric of interoperability, exploratory thoughts that motivate discussion in the standards-making organizations include the nuances of user experience at the various touch points, from an architectural and procedural perspective. The environmental footprint impacts of the mobile communication infrastructure and practices are an arena that benefits from the energy consumption efficiencies afforded by standardization initiatives, such as in ETSI2.12.12.1[†] and 3GPP.[‡]

---

* http://collaborate.nist.gov/twiki-sggrid/pub/SmartGrid/PAP02Wireless/NISTIR7761.pdf.

[†] ETSI ES 201 554. Measurement method for energy efficiency of core network equipment. (http://www.etsi.org/deliver/etsi_es/201500_201599/201554/01.01.01_50/es_201554v010101m.pdf)

[‡] 3GPP TR 36.927. Study on potential solutions for energy saving. (http://www.3gpp.org/DynaReport/36927.htm)

The standardization process is an integral component of information evolution. It embodies an exercise in the early and incremental resolution of uncertainty, as new ideas are examined and articulated in the form of specifications. It is the contextual nature of the standards process that transcends the specifics that are enumerated in the language of the specifications. This attribute therefore demands the appropriate interpretation of the intent of the specification or specification framework, for a corresponding design of a product or service. The binding veneer of standardization transcends the regional regulatory nuances around the globe. It enables a consistency of behaviors, in product and service implementations, while allowing customization outside the scope of the standardized specifications to suit any regional regulatory constraints or market preferences. The common aspects of architectures, protocols, and procedures, along the theme of oneness, within the standardized specifications are significant in minimizing design costs and operational costs and in market expansion potential, in a global economy.

The relatively nascent era of next-generation mobile computing and communications portends radical transformation of our personal space. It is one where a redistribution and reconstruction of the Internet into an untethered fabric of decentralized interfaces and systems is immanent. M2M communications—where devices connect to assorted communication systems and the physical world—is an arena that enormously benefits from the standardization process, through a systematic handling of the signaling and traffic procedures to promote uniformity in diversity, which in turn manages both scale and complexity. The data traffic generated by M2M applications can be quite different than traffic from the human-centric applications for which today's communication networks were designed. M2M applications typically involve large numbers of devices, small amounts of data per device, heavy signaling loads, and high peak data traffic during events such as power outages and earthquakes. The unique characteristics of M2M applications and their data traffic have prompted a number of standards bodies to develop M2M-specific enhancements to their communications network standards.

The flexibility in the confluence of mobility and packet-oriented multimedia services is paramount in its ubiquitous presence in the evolution of cloud computing. Standardization is both a catalyst and immanent in the evolution of a mobile service–inspired cloud framework. From the wireless perspective, underlying enabling technologies, such as software-defined radio* (SDR) and remote radio head (RRH), are among the emerging actors in the story of cloud-oriented mobile services. An identification of logical interfaces, between intrinsically orthogonal functions, for distribution and autonomy, with a skin of coordination, affords both uniformity and customization. The realizations then naturally separates software and hardware elements of functionality, from a system lens, while focus continues to shift toward software renditions of the overall functionality for flexible and dynamic configurations in a distributed connectivity and service environment—directions in collaborative processing, with standardized plug-and-play capabilities, lending themselves to scale and autonomy.

The base station element in the wireless mobile ecosystem is a relatively high capital investment, where resource allocation occurs depending on the throughput demands imposed by the mobile devices it serves for access to services. The radio-link protocols and procedures are governed by a variety of standards, for both near- and far-range access. As the evolution toward a cloud-oriented mobile information exchange system continues, the distribution of base stations through interconnectivity, such as via the X2 interface in the LTE EPS is pivotal. These directions encourage a placement of the base station closer to the serving gateways, which are typically

---

* Mitola III, L., "Technical challenges in the globalization of software radio," *IEEE Communication Magazine*, vol. 37, no. 2, pp. 84–89, February 1999.

colocated at the data centers, where the RRH would be distributed geographically over a coverage footprint, with variable sizes to manage and balance the demand for coverage and capacity. The wireless technologies within the LTE EPS continue to evolve through the standards process in these directions, with capabilities such as MIMO, with cooperation among multiple base stations, realized through coordinated multipoint (CoMP) transmission and reception. These trends in LTE advanced and beyond are aligned with a cloud model and offer significant enhancements in capacity and resource utilization. The cloud model paves a path toward virtualization over the various segments of mobile communications that span connectivity and service.

Flexibility and adaptability in the exchange of information across the different components of connectivity and service system are paramount in the realization of both M2M and cloud communications. Service-oriented API definitions, for devices and networks, provide a tool chest of such information exchange capabilities, within the connectivity and service layers. For example, in the various flavors of location-oriented applications, the location information (e.g., map application) could be overlaid with the corresponding area attraction information, such as cafes, restaurants, and entertainment. The OMA API framework is engaged in the specification of both network and device APIs. The exposure of network resources and device capabilities is a requisite for promoting innovation applications and services. APIs for cloud computing may be categorized in terms of infrastructure, platform, and software, where each of these are rendered as a service—infrastructure as a service (IaaS), platform as a service (PaaS), and software as a service (SaaS). The IaaS components include infrastructure provisioning, such as virtual machines, and configuration. The PaaS components include interfaces to specific cloud capabilities, such as charging, user profiles, device profiles, content, storage, security, network access, and maps. The SaaS components include procedures, such as ERP, CRM, and customer service, to extend and interface with web applications.

The API framework provides a decoupling of the execution environment in the mobile device or mobile network, from the underlying mobility protocols and procedures, to allow robust interactions between applications and the mobile device or mobile network. The decoupling insulates the applications for a graceful behavior and enhanced resilience—avoidance of a disruption of user expectations and experience—in the event of mobile connectivity degradation, or change.

The publication of a specification is just the beginning of the next step of evolution or an incremental enhancement of the associated features and capabilities. This entails ongoing work to align the evolution of the specifications with the pace of technology innovation and the winds of market directions—regional or global. The forward-looking nature of specifications implies that there is a natural time gap between publication and a realization of implementations. The realization phase involves the engagement of industry forums or organizations, which identify specific profiles of published specifications, to meet the demands of assorted business models and commercial interests. Industry forums, across the various segments of the mobile communication system, serve as collaborative intermediaries between the standardization process and the various product and service implementation for seamless interoperability across a multitude of design realizations. Conformance to standardized specifications is validated in the industry through testing for interoperability; certification of components within the system, such as mobile devices; and statements of compliance. Cooperation across the different industry segments that impact the value chain, regulations, and the standards process on a global scale generates enormous value in terms of optimizing both capital and operational costs, for all parties engaged in product and service offerings, in the evolution of mobile communications.

The publication of forward-looking specifications spurs innovation in a variety of forms, including the creation patent portfolios that in turn promote the licensing and sharing of compelling designs that are attractive for different business models. The resulting dissemination of knowledge and imagination inspires continuing research that augments the momentum of evolutionary directions and in turn opens avenues for the development of specifications that embrace the emerging ideas. These are among the underlying ingredients that propel the spirit of open standardization processes—hinging on cooperation, collaboration, and consensus building—at the edge of information chaos and evolution. A vista—that creates and embellishes the technology-mediated world around us—fertile in ever-changing ways.

2.4 Leadership

The information age symbolizes a techno-economic global epitome. It embodies an era, where change and chaos incubate complexity. The value and uniqueness of products and services hinge on an awareness and utilization of information interdependence. Carlota Perez, a noted scholar on technology and socioeconomic interdependence, postulated the notion of a "techno-economic" paradigm.[*][†] The information mobility era epitomizes incessant change and equilibrium disruption. The absence of equilibrium in technologies, human behaviors, and markets demands strategic perspectives. A strategy of adaptation is one among potential options to manage the disruptive nature of evolution. This approach considers the detection and a rapid reaction to the changing and new technology capabilities. While this option is effective on the fringes, it lacks the vision of sustainability, which demands a different way of thinking, while carefully examining the emergence of technologies.

A more potent option is to take abstract the essence of the emerging technologies, while being flexible on the particulars to synthesize a new vision. The vision can then be applied to craft influential strategies to shape new opportunities and markets, through new technology directions. This is a departure from conventional wisdom, where sameness and difference are creatively blended. Google's advertising scheme and LinkedIn's social networking service are emblematic different ways of thinking in technology, while catering to the universality of human nature. These directions embody a bold vision, as compared to enhancements on the fringes. These strategies are vital in the creation of next-generation mobile specifications, through a creative adoption and introduction of new technology capabilities that shape new ecosystems on a global scale.

Prevailing perceptions of risk and reward have a tendency to be modulated by existing technologies and market demands. On the other hand, forward-looking considerations that are rooted in a robust vision maximize the potential for market creation through innovation in products and services. It is in the realm of forward-looking concepts and capabilities, where the strategic elements of leadership in the standardization process are notable. A reduction of the perceived risk, and promotion of the perceived opportunity through analysis and discourse on the technology capabilities and its potential applicability, is a leadership endeavor within the scope of standards. Loosely described, big-picture views that describe and elucidate the context and implications of forward-looking capabilities and applicability are the seeds that motivate the adoption of a related topic or topics for further definition and standardization. Strategies that shape the standardization of new

---

[*] Drechsler, W., Kattel, R., and Reinert, E.S. (Eds.), *Techno-Economic Paradigms: Essays in Honour of Carlota Perez*. London, U.K.: Anthem, 2009.

[†] Hagel III, J., Brown, J.S., and Davison, L., *Shaping Strategy in a World of Constant Disruption*. Boston, MA: Harvard Business School Publishing Corporation, 2008.

mobile connectivity and service technologies provide unique and unprecedented opportunities for economic expansion through widespread and decentralized innovation.

Among the prominent leadership dimensions with respect to technology strategy are the following:

Identification of market potential and related motivators

Identification of long-term market implications and the nature of potential changes

Identification of a high-level vision that accommodates evolutionary change, while preserving the vision

Identification of social, cultural, and business opportunities that accrue from a proposed forward-looking vision

Identification of communication methodologies within and across standards organizations and the industry

The IETF symbolizes openness as a nonmembership or industry-funded standards organization. Established in 1986, it was an evolution of Advanced Research Project Agency* (ARPA), with the Internet Change Control Board (ICCB) to help manage the Internet program. Among the guiding tenets, David Clark noted: "We reject kings, presidents and voting. We believe in rough consensus and running code." The nature of this openness is instrumental in attracting a global participation from enthusiasts, academia, research organizations, and industry. The advent of the third-generation mobile era marked the entry of wireless connectivity and service aspects within the IETF working groups. This direction highlighted an architectural revolution toward packet-switched IP transport in mobile communications.

Since the advent of the Industrial Revolution era, during the eighteenth and nineteenth centuries and its maturation, during the earlier portion of the twentieth century, where manufacturing was propelled by the utilities of electricity, railroads, automobiles, telephones, television, etc., the Internet has emerged as the single most profoundly transformational technology, in the history of mankind toward the end of the closing decades of the twentieth century. A large part of the commercial success of the Internet is attributable to the behind-the-scene standardization initiative of the IETF. It is an exemplary testimony of the virtually boundless possibilities that appear, when the process is open and voluntary. It is an illustration of the enormous potential that is unleashed when institutionalized constraints that often tend to predetermine or preordain outcomes are avoided. Institutionalized constraints have nontechnology-driven, business-oriented ingredients that may not motivate a broad vision that inspires widely interoperable approaches that are most likely to benefit these institutions in a more sustainable and forward-looking fashion. This is direct consequence of examining the market trends and behaviors to guide the standardization process, in a business model neutral manner. It also opens the possibilities for inferring the texture of the market evolution, such that the standardization process may identify imaginative ideas to align with the market evolution as well as to attractively shape future market directions. The IETF operates under the auspices of the Internet Society (ISOC), which is a nonprofit organization that serves to evangelize the evolution of the Internet. It is from the spirit of openness that the IETF specifications derive their robustness and universality, in the realization of Internet-related products and services.

Liaison relationships between the IETF and the prominent global SDOs, such as 3GPP, 3GPP2, and OMA, among others, are a vital ongoing process in preserving the universality of

---

* IETF RFC902 ARPA-Internet Protocol Policy.

the procedures and protocols at the IP layer and above. The protocols and procedures, while preserving universality, through an IETF endorsement, are adapted as needed to suit the nuances of wireless connectivity and service behaviors. IETF specifications are not bound to legislative regulations, as compared to the legally endorsed and business-sponsored SDOs, such as 3GPP, 3GPP2, and OMA. The openness and universality inherent in the IETF standards are a valuable reference for adoption by the prominent global SDOs. The universal building blocks, made available through the IETF standards, serve as a catalyst for a widespread implementation of the specification developed by the prominent global SDOs.

The International Telecommunications Union-Telecommunications Standardization Sector (ITU-T) has a well-developed relationship with the IETF for mutually beneficial collaborative undertakings, especially with respect to the packet-switched nature of mobile next-generation networks, with IP as the unifying protocol. Cross-pollination across the prominent global SDOs, under the umbrella of the ITU-T, creates an awareness and specification of protocol extensions or new protocols to meet the changing and evolving requirements for a mobile Internet. In the spirit of openness, the submission of proposals is not mandated or policed, but allowed as a consequence of both individual and collective thinking. Interests in specific arenas of connectivity or service technologies that are spawned one community (ITU-T or IETF) are encouraged for presentation in the other, if impacts or benefits are anticipated. This process is productive across other standards-producing organizations as well, in the spirit of collaboration, awareness creation, specification improvement, innovation, and leadership. Only updates, extensions, or modifications to protocols and procedures that affect the network layer are within the scope of the IETF standards, which includes cross-layer aspects—especially with the convergence of wired and wireless IP access networks, for the creation, delivery, and consumption of IP multimedia services.

The leadership role, within SDOs, embraces a variety of attributes that are characteristic and relevant in a variety of human endeavors. Among the notable descriptors, in the process of standardization, are catalytic, transformational, interpersonal, participative, guiding, influential, and inspiring. These are quintessential descriptors in the making of consensus-building skills and effectiveness, where a leader deftly navigates through the fog of diverse and often conflicting views on a global stage, colored by cross-cultural nuances and motley business models. Styles of leadership that allow conflicts to play out, without reaction or interference, have been among the effective strategies, to better understand the conflicting views, toward a natural resolution—typically lend themselves to the production of better specifications. Such a style reflects a leader's capacity to elicit a spirit of cooperation through a selective noninterfering strategy. In this context, the leadership could be exhibited either in a participatory role (a contributor to a specific subject proposal being discussed) or in a moderator's role (a chair or a convener of a specification group).

A system-level view of the mobile communication system, together with the various contextual descriptors, is paramount in a leadership role. This is the domain of broad thinking, with an awareness of the complexities of interdependence, where the trade-offs and impact to specific topics are understood and explained with clarity. The carving of logical architectural models and the identification of pertinent interfaces for interoperability, independent of designs, usage, or deployment scenarios, are vital in the directions toward viable standardized specifications. The independence of specifications requires insights at a level, where the system and its envisioned component behaviors are understood and articulated in the progress of discussions toward the establishment of specifications. This requires the ability to both abstract and generalize the specifics into concepts, to transcend the boundaries of the components of the system. It is the path that reveals the essential ingredients or the nature of the system. These directions serve as the pillars in the foundation of elegant and simple architectural models. The twin attributes of elegance and simplicity serve as significant themes for a realization of innovative, interoperable connectivity and service realizations, while not constraining imagination in the making of standards compatible designs.

The end-to-end functionality and performance of the system are the primary objectives in the formulation of architectural models, interfaces, protocols, and procedures. The clear and orthogonal decomposition of the elemental functional ingredients and their interfaces is vital for robustness, design, and choices for implementation. These attributes facilitate a systematic progress of specification development that comprises Stage 1 (system-level and functional requirements), Stage 2 (architectural models, interfaces, and protocols), and Stage 3 (procedures). A proper crystallization of ideas, within each of these phases, ensures the quality and integrity of the specifications. The vision to anticipate some of the critical implications for Stage 3 phase, during Stage 1 and Stage 2 phases of specification development, is an integral aspect of the leadership role in the standardization process.

The nature of the standardization process, in realizing a vision aligned not only with the existing market patterns, but also one that is concerned with market evolution embedded in forward-looking insights, demands a loosely coupled, decentralized collaborative model. Decentralization, with loose coordination, is vital, to adapt to complex, assorted, and changing market demands. It is in this paradigm that the leadership role, in the standardization process, is that of a catalyst,* where influence is incubated, without rigid, top-down hierarchically oriented directives that tend to stifle innovative thought creation and adaptation, in a dynamic environment. In a decentralized paradigm, overall directions are inferred and applied through loose coordination, which then serves to modulate and enhance the overall directions iteratively.

In the various costumes of mobile evolution, catalysts promote change and openness to maximize the landscape of thought and awareness to allow the introduction of innovative ideas for an adaptation to new challenges. They inspire the standardization process to accept and explore the nature of market directions, while at the same time providing guidance to influence the fabric of market evolution.

Coordination is vital in the changing and evolving complexities of interdependence, in the decentralized nature of mobile communications. The scope of decentralization is vast and spans the mobile communications system both from a factual (the mechanistic aspects of protocol, procedure, architecture, network, and device) and from an experiential (the attractiveness of form, function, and service) perspective. This is in stark contrast to the hierarchically oriented, centralized mobile communication systems of yesteryear in a predominantly voice-centric world. Decentralization and distribution are a natural consequence of evolutionary directions in mobile technology to adapt to a rich, contextual arena of the multimodal nature of human communications and experience. This shift is one where a common fabric of communications—textured with multimedia—contains virtually unbounded possibilities for choices and customization at the level of the individual, while being coordinated for consistency and reliability. The starfish exemplifies a rendition, in nature, of a decentralized nervous system and intelligence, where coordination of autonomous limbs—different components of the whole—while providing a common purpose for movement and existence also ensures a high level of survivability. The behaviors and capabilities of this natural marvel were explored in the Great Barrier Reef in Australia, where an abrupt rise in the population of the starfish was noticed in their coral habitat. To preserve

---

* Brafman, O. and Rod, B.A., *The Starfish and the Spider*. New York: Decentralized Revolution, LLC, 2006.

the coral reef, mechanical methods, such as knives, were used to eliminate* the starfish. These methods were futile, since they resulted in exactly the opposite effect of creating more starfish, since the dismembered limbs had sufficient autonomy to spawn new starfish. Further research revealed that the environmental conditions, such as managing the elevated water temperature and pollution, were effective ways to control the starfish population. Insights from this behavior divulge the characteristics, where the conditions that require coordination are those where the role of a catalyst inspires effective leadership, in the ever-changing horizons of the mobile information age.

The theme of coordination, across seemingly disparate ingredients of mobile evolution—technology, business, and the human experience—is elemental in the role of leadership through the various aspects of standardization in the age of information. Carl Rogers,[†] points to a changing and evolving future that embraces the directions of a coordination-centric leadership. These are rooted in a deep awareness and understanding, within the context of the human dimension. An assertion of the significance of interdependence, in the vanishing boundaries that are inherent in a mobile-inspired virtual world. These interdependent and interpersonal elements of leadership style are reflective of evangelism in the evolution of mobile systems. It is a process of selling a dream[‡]—an assortment idea selectively crafted to address the new challenges and frontiers of change as personalization and experience emerge as the dominant themes in mobile communications.

Evangelism in technology embodies an array of qualities—enthusiasm, zeal, insight, trust, catalyst, creativity, empathy, fervor, vision, and influence. These attributes are woven in an array of costumes of advocacy, consensus building, collaboration, cooperation, technology direction, strategic thinking, and thought leadership. It is a presence that inspires people to act in positive and beneficial ways toward a realization of both specific and broad objectives that transcend conflict and ideological perspectives. It reflects a commitment to the shaping and sustainability of business objectives and the related products and services. In the colorful and manifold terrain of global standardization, evangelism is cardinal. Intrinsically organic, technology evangelism sows and harvests fields of innovative thought through the cycles of mobile communication evolution. It invites distinctive styles (loosely structured and autonomous) that explore the various contextual themes (social, human, and artistic) associated with the implications of the standardization process in changing and emerging patterns of technology-mediated communications.

Self-direction and a drive to inspire perspectives through a provision of clarity and understanding to influence collective views are the hallmarks of evangelism in leadership that moves the standardization process in the making of next-generation mobile specifications. The style encompasses mentoring, coaching, and an appropriate delegation of details—one that inspires and allows. It is a substantive element, in a complex, changing, and interdependent world of information creation and consumption. The ability to rapidly dissect and distill information, across the boundaries of assorted subject matter, through a revelation of the underlying concepts, motivators, and trade-offs is integral to technology evangelism, within the leadership role. It persuades and embraces collective views, while adapting them to specific or customized objectives. The gestalt

---

* "Controlling crown of thorns starfish," CRC Reef Research Centre. http://www.reef.crc.org.au/publications/explore/feat45.html#t5-2.
† Rogers, C., *A Way of Being*. New York: Houghton Mifflin Company, 1980.
‡ Kawasaki, G., *Selling the Dream*. New York: Harper Collins Publishers, 1992.

of information—boundaryless and interdependent in nature—experienced through the lens of mobile communications, demands evangelism that crystallizes and transforms new ideas and thought, in the face of incessant change, akin to the flywheel effect.*

Conformity in variety, while unveiling conceptual and strategic frontiers, is a dominant signature of the mobile connectivity and service standardization process.

---

* Collins, J., *Good to Great*. New York: HarperCollins, 2001.

Chapter 3

Connectivity

Everything that counts cannot be counted.

— **Albert Einstein**

From the beginnings of existence—through the eyes of recorded history—to the present, the inherent need of human beings to connect with one another is both vividly and diversely depicted. Connectedness—enabling discovery and experience of our existence in unique ways, while simultaneously revealing our common heritage—is an ethos of connectivity.

This natural inclination to connect has a rich texture that may be characterized simultaneously by sameness and distinctiveness. The emerging mobility paradigm, virtual connectedness, continues to foster this innate desire in humans to experience the flavors of sameness and distinctiveness through the lens of technology innovation—a global pursuit. It is one that spawns endeavors to promote the introduction and adoption of technologies that mediate connectedness, in a complex and dynamic world, where the time–space separation is frequent, across people, places, and events.

Mobile communications—deeply rooted in the ethos of connectedness—is now pivotal in human affairs. A vehicle promoting unbounded frontiers for usage spurred innovation and information dissemination.

3.1 Moments and Memories

Connections, among people, across localities are depicted in a variety of stories throughout history. Instinctive and insatiable, it is an appetite fueled by a pastoralist heritage. Mobile broadband has allowed an abundance of choices, where connectivity barriers are transcended with time–space collapsed experience. Wanderings have shaped human behaviors and cultures, through history, well before the age of information ushered in following the advent of the Industrial Revolution. For centuries, from a distant past, nomadic behaviors have been practiced by people for a variety of innate motivating factors—ranging from relationships, trade, wars, education, exploration, weather patterns, etc.

Through the ages, curiosity, a search for purpose, has compelled both individuals and groups to surmount the barriers of time and space. Seasonal weather patterns encouraged herdsmen in Central Eurasia* to move hundreds of miles in search of temperate climes. Stories etched in the passages of history, such as the *Travels of Marco Polo* and *The Return of Martin Guerre,*[†] depict this urge in humans well before the dawn of the Internet and mobile communications. Notions of an individual and a collective kinship, enriched through colorful experiences, played out in temporal and spatial dimensions in a celebration of life's journey.

Experiences shape behaviors, mold perceptions, and crafts expectations. Unfolding moments encapsulate experiences, in intensities carved in emotional context or stowed away in memory for playback, displaced in time and space—sometimes spontaneously, sometimes deliberately recalled. Anita Loos, an American screenwriter, playwright, and author, noted, "Memory is more indelible than ink"—a sentiment that captures the nature of human memory and one that is rich in emotional context, a feeling, a time–space collapsed experience.

It is to this experience that connectedness becomes a compelling necessity in the human saga. How does the Internet and mobile communications, in conjunction with the various renditions of information and communications technology (ICT), both allow and complement this experience? The ensuing sections depict the nature and fabric of an evolving next-generation mobile communications landscape. These directions allow a weaving of attractive experiences and a pattern that invites visionary cross-functional ideas and thought leadership to both transcend and leverage the tools crystallized through technology innovation, in conjunction with the cornerstone of human factors.

3.2 Metamorphosis: Voice to Data

Voice communications—a widely adopted mode of human communications—is clearly reflected by the global adoption of voice services over mobile phone. It is beyond cultural and economic boundaries, spontaneously appealing, and inherently organic.

The primal, instinctive communications—encapsulated in sound and synthesized in voice—remains as a significant modality to connect beyond the boundaries of time and space. Acoustic transport for proximity voice communications is a timeless enabler. Wireless transport for long distance voice communications is a mobility enabler. Electromagnetic wave propagation (radio communications), characterized mathematically by James Clerk Maxwell, served as a precursor to mobile voice communications. An early realization of mobile voice communications—by Guglielmo Marconi[‡]—was a radio telegraph transmission from a ship in New York Harbor to the Twin Lights in Highlands, New Jersey. The evolution of radio telephony toward cellular telephony to accomplish better quality and coverage for that coveted form of mobile communications—voice—continued through the latter part of the twentieth century.

The dawn of the Internet and its growing popularity since the latter part of the twentieth century ushered in new usage paradigms (rich in context) of voice and information for virtually unlimited possibilities and experiences in connectedness. These new and yet to emerge uncharted

---

* nomadic_cultures_ref-1.pdf.
[†] Davis, N.Z., *The Return of Martin Guerre*. Cambridge, MA: Harvard University Press, 1983.
[‡] Greenstein, L., *100 Years of Radio*, Speech at WINLAB Marconi Day Commemoration, Red Bank, NJ, September 30, 1999.

paradigms form a compelling basis for the evolution of next-generation mobile communications system—characterized loosely in terms of generational chronologies, such as fourth generation (4G) and beyond. Voice then simply becomes another manifestation of information, in an abundant ocean of embellishments, to enrich and energize human communications. A virtual, data-centric world is woven in a language of connectedness—powered with the freedom to move in time and space.

Computer-mediated communications, shaping the world of ICT, has arrived and is emerging in directions with endless possibilities—many yet to be unveiled. An unpredictability embedded with opportunities is fraught with challenges, as Machiavelli mused, "It must be remembered that there is nothing more difficult to plan, more doubtful of success, nor more dangerous to manage, than the creation of a new system." Yet these barriers are pale in comparison with the possibilities incubated in the appeal of connectedness—a universal persuasion. The evolution of the mobile communications landscape, from the beginnings of cellular telephony (voice-centric) toward a mobile Internet (packaged and transported as data) through the lens of innovation and standardization, continues unabated.

The exchange of information—an inherent social desire—provides an experience of both connectivity and content. While the experience of connectivity holds the fabric of human communications, the experience of content evokes thoughts and emotions that enrich connectedness—a feeling of presence through the chasms of time and space. Amalgamation—blending styles and perceptions—is espousing uniqueness in universality.

Content is central to the theme of mobile broadband, which in turn implies enormous data traffic volumes in the ICT landscape—changes that demand seminal changes in the vision, technologies, and the strategies to navigate the uncharted seas in the age of information. What are the technologies and strategies needed to meet an unprecedented demand for information flows across a myriad of mobile devices and usage patterns across the globe? The approaches to meet these demands hinge on an understanding of the complex implications of the shift from voice to data transport. The approaches are multifaceted (beyond the limitations of technology and business silos) and are rooted in unencumbered thought patterns.

Figure 3.1 shows a global proposition* toward connectivity evolution, where mobility and wireless are pervasive, in a collage of growing adoption patterns.

The enabling mobile system architectures to promote data-centric communications are distinct relative to those inherited from the fabric of voice-centric communications. The implications of this distinction are profound in nature. It is these implications that contain both challenges and opportunities to navigate and leverage a variety of ideas that transcend the traditional barriers—silos of thought. Information propelled through the various modalities of communications of people and things is a naturally growing enabler in the advancement of knowledge and applications across a globally networked society—a virtually unbounded assortment of human initiatives and endeavors. It is a dynamic interdependence among people, places, and things with the potential to harness and distribute ubiquitous well-being, limited only by imagination. This naturally implies new ways of exploring and innovating connectivity across people and things with high-capacity information access or broadband access and mobility. Agility of information processing, such as in the cloud, embellished with individualized context is among the attributes that naturally augment mobile connectivity. The freedom of movement, individualized empowerment,

---

* http://www.gsma.com/connectedliving/connected-life-the-impact-of-the-connected-life-over-the-next-five-years/.

Evolving scope of socioeconomic impact of connected life
(projections over the next 5 years)

One million
The number of lives mHealth will save in sub-Saharan Africa over the next 5 years

$400 billion
The amount saved in 2017 from the annual healthcare bill in developed countries as a result of mobile healthcare solutions

40 million
The number of people in developing countries, equivalent to the population of Kenya, that can be fed each year due to fleet telematics preventing food wastage

One in nine
The number of lives saved in road accidents in developed countries over the next 5 years due to mobile enabled in-car emergency services

A week back every year
Smart commute interventions in developing world cities will give commuters back a whole week's worth of time every year

1.2 billion trees
In developed world cities, smart metering will reduce carbon emissions by 27 million tonnes—equivalent to planting more than 1.2 billion trees

180 million
The number of children in developing countries who will have the opportunity to stay in school between now and 2017

Figure 3.1 A view of the global connectivity evolution landscape.

and a choice-driven lifestyle are among the universally appealing elements of mobile connectivity. This is reflective of an attractive confluence—a promotion of interdependent well-being and transformation across individuals and societies. The seeds of change in this shifting paradigm are foundational in the emergent ICT era, one where humans and things connect anytime, anywhere in a dizzying array of lifestyle-enhancing modalities. The Internet of Things (IoT) is an ongoing realization of a seamless movement of information through a myriad of devices for diverse multimedia services such as telemetry, telemedicine, education, etc.

Usage scenarios—the guideposts to emerging trends—are manifold. Multimedia service experiences in the mobile paradigm—creation and consumption in terms of global mobile data traffic volume—are estimated to exceed 15 EB* per month by 2018. The primary drivers that underscore the growth of global mobile data traffic are the proliferation of video content over mobile devices, together with the widespread adoption of mobile devices. A view of this prolific growth trend is shown in Figure 3.2.

A large portion of this mobile data volume—the video experience—is about 69.1%, as depicted in Figure 3.3.

The twin vehicles—smartphones and portables[†]—are carving new and uncharted pathways for information creation and consumption. The drivers of these trends are the intuitive and ergonomic interfaces embellished with high-quality displays, with attractive screen form factors and high-quality video supported by high-speed mobile broadband access. Studies project that these vehicles of data access, collectively referred to as "smart devices," will be a significant component of mobile data traffic—96 % of all mobile data traffic by 2018*. Figure 3.4 illustrates a view of the

---

* Cisco systems—http://www.cisco.com/en/US/solutions/collateral/ns341/ns525/ns537/ns705/ns827/white_paper_c11-520862.html.
† Wearables, sensors and actuators.

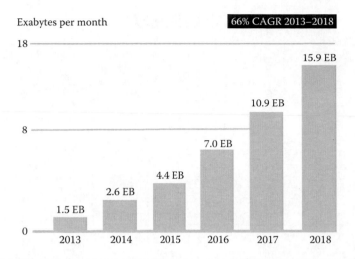

Figure 3.2   Mobile data traffic growth pattern. (From Cisco VNI Mobile, 2014.)

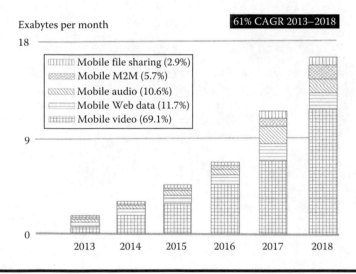

Figure 3.3   Mobile video significance in mobile data growth pattern. (From Cisco VNI Mobile, 2014.)

components of mobile data traffic projections in the mobile information age, providing perspectives on the directions in the emerging data-centric trends.

A shift toward a data-centric world catering to the explosive demand for the creation and consumption of information, mobile broadband is an idea whose time has come. Spectrum* availability and harmonization are critical ingredients for a global adoption of mobile broadband, economies of scale, and innovation. Long-Term Evolution (LTE) and LTE-Advanced

---

* A range of electromagnetic frequencies for a wireless transport of information.

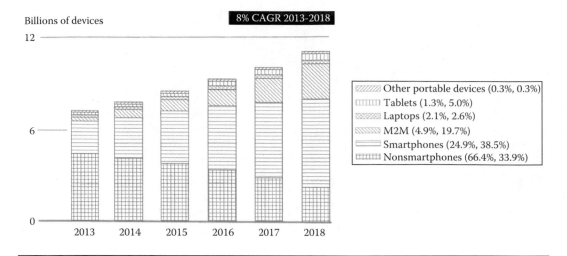

Figure 3.4 **Global mobile devices and connections growth pattern. (From Cisco VNI Mobile, 2014.)**

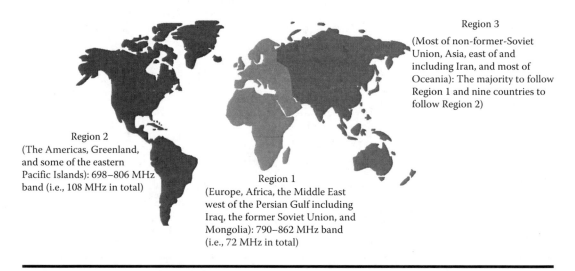

Figure 3.5 **Regional classification of mobile broadband spectrum.**

(LTE-A) are evolutionary technologies with tremendous global interest and are pivotal in the emergence of mobile broadband toward 4G and beyond. On a global basis, the International Telecommunications Union (ITU) has identified three regions* for spectrum harmonization, as illustrated in Figure 3.5.

The widespread appeal of ICT is a testimony of an intrinsic demand for connectedness, manifested in a variety of modes of information—presence, voice, video, and data. Relationships,

---

* http://life.itu.int/radioclub/rr/art05.htm#Reg.

powered through connectedness, are no longer confined to the physical boundaries of meeting places—local or global. Networks of connectedness, characterized by individuals or groups bound by common ideas and perspectives, are not only preserved beyond the circles of places and events but remain vibrant through the artifacts of virtual connections in the age of information. The ICT evolution—a framework for a global locality—continues inexorably in a changing world without boundaries.

Since the dawn of the Industrial Revolution (early portions of the twentieth century), relationships predominantly thrived through the vehicles of transport (automobiles, trains, planes, and ships) to surmount the barriers of time and space. These shifting trends took people away from small close-knit communities to long-distance relationships for both business and personal pursuits—categories of social behaviors native to the human condition. Lifestyle enhancement opportunities, bolstered by technology innovation, allowed for these attractive trends to be adopted far and wide with a growing appeal. In the mix of these profound winds of change, complexities continue to compound as the possibilities to satisfy individual and community needs flourish with an endless potential. The connections tend to become more fluid, with choice-driven and variable degrees of coupling, which in turn fostered new possibilities.

The demands for individual choices and flexibility grew out of new and emerging possibilities for connectedness—empowered by technology innovation and motivated by creative thinking. The compounding complexity to satisfy social demands—carved in the boundaries of groups and communities—together with the technology augmented possibilities for individuals to explore and enact their life's purpose, allowed for a proliferation of individual connectedness. A free-spirited paradigm—akin to the natural world—is embedded in the ageless notion of connectivity in the human saga.

Networks (social networks) such as LinkedIn* sprout islands of connectivity. It is an exemplification of unbounded relationships—flexible, agile, choice driven, and purposeful—where information of sorts traverses wired and wireless highways to suit an insatiable human desire for self-expression draped in a sea of connectedness, and a microcosm that influences widespread behavioral and cultural trends.

In a compelling narrative *The Strength of Weak Ties*,† Granovetter provides a glimpse of the macro implications of micro-level interactions—a principle that fosters a natural diffusion of information and influence in a mobile setting. Weak ties, informal in nature, provide a foundation for diverse perspectives, beyond the well-advertised and prominent ties, which are often widely recognized, endorsed, and trusted. The story of multinational banks, where bankers relied on information provided by weak ties, in studies‡ by Mizruchi and Stearns, found that the well-known pathways of information were limited in terms of insightful inputs, while the informal pathways—characterized by weak ties—added different dimensions that often provided the impetus to successfully close deals with clients. Connectedness at the micro level is implicit, with a high level of personalization and context in the untethered world of mobile communications. Sparse connections elicit diverse viewpoints.

The connectivity layer for mobile broadband access is heterogeneous—an assortment of wireless technologies. A series of concatenated connectivity segments link the endpoints of a

---

* http://www.linkedin.com.
† http://www.stanford.edu/dept/soc/people/mgranovetter/documents/granstrengthweakties.pdf.
‡ http://www.scribd.com/doc/24760976/The-Impact-of-Social-Structure-on-Economic-Outcomes-Mark-Granovetter.

connection. The connectivity layer manages the setup, mobility, maintenance, update, and tear-down of the connectivity segments. The concatenated connectivity segments traverse a variety of wired/wireless paths between the endpoints. The transfer of information between the endpoints engaged in communications or information access is orchestrated by the connectivity layer, through architectural frameworks, functional procedures, interfaces, and signaling messages. Mobility is enabled through a wireless path—the link between the end-user mobile device and the wired access network—in an array of concatenated segments between the endpoints. The IoT—humans and objects—is linked together organically, embellished in universality and uniqueness. Linkages are augmented with a dizzying array of possibilities through the ingredients of diverse but unifying networking, computing, storing, and sensing capabilities. Ambient intelligence—a pervasive connectivity—is a seamless connectivity experience. In this quest, the wireless path—a cornerstone for a mobile experience—along with the end-to-end connectivity chain forms the highways for high-quality, high-bit-rate, low-cost transport of information. It is in this quest that the forward-looking, next-generation mobile broadband systems are being explored on a global scale to provide attractive, lifestyle-enhancing information services for the mobile end user anytime, anywhere. Mobile Internet, orchestrated by the connectivity layer, is ubiquitous and heterogeneous.

Heterogeneous wireless paths interface with the common wired paths for connectivity across endpoints. The common wired paths are established through the de facto Internet Protocol (IP). Distribution is intrinsic to the working of IP, where the survivability of these common wired paths (IP networks) is the dominant guiding principle. This is a shift from the rigid design principles of the traditional voice-only communications networks, which are inherently centralized. The continuing demand for information, exemplified in the exponential rise of mobile multimedia communications, points to necessary shifts in the guiding principles for embracing the dawn of a new ICT era—one where voice is simply one component of human expression and experience. The shift toward multimedia types of communication modalities is realized through a decoupling of the connectivity layer and the service layer.

The new directions, implied by the shifting trends, quintessential in the age of information, point to a dominant theme—distribution. Information access and flow demand a distributed connectivity layer. Decentralization and distribution of the connectivity layer provide the primitives necessary for managing the complexities and the robustness of the wireless and the wired segments between the mobile device and a remote destination. Autonomic capabilities in the connectivity layer—self-configuration, self-healing, and self-optimization—are among the myriad of significant elements enabling information pathways. The diversity and complexity of information access and flow demands a distributed connectivity layer. In the natural world, notions of distributed systems are manifested in the depths of the oceans: blue Linckia* belonging to the family of echinoderms† with complex nervous systems, without a centralized control function—the brain. Their nervous system is a network of interlaced nerves. The evolutionary characteristics of the mobile broadband connectivity layer are akin to the distributed nervous system of these amazing creatures that embrace autonomic principles.

Distribution of connectivity, a mobile story rich in enabling capabilities, is the center stage of ICT, enabling a proliferation of information without boundaries (an indispensable archetype), and one that is transformational and transforming in the human story.

---

* http://www.susanscott.net/OceanWatch2001/may25-01.html.
† http://www.sheppardsoftware.com/content/animals/animals/invertebrates/starfish.htm.

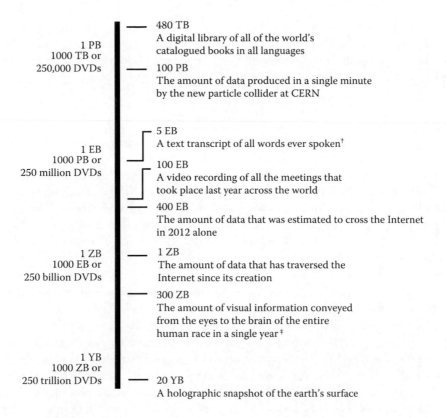

1 PB
1000 TB or
250,000 DVDs

— 480 TB
A digital library of all of the world's
catalogued books in all languages

— 100 PB
The amount of data produced in a single minute
by the new particle collider at CERN

1 EB
1000 PB or
250 million DVDs

5 EB
A text transcript of all words ever spoken[†]

100 EB
A video recording of all the meetings that
took place last year across the world

400 EB
The amount of data that was estimated to cross the Internet
in 2012 alone

1 ZB
1000 EB or
250 billion DVDs

— 1 ZB
The amount of data that has traversed the
Internet since its creation

— 300 ZB
The amount of visual information conveyed
from the eyes to the brain of the entire
human race in a single year[‡]

1 YB
1000 ZB or
250 trillion DVDs

— 20 YB
A holographic snapshot of the earth's surface

† Roy Williams, "Data Powers of Ten," 2000
‡ Based on a 2006 estimate by the University of Pennsylvania School of Medicine that the retina
transmits information to the brain at 10 Mbps.

All other figures are Cisco estimates.

Figure 3.6 A perspective of measures of information traffic volumes. (From Cisco systems, http://www.cisco.com/cdc_content_elements/networking_solutions/service_provider/visual_networking_ip_traffic_chart.html.)

A testimony of the continuing metamorphosis toward a data-centric world is exemplified in the projections for the growing global Internet traffic—a staggering volume* of 1.6 ZB—in the vicinity of the 2018 timeframe. A perspective of the measures of traffic volume is illustrated in Figure 3.6.

3.3 Transport

Ubiquity, a vital artifact, enables connectivity without boundaries. The cornerstone of viability in the next-generation mobile ecosystem is encapsulated in a single word, *interoperability*. It is the ingredient that promotes seamless connectedness in the landscape of mobile communications. It is elemental (richness and diversity) in a field of possibilities, aligned with attributes that

* Cisco Systems. http://www.cisco.com/c/en/us/solutions/collateral/service-provider/visual-networking-index-vni/VNI_Hyperconnectivity_WP.html.

are inherent in the tapestry of human communications, and a cornerstone of the viability of the next-generation mobile ecosystem. It is the ingredient that promotes seamless connectedness, in the usage of mobile communications, while allowing an experience of the richness and diversity that shapes the appeal of mobile multimedia services.

3.3.1 Entities

Mobile networks are an integral component enabling human communications. Mobile devices, keepsake artifacts that adapt to the demands of behaviors, choices, and styles, are a repository of personalized context. Mobile devices and networks collaborate to provide connectivity for human communications—a mobile broadband system. The motivation for these communications is the innate desire among humans for social interactions—one that provides a framework for a network of communities as well as for the dissemination and acquisition of information.

A network of communities, enabled and stitched together by mobile broadband systems, is a stage for communications that both leverages and influences the multifaceted connectivity technologies that are a fabric of mobile broadband systems—a pervasive interdependence. This mutual dependence is one that is rooted in the realm of the tangible (seamless connectivity) and the intangible (evolution of enabling connectivity technologies). This symphony evokes a complementary metaphor: "Not everything that counts can be counted"—Albert Einstein.

The social dimension that is elemental in the creation and sustenance of communities has variable boundaries, local and global, where the former has a texture of neighborhood localities while the latter is universal. Within these categories, the social interactions may be between individuals or groups. The flexibility of boundaries has been enabled through technology-mediated communications such as mobile broadband systems—a multimedia data paradigm. Ubiquitous connectivity is the essence of attractive possibilities, enabling innovation and a rich experience in the saga of human communications.

While ubiquitous connectivity is powered by mobile broadband systems, the archetypes of human communication modalities, entrenched in temporal and spatial boundaries over the ages, color the usage, evolution, and innovation of the technologies behind the scenes. The underpinnings of mobile broadband systems are rooted in the underlying technologies—a framework that is invisible to the user. Yet this framework must be cognizant of the age-old precepts that have governed the affairs of human interaction and concern.

3.3.2 Pathways

The notion of boundaries has faded in the milieu of communities. Explicit boundaries, shaped by proximity orientations such as clubs, restaurants, theaters, cafes, resorts, and the like, comfortably coexist with virtual boundaries, shaped by technology-mediated capabilities. In either case, social interactions, motivated by ideas, interests, choices, and styles, are a seminal fabric and are boundless (strewn in imagination). Technology-mediated capabilities—communications—allow unprecedented possibilities in the realization of social interactions and access to information. Multifaceted in nature, technology innovation and standardization provide a stage where evolution of communities enacts the stories and scenes through the language of social interactions.

The stage of technology-mediated capabilities has a foundation of connectivity—one that is rendered in the form of networks of communication pathways. Mobile connectivity is a pathway constructed with gateways and functions. These gateways and functions provide the links in communication network infrastructure that are a stage in the actualization of virtual communities.

A mobile communication network infrastructure (the stage) that enables social interactions and access to information has morphologies that resemble those of communities. Technology-mediated capabilities have inherited some of the models prevalent in communities in the social context—information transport and transactions. The stage and the actors (communities) engage and influence each another.

3.3.3 Symphony of Choice

Digital in nature, virtual connectivity is empowered by a variety of mobile communication systems and comprised of network infrastructures and mobile devices. This stage in the lineage of mobile broadband of mobile communication systems is also referred to as 4G and is intrinsically a data connectivity paradigm.

LTE is a rendition of mobile broadband connectivity and a symphony of architectural models, protocols, and procedures, which collectively characterize the Evolved Packet System (EPS). With the proliferation of ubiquitous Internet connectivity, the LTE framework for digital connectivity brings mobility in the Internet stage.

The architectural model in the LTE framework provides a blueprint for enabling ubiquitous connectivity, with mobility as the central theme. It embodies the entities and interfaces for connectivity between a mobile device user and a remote entity (another user), or an information source. The pathways of connectivity enable social connectedness, through the stories and scenes that are brought to mobile users, via these pathways invisible to the mobile users. For them, it is the experience that is compelling and memorable.

Protocols provide a structured environment for the creation, consumption, and clearing of pathways, through junctions of gateways, for the passage of information between the mobile device and the mobile network infrastructure. Procedures influence the creation, consumption, and clearing of pathways to suit the movement of information.

The symphony of choices—architectures, protocols, and procedures—is shaped and specified through the lens of standardization, innovation, business models, regulatory policies, and market trends. The virtually unlimited choices are rendered within the context of interoperability through the standards process for global applicability with consistent behaviors. The rich and virtually unbounded possibilities are shaped through a limited set of guiding principles. What are the guiding principles (tenets or conceptual principles) that govern the rendering of rich and diverse possibilities for implementation and deployment of mobile connectivity?

3.3.4 Behind the Scenes

The conceptual primitives and the unchanging tenets are few. These are elemental in nature (abstractions) that allow a field of creative thinking. The abstractions are generalized concepts that are embedded in the field of possibilities. Simultaneously, these transcend the barriers of possibilities—architectural model, protocols, and procedures. They enable the field of possibilities to evolve, unencumbered by the existing field of products and services. Their applicability is cross functional.

"Only connect," an epigram from E.M. Forster's book *Howards End*, depicts the applicability of conceptual primitives, woven in the fabric of the human condition. Connectedness is a conceptual primitive in a field of possibilities (communications, interactions) in the socioinformational landscape. The Internet, with the mobile paradigm at the epicenter, allows the application of the elemental notions to sustain and evolve a compelling communication fabric.

History, derived from the Greek* word "historia" meaning "knowledge from inquiry," provides clues to these hidden conceptual primitives played through the various stories, time and again, delineated in time and space. These gems identified through inquiry and thought pervade across the apparent boundaries in an ocean of experiences, rendered in connectedness and interdependence.

Partitioning, categorization, and formalized structures are some of the facets of recognizable and repeatable patterns across the field of possibilities—architectures, protocols, and procedures. It is in this field of possibilities that the evolution of connectedness is shaped in the domain of mobile communication systems. How does information traverse the links between mobile devices and mobile networks? How do the gateways along these links identify, interpret, and convey information between a source and a destination? Information preservation under mobility conditions in an adaptable manner is a requisite for an untethered service experience. Identification of conceptual primitives over changing market trends is a compelling persuasion in the exploration of new possibilities and implementation strategies. What are the choices of these conceptual primitives in the standardization of architectures, protocols, and procedures? Why must the standardized enabling capabilities be decoupled from designs? How is interoperability achieved? How is elegance (simplicity and robustness) promoted with the use of conceptual primitives? Why are the conceptual primitives significant in the formulation of effective and forward-looking strategies? These unbounded nuggets are prescient in nature.

The answers to these queries are in the conceptual primitives that are critical in unveiling insights for evolutionary directions in a mobility framework of architectures, protocols, and procedures. Pivotal to the field of possibilities shaped by architectures, protocols, and procedures is an inquiry into the conceptual primitives that are invisible scaffolding. These conceptual primitives are explored in the context of LTE—a globally established initiative for the evolution of mobile broadband communications systems. The conceptual primitives are foundational in the crafting of robust architectures, protocols, and procedures in the LTE system. Enablers of mobile connectedness are formulated in standardization, innovation, and openness in a landscape where the constant is change. These enablers are identified and explored to reveal the nature of the hidden conceptual primitives in the context of the LTE system. The implications traverse a plethora of scenarios through life's journey in the emerging interdependence between technology and the human condition.

3.4 Wireless Framework

The dichotomous nature of the world around us provides clues about the formless, invisible, intangible energies behind the world of the manifest. This is an epiphany in the world of technology, where tangible products and services have elevated the lifestyles around the globe. Imagination (revealing new possibilities), on the other hand, delineates boundless frontiers. Ideas thrive in imagination to propel innovation. The emerging, changing arena of mobile broadband communications hinges on technological innovation and understanding of a plethora of concepts. Concepts are the essential building blocks that are framed in ideas, applicable across different manifestations, different services, and different products. Products and services change, while the concepts that brought them to be remain unchanged. Concepts are portable and subjectless powerful

---

* Joseph, B. and Janda, R. (Eds.). *The Handbook of Historical Linguistics*. Malden, MA: Blackwell Publishing (published December 30, 2004), p. 163, 2008.

abstractions that defy boundaries of specific products or services. Mobile connectivity, which hinges on the challenges associated with the probabilistic nature of the wireless information link, ushers new and unbounded horizons for human and machine connectivity.

The milieu of concepts (building blocks) is the essence in the emerging frontiers of technology and innovation. These unchanging primitives in the next-generation mobile broadband access technologies are the building blocks of the LTE wireless access and the complementary packet-oriented (IP) system architecture.

3.4.1 What Is the Heritage?

Traditional cellular systems (voice-centric) in nature did not require wide wireless highways of transport. Voice communications—a primal human communication modality—over space and time, enabled by technology innovation, satisfied the instinctive social need among people across all walks of life. The innate social need is rooted in the essence of human behavioral patterns that promotes new frontiers in emerging technologies. Access-technology enhancements, where voice is a specific type of information in a generalized data transport model, introduce both challenges and opportunities for adoption, while the underlying ideas and creative thought are a catalyst in continuing innovation.

Through the ages, the collective knowledge and wisdom were preserved and propagated by scribes—a practice limited to the literate, passionate few since ancient times. The scribes were masters at both creating text and copying for distribution among the people. In the fifteenth century, with the invention of the movable type, technology innovation automated a part of the scribe's role—copying. The essential aspects of the role of the scribe not only remained unchanged but became more significant as a result of technology innovation, allowing a much wider and rapid distribution of their distinctive works. The tedium of copying was replaced by the movable type gradually, as the application of new capabilities and adaptation to a social need for distribution continued to be augmented, which birthed in the crucible of technology evolution. The scribe's copying role diminished in significance, with social adaptation lagging the capabilities afforded by unprecedented technology innovation. The innovation did not replace the creative and literary mind of the scribe.

The invention of the movable type reveals a universal pattern—the separation of an innovation from the motivators for the innovation. The latter is the source that sparks ideas that create and embellish the innovation to suit a social demand, directly (at the user level) or indirectly (at the access level), in the world of mobile broadband communications.

The advent of data, providing a generalized transport of multimedia information, poses both technological and social challenges. Akin to the impacts of the invention of the movable type, these challenges point to a lag in the adoption of technology evolution, driven by existing dominant social behaviors, where mobile voice communications have been traditional. Scribes preserved their value. In the mobile world, creative thought and innovation in information transport is an implicit value preserved through the transformational nature of technology evolution.

The following are some questions to ponder: Why do varieties of radio-access technologies (RATs) exist? Is it market preference? Is it usage scenarios? Is it open-access differences? Is it pricing? Is it usage convenience? The expansive nature of human thought and behavior suggests that exploration and innovation reveal insights through an interdependent perspective and understanding of these multifaceted questions over evolutionary horizons.

The LTE initiative is an evolution of mobile communications that embodies the all-IP mobile broadband vision. Global participation and collaboration are a significant impetus behind this initiative. Standardization, enabling open interfaces and interoperable procedures is a cornerstone in this venture beyond the incremental steps that have been established since the origins of the third generation (3G) of mobile packet data (Internet mobility–oriented) access. LTE is the next generation of the Universal Mobile Terrestrial System (UMTS) family of technologies beyond high-speed packet access (HSPA). The HSPA vintage of mobile packet access technology has a relatively similar chronology, with respect to 1xEVolution Data Optimized (1xEV-DO), which is part of the cdma2000 family of 3G technologies.

A harmonization of packet-oriented access technologies, under the umbrella of LTE, is a promising direction toward a viable and ubiquitous mobile broadband ecosystem—one where the ubiquity of the network access layer bound by IP provides a common core network framework for converged access (wired and wireless access technologies) while allowing market-driven heterogeneity of access at the radio layer (Wi-Fi, etc.). The common core network framework accommodates a variety of flexible radio coverage footprints (macro cells, small cells). A management of the different and overlapping global spectrum allocations for mobile broadband usage provides economies of scale through standardization and collaboration that serves the common objectives of achieving universal mobile broadband connectivity. These endeavors minimize or avoid chipset variants for mobile device implementations, resulting in a wider adoption across markets with diverse business strategies and models. Globalization trends have ushered in an era of widespread travel—user choice and lifestyle driven—where the role of mobile broadband communications is imbued with a significant evolutionary value potential. This implies a seamless mobile broadband service across markets, geographical boundaries, and time zones. Roaming is where a uniform and contextual mobile broadband service experience is indispensable. Standardization and advocacy play a crucial role in accomplishing these multifaceted objectives.

The wireless interface (LTE environment) is a departure from the ancestral 3G UMTS access technologies, such as HSPA. This departure is characterized by a packet-oriented switching model, which embraces resource sharing and decentralization. As the adoption continues, islands of LTE wireless access are expected to grow in greenfield areas as well as within the existing 3G wireless access areas, during the migration toward mobile broadband (4G) access and beyond. Migration is enabled with flexible and incremental choices, where reverting to a legacy (circuit-switched-oriented technologies) 3G type of wireless access is allowed. The notion of preserving operational compatibility with legacy technologies serves as a bridge, as new horizons of forward-looking capabilities are explored and mined. This strategy promotes a graceful migration of operational systems, guided by business models and market evolution. With the packet-oriented switching model of the LTE ecosystem, the paradigm shift is foundational in allowing a virtually unbounded potential for innovation and differentiation in terms of usage scenarios, business models, and market trends, driven by pricing, human factors, and services. These opportunities promote market directions toward harmonization and interoperability for single wide-area radio-access segment, with LTE as an evolutionary technology of choice, on a global scale.

LTE utilizes orthogonal frequency division multiple access (OFDMA) on the downlink and single-carrier frequency–division multiple access (SC-FDMA) on the uplink. The difference between the uplink and the downlink schemes lies in the multiplexing technique. This is motivated by the need for minimizing the peak-to-average power ratio (PAPR) on the uplink—inherent in the nature of SC-FDMA—for optimizing mobile device designs and battery-life longevity. This enhanced radio-access subsystem interfaces with a flattened packet access network subsystem, referred to as the Evolved Packet Core (EPC). The synergistic combination of these two subsystems provide mobile connectivity, and the integrated system is collectively referred to as the Evolved Packet System (EPS). The LTE radio-access subsystem, with adaptable channels or highways for the transport of information, in concert with the EPC provides different choices for mobile broadband access, where the choice is driven by mobility, pricing, availability, and service usage scenarios.

3.4.2 Motivating Factors

A compelling and unprecedented trend—global mobile broadband adoption and demand revealed in the experiences through the deployment of 3G—marks the foundations of the LTE initiative. Evidence of this trend—sustained via a rise in mobile broadband connections from a few thousand in 2006 to nearly 100 million connections in early 2009—has been accompanied by an enormous appetite for higher data rates and data capacity. This appetite and the attractiveness of deployment flexibility, lower capital, and operating costs have been pivotal motivators for the development of a packetized connectivity fabric characterized by the nature of the LTE system architecture and features. In turn, this intrinsic appeal serves as the fuel for continuing innovation and adoption that are catalysts for the study and standardization of new features. The essence of these features is characterized in terms of meeting the growing market demand of capacity and coverage. For example, features such as spatial multiplexing, through Multiple-Input Multiple-Output (MIMO) antenna technologies, allow the multiplexing of information streams that increase the information rate for a mobile device user or the system capacity for multiple users. Coordinated information transmission over multiple radio-frequency links, between the mobile device and the network attachment points—coordination across multiple base stations, for example, coordinated multiple point transmission (CoMP)—is among the prominent features in the 4G of mobile communications. Topological variations in terms of coverage footprints, assorted RATs, and aggregation of individual radio-frequency carriers, for wider transport channels and coordination across multiple points of attachment for serving a mobile device, are collectively conceived within the framework of heterogeneous networks in the 4G version of LTE—LTE-A. In other words, the framework of heterogeneous networks promotes the optimization of radio-frequency resources across, space, time, and frequency domains together while allowing assorted power requirements and mitigating interference. These enhancements are geared to incrementally enhance the performance of mobile connectivity while bolstering the growing demand for the related capacity and coverage.

3.4.3 Conceptual Underpinnings

The probabilistic nature of a radio link poses a variety of challenges in terms of the integrity of the lanes of information transport between the mobile device and a wireless access network, to which the mobile device is attached at any given time. Nondeterminism creates both inherent value and a field of enormous possibilities (richness in unpredictability). Mobility implications are in stark contrast to the relatively predictable nature of wired systems—a rigid, motionless model of tethered connectivity.

The unpredictability inherent in a radio link translates to the benefit of mobility but demands an identification and understanding of the underlying patterns and behaviors. Benoit Mandelbrot's book *The Fractal Geometry of Nature* is an expose on the reduction of the world around us into repeatable patterns known as fractals—composing a whole. The repeatable patterns are abstracted as mathematical objects in the Mandelbrot set. When applied recursively, these repeatable patterns produce a complex whole—characterized by randomness—inherent in nature. These patterns are applicable in a diverse array of topics, such as poetry, art, geography, music, design, aesthetics, and industrial applications, and present a notion based on self-similarity, which allows complex phenomena to be explained through the use of smaller repeating patterns. In other words, an astute application of unchanging concepts—repeating patterns—produces a complex whole, which is shaped and designed through the identification and use of the unchanging concepts. The radio link—complex in the whole—can be identified, examined, and applied through the use of the underlying concepts, which serve as repeated patterns in a mobile broadband system.

The stochastic nature of the radio link produces a level of randomness or chaos, where both spatial and temporal dimensions produce a variety of complexities. These complexities are reflected as changes from a set of initial conditions that may not be sufficient to accurately predict the state of the radio link, with changes in the spatial and temporal conditions associated with the radio link. Adaptation to these dynamic conditions requires analysis and application through the use of conceptual primitives. Electromagnetic interference impact on a radio link changes with a variety of factors, such as the strength of an intended signal and its proximity to other signals, which appear as interferers to the intended signals. Mobility, over changing physical topologies (natural or artificial structures), adds another level of unpredictability. Effective countermeasures can be leveraged to enhance the integrity of the radio link through the use of elemental concepts (repeating patterns). Indeed, it is a reflection of nature's arrangement of primal ingredients that reveal enormous value potential in simplicity, decentralization, and interdependence within a complex whole, which is composed of self-similar repeating patterns, namely, conceptual nuggets.*

The computationally intensive nature of iterating self-similar, repeating patterns to produce information about the whole has been enabled through technological advances in computing capabilities—continuing development and exemplification of chaos theory. The concepts are constant in change (orchestrated in the universe), while technology advances are propelled through the understanding and application of concepts.

The probabilistic nature of radio links is considered in the evaluation of the conformance of radio elements (mobile device and the base station) in the wireless connectivity system, to the associated core requirements and the related measurement uncertainty specifications. The notion of the shared risk principle is applied for consistent behaviors from an industry practice perspective. The principle entails a relaxation of radio parameter values (e.g., maximum radio transmit power, radio receiver sensitivity) on a case-by-case basis considering the critical factor for system performance and measurement uncertainties.[†]

Advancement in the radio-link capabilities is one of the two significant pillars in the evolution of the LTE system—the other being advancement in the mobile system architecture. At the radio-link level, reduced cost per bit, flexibility in the use of bandwidth, quality of service (QoS), flexible coverage footprints, and increased capacity/data rate are among the significant distinguishing 4G features. Concepts are vital in identifying and establishing the guiding principles and trade-offs for shaping behaviors in the radio-link layer as well as in the mobile system architecture.

The LTE initiative embraces a global vision for wide, high-speed highways in the quest for mobile access to information and services. For access providers, it holds the potential for higher information highway capacity and speed while lowering operational costs. From a user perspective, these enhanced information highways are invisible—it is the access experience that is indispensable. The agents of user experience, IP multimedia applications, are a myriad of applications (service ingredients) crafted by communities of designers and developers. For these communities, wider, high-speed information highways bode well, in terms of providing more opportunities for

---

* Kellert, S.H., *In the Wake of Chaos: Unpredictable Order in Dynamical Systems*. Chicago, IL: University of Chicago Press, p. 32, 1993.
† ITU-R M.1545.

service innovation. These broad yet compelling motivators propel the LTE system to the epicenter of the evolution of next-generation mobile broadband experience, shaping new horizons in the ICT milieu.

3.4.4 Nature of LTE Radio Access

A packet switched paradigm. Flexibility in the usage of available spectrum (a scarce resource) is the fundamental motivator for design requirements that allow variable bandwidth combinations for an aggregate wider band. Radio-access bandwidth segment combinations enable wider bandwidths for access to information. The flexible building-block segments allow a variety of deployments choices, enabling dynamic bandwidths. The component building-block segments may consist of contiguous or noncontiguous bandwidths. This flexibility translates into aggregated low-frequency bands for large coverage areas or aggregated high-frequency bands for higher capacity with small cell deployments in dense usage areas.

The transition of analog TV (ATV) to digital TV (DTV) has yielded additional spectrum in the lower frequencies—700 MHz in the North America—for mobile broadband access. This transition has yielded about 400 MHz of spectrum, referred to as the digital dividend. The benefit of the use of the lower range of frequencies is that a wide coverage is enabled for access. The regulatory policies are motivated to allow both high-range frequencies and low-range frequencies, albeit with varying levels of fragmentation, in the form of diverse interference impacts, spectrum holdings, and the related deployment models that demand innovative coexistence strategies. This promotes opportunities for access providers to organize the access networks to serve both densely and sparsely populated regions—metropolitan and rural. On a broader note, the spectrum of frequencies for both terrestrial and satellite mobile broadband span the range from vicinity of 700 MHz to 3.5 GHz. Among the distinguishing characteristics of the LTE, RAT is the ability to enable a flexible and efficient usage of spectrum—frequency division duplex (FDD) and time division duplex (TDD)—through the use of frequency versus temporal scheduling techniques, for the same type of modulation.

Intrinsically, the quality of a wireless or radio-frequency channel varies instantaneously, in terms of time, frequency, and space. These variations also include those affected by the multiple paths along which radio-frequency energy waves are propagated. Typical methods for mitigating these variations in the quality of a radio-frequency channel, for a steady rate of movement of information over the radio-frequency channel, utilize time–space diversified transmission of radio-frequency energy waves that are the fabric of a radio-frequency channel. In turn, the impacts of the instantaneous radio-channel quality variations are minimized in terms of a mobile service experience. From a service experience perspective (multimedia packet-switched information), the instantaneous radio-channel quality variations are imperceptible. The essence of the LTE RAT hinges on a mitigation of these inherent radio-channel quality variations, in both the time and frequency domains, through the use of OFDM. The principle consists of the use of multiple and simultaneous transmission of narrowband radio-frequency carriers. These transmissions, in conjunction with a cyclic prefix (CP), allow transmissions to be immune to temporal variations in the radio-channel conditions. Consequently, the mobile device receiver (the downlink) baseband processing is simplified since complex channel equalization procedures are not necessary. In turn, benefits accrue in terms of the mobile device cost reduction and an increase in the battery-life longevity. These principles are attractive, especially with respect to the wide radio-channel bandwidths (single or aggregated frequency bands) allowed in LTE. From a mobile device transmitter perspective (the uplink), power consumption efficiency is a critical attribute for battery-life longevity, lower cost, and coverage improvements. The mobile device transmission efficiency enhancement in the uplink is realized through a reduction of spikes in the transmission power and reduction of the peak-to-average transmission power.

The fundamental radio-frequency resource within the LTE access technology corresponds to a bandwidth of 180 kHz that spans a 1 ms subframe, which contains two resource blocks (RBs) of data. Flexibility, in terms of realizing a variety of information transfer rates, can be accomplished by aggregating this fundamental radio-frequency resource, in conjunction with different orders of modulation and radio-channel coding rates. LTE access technology accommodates radio-frequency operation in both FDD and TDD modalities of information transmission and reception. The FDD mode allows both full-duplex and half-duplex information transactions, where the half-duplex arrangement consists of a separation of the transmission and reception of information in both frequency and time. This separation in time and frequency of information transaction simplifies the mobile device complexity in terms of a relaxation of the duplex filter performance requirements, which in turns lowers the mobile device cost. In a spectrum regime, where the availability of spectrum is distributed across different stakeholders (service providers), the usefulness of FDD frequency bands is leveraged through the use of viable duplex filter capabilities to support variations in the duplex gap—separation between the transmission and the reception frequency bands. This inherent flexibility of the LTE access technology is particularly attractive, where the relatively arbitrary nature spectrum allocations are driven by the complexities of a dynamic business, market, regulatory, and spectrum refarming environment.

The use of common radio resource frame structures, with the LTE access technology, lends itself to flexibility, independent of frequency bandwidths and spectrum allocations. Further, a high level of radio resource frame structure commonality is preserved across the two different duplexing modes—FDD and TDD. These ideas are reflective of mining foundational concepts, unchanging within a fabric of forward-looking technologies.

The utilization of OFDM technology within the LTE radio interface enables its innate resilience to interference impacts and bandwidth flexibility for enhanced information throughput. This is a prerequisite for a data-centric mobile world. These attributes have been instrumental in the adoption of OFDM technology in 4G and beyond radio-link technologies. Furthermore, industry experience in the use of OFDM techniques, in recent wireless technology implementations (e.g., Wi-Fi), have established its value as an interference-resilient modulation technique for high data rate transport. Much like the blue Linckia,* a powerful metaphor from the natural world, OFDM embodies the themes of distribution and decentralization of information transport.

In the move toward seamless access (a mobile Internet vision), OFDM has emerged as the radio-access modulation of choice and an elemental building block within a wireless mobile connectivity system. Diverse access technologies share different flavors of an OFDM framework. On the other hand, nonorthogonal multiple access (NOMA) methods[†], with interference cancellation considerations hold promise in the quest for improved radio-frequency spectrum efficiencies in the exploration of future radio-access capabilities, for mobile connectivity.

---

* Brafman, O. and Beckstrom, R.A., *The Starfish and the Spider: The Unstoppable Power of Leaderless Organizations.* New York: Portfolio, 2006.

† Saito, Y., Kishiyama, Y., Benjebbour, A., Nakamura, T., Anxin Li; and Higuchi, K., "Non-Orthogonal Multiple Access (NOMA) for Cellular Future Radio Access", Vehicular Technology Conference (VTC Spring), 2013 IEEE 77th DOI: 10.1109/VTCSpring.2013.6692652, 2013, 1–5.

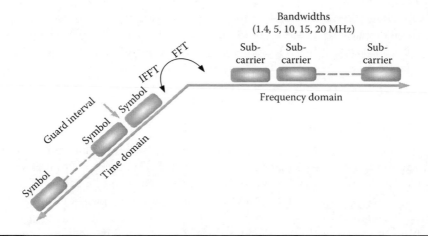

Figure 3.7   Time and frequency domain representation of an OFDM signal.

A defining characteristic of OFDM is the transport of multiple symbols of information, assembled in the form of quadrature phase shift keying (QPSK), 16-QAM (quadrature amplitude modulation), or 64-QAM information modulation schemes. These symbols are contained within each resource element of QPSK, 16-QAM, or 64-QAM output. These symbols are transported in an OFDM symbol represented by multiple orthogonal frequency carriers referred to as subcarriers. The spacing of these subcarriers is selected to minimize intersubcarrier interference at a mobile device receiver. The frequency and time domain representation of an OFDM signal is depicted in Figure 3.7. The fast Fourier transform (FFT) mathematical procedure transforms the time-domain function into a corresponding frequency-domain function. The inverse FFT (IFFT) performs the transformation in the reverse direction, from the frequency-domain function to the time-domain function.

The modulation technology on the downlink is OFDMA, which is essentially OFDM with multiple accesses, and implies that downlink data can be accessed by multiple mobile devices. CPs are used with the OFDMA in the downlink and SC-FDMA in the uplink, for avoiding susceptibility to intersymbol interference (ISI), resulting from multipath reflection delay spread of wireless signals. The orthogonal nature of the component information carrier signals of an OFDM signal transmission avoids mutual interference among the components. The closely spaced component carrier signals are modulated with low data rate baseband information signals that are transported over the carrier signals. This is accomplished by aligning the spacing of the component carriers, with the symbol period, which is the interval. If the CP is not longer than the duration of the multipath reflection delay spread ISI is not mitigated, while if the CP is too long then the increased overhead reduces the capacity of information throughput. The generation of the CP is delineated in the physical channel and modulation specifications.* The spacing of the component carriers, with the periodicity of a symbol, results in an integral number of symbol cycles during the demodulation process. This implies a zero cumulative sum of mutual interference. The baseband information transported, distributed over the component carriers, is retrieved in the demodulation process.

---

* 3GPP TS36.211.

Error correction techniques enable information reconstruction possibilities, such as where multipath signal propagation effects result in the loss of some of the component carriers. Since the information is transported at low rates, across multiple component carrier signals, the impacts of intersymbol interference (ISI) and reflections can be surmounted to preserve the integrity of the information.

3.4.4.1 OFDM CP

A group of consecutive OFDM symbols constitute an OFDM frame. Downlink subframes and uplink subframes represent the composition of an OFDM frame. The subframes are laden with information experienced by a mobile user. A conceptual depiction of the Evolved Universal Radio Access Network (E-UTRAN) distribution of radio resources (frequency and time domains) and mobile device users in the LTE ecosystem is illustrated in Figure 3.8.

For each downlink transmission, a mobile receiver detects multiple copies of the transmitted signal, via different paths, through environmental signal reflection and scattering, with the corresponding differences in the arrival times at the receiver. This delay spread, resulting from multipath fading effects, raises interference impacts on signal reception. A significant evolutionary aspect in the LTE radio interface, relative to 3G renditions in the UMTS family, is that the data streams are distributed and transported simultaneously over multiple low data rate subfrequency carriers. This strategy allows an overall high data rate signal transmission while mitigating the adverse impact of multipath reflection delay spread through the use of low data rate, subfrequency carriers. Additional subfrequency carriers can be used for higher overall transmission data rates, independent of the interference impact of delay spread, since each subfrequency carrier has the same, fixed low data rate.

On the uplink, the modulation technology is SC-FDMA. The reason for this choice for the uplink is the inherently high PAPR in the OFDM scheme, which is detrimental for the longevity of battery life in a mobile device. The single frequency carrier operation in SC-FDMA—also referred to as discrete Fourier transform (DFT)-spread OFDM—reduces the PAPR, relative to the OFDM scheme used on the downlink, which then leads to a reduction in mobile device complexity, together with enhanced power-consumption efficiencies.

Signal transmission from the mobile device is power intensive with the power amplifier (PA) being a significant power consumer. The PA element in the mobile infuses the necessary energy for signal transmissions to be detected by the network. Since the PA is arguably the largest energy

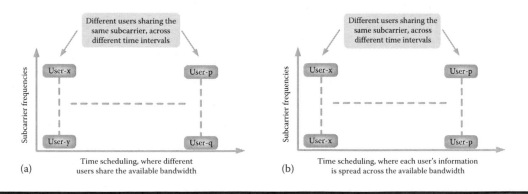

Figure 3.8   Conceptual view of OFDMA and SC-FDMA in the E-UTRAN interface. (a) E-UTRAN downlink (OFDMA scheme). (b) E-UTRAN uplink (SC-FDMA scheme).

consuming element in the mobile device, a high PA operational efficiency is of paramount significance. This efficiency is impacted by the type of modulation format and the signal format. Information bearing signals that have a high PAPR are not naturally amenable to the linear operation of the PA. The peak value of the amplification of the information bearing signal determines the power consumption of the PA, which is subject to the silicon implementation constraints, in terms of process and temperature conditions. To minimize the PA power consumption, it is of critical importance that the PA is operating as efficiently as possible. This efficiency is impacted by the type of modulation format and the signal format.

Since the data transmission rates are no different between a peak power signal transmission and an average power signal transmission, it is essential to keep the PAPR low, which is the case with SC-FDMA, as compared with the OFDMA modulation scheme, which enables higher data rates but incurs higher power consumption. The SC-FDMA modulation format leverages the low PAPR nature of single-carrier systems, together with resilience to multipath interference and the flexible use of subcarrier components for information transport, inherent in OFDM systems. Information bearing signals that have a high PAPR are not naturally amenable to the linear operation of the PA, which limits the efficient operation of the RF amplifier. Consequently, a modulation format that has an almost constant power of transmission is desirable other than OFDM, which inherently has a high PAPR. A high PAPR is less problematic for the base station since battery life limitations are not relevant, although optimizing PA power consumption is important for proper operation. From a mobile device perspective, a nearly constant transmission power modulation format is very desirable to conserve power. This is the motivation for the use of SC-FDMA in the uplink LTE radio access.

The use of noncontiguous subcarriers for the uplink SC-FDMA transmission accommodates a flexible uplink frequency selective scheduling for enhancements in the radio-access system performance.

From a mobile device performance perspective, the longevity of battery life is critical for an enhanced untethered experience. The processing demands on power consumption are required to be astutely optimized for prolonging the stored energy availability. Mobile device designs require a balance between battery life and PA power consumption, where the two metrics are inversely related. Strong channel coding, equalization techniques at the receiver, and predistortion schemes at the transmitter mitigate the impacts of nonlinear distortion introduced by the PA. These schemes reduce the signal-to-noise ratio (SNR) loss for a given bit error rate (BER). A low PAPR translates to a relatively constant uplink transmission power, which enables highly efficient PA designs, within the battery-power constraints of the mobile device.

The standardized frequency bandwidths for the LTE radio link in megahertz (MHz) are 1.4, 3, 5, 10, 15, and 20, with wider bandwidth attainable through bandwidth aggregation in LTE-A (4G). The subcarriers in the OFDM signal are spaced at 15 kHz apart, which translates to a symbol interval of $1/15000 = 66.7$ μs. This implies that each component carrier transports information at the rate of 15 kilo symbols per second or 0.015 (Msps) mega symbols per second. The relationship between the data rate (bit rate) and the baud rate, which assumes 1 bit per symbol, is data rate = baud rate * $\log_2 Q$. For a 64 QAM, $Q = 64$. In the case of a 64 quadrature amplitude modulation (QAM), the number of bits per symbol is 6 since 2^6 allows for 64 different symbols with a 6-bit representation. The number of subcarriers (channels) in a 20 MHz bandwidth is 12 subcarriers/RB * $100 = 1200$ subcarriers. Therefore, the maximum symbol rate, over the 1200 channels, would be 0.015 Msps * $1200 = 18$ Msps. For the 64 QAM case, the overall bit rate, including overhead, is baud rate * $\log_2 64 = 18 * 6 = 108$ Msps. The overhead includes signaling and coding, and gains in the rate are afforded by techniques such as spatial multiplexing.

Table 3.1   Channel Bandwidth and RBs

Channel bandwidth (MHz)	1.4	3.0	5.0	10	15	20
Number of RBs	6	15	25	50	75	100

The initial vintage of LTE systems enabled peak data rates for the downlink access in the vicinity of 100 Mbps in a 20 MHz channel. This is subject to further augmentation through the use of a higher-order MIMO antenna configurations (e.g., 2 * 2 MIMO). The latency targets are estimated to be in the sub-100 ms vicinity for the setup of an access network connection. The data transfer latency over the radio-access network portion of this connection is in the sub-10ms vicinity. Further latency improvements toward sub-5ms are feasible, with low levels of hybrid automatic repeat request (HARQ) block error rate (BLER). The physical layer of the radio-access interface is organized in a bandwidth agnostic fashion, allowing a variety of cumulative bandwidths, up to 20 MHz and beyond. The radio resources are composed of RBs, where each RB consists of 12 subcarriers, with 15 kHz spacing between each RB, and time duration of 1 ms for each RB. A dynamic allocation[*][†] of RBs for users harnesses multiuser diversity[‡] gain—both in the frequency and time domains, where the RB allocation is influenced by the channel condition. Performance improvements are realizable through an allocation of carriers, where the best channel conditions are being experienced, under different mobility scenarios. This capability leverages adaptive modulation and coding (AMC), in conjunction with hybrid automatic repeat request (HARQ), conceptualized in a subsequent section. For example, Table 3.1 illustrates the relationship between the channel bandwidth and the associated number of RBs.

The radio-frequency resource—physical resource or physical layer—can be represented as a time-frequency grid, with specific attributes in the frequency domain and time domain. In the frequency domain, the spacing between each of the multiple orthogonal subcarriers is $\Delta f = 15$ kHz. The OFDM symbol duration time is $1/\Delta f + CP$. The CP enables the preservation of orthogonality among the subcarriers, for a radio channel dispersed in the time domain. They serve as a guard time interval between the subcarriers to avoid ISI, such as from a multipath fading condition. There are two types of CPs in the time domain—the normal CP (4.7 μs) and the extended CP (16.7 μs).

The signal power in an OFDM signal can reach a peak of the average power of each subcarrier multiplied by the number of subcarriers. It is the instantaneous voltage addition that results in a power peak. The peaks of OFDM signal power raises the level of intermodulation distortion, which in turn raises the information BER, while also adversely impacting the behavior (nonlinearity) of the signal PA resulting in potential signal distortion. It then implies that a minimization of PAPR, through an avoidance of power peaks, promotes improvements in the SNR and in the behavior of the signal PA for signal transmission. For mobile devices, a reduction of PAPR allows for enhanced signal transmission efficiency while optimizing energy consumption to promote a longer battery life.

The fabric of radio-frequency resources that constitute the highways of connectivity between mobile devices and the radio-access network is defined such that the bandwidths can be flexibly

---

[*] 3GPP TS 36.211.

[†] 3GPP TS 36.213.

[‡] Jinping, N., Daewon, L., Xiaofeng, R., Geoffrey, Y.L., Fellow IEEE, and Tao, S., "Scheduling exploiting frequency and multiuser diversity in LTE downlink systems," *IEEE Transactions on Wireless Communications*, vol. 12, no. 4, pp. 1843–1849, April 2013.

arranged to enable various spectrum allocations, associated with system deployment scenarios. The building blocks of entities, consisting of radio-frequency resources, dispersed in time and space that weave the fabric of wireless information highways between the mobile device and the E-UTRAN, are illustrated in Figure 3.9.

The product of OFDM symbols in the time domain and the OFDM subcarriers in the frequency domain constitute a radio RB. A single RB spans 180 kHz (12 subcarriers) in the

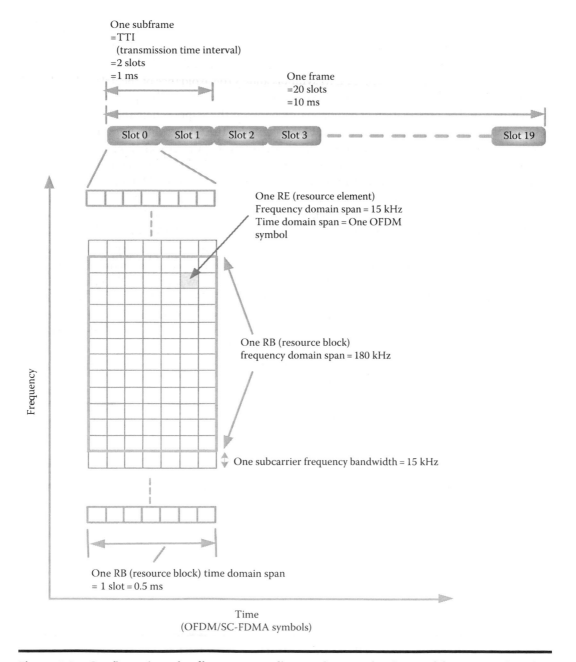

Figure 3.9 Configuration of radio resources dispersed across the time and frequency domains.

A selected modulation choice depends on the SNR conditions. The lower-order modulation formats, such as QPSK, do not require a high SNR but have a lower information rate (throughput). The high-order modulation techniques are applicable under good SNR conditions.

Higher-order modulation schemes, such as 256-QAM, provide capacity improvements for small cell deployments, under good radio link conditions, and low error vector magnitude (EVM) values in the PA. Improvements in the linearity of the PA operation, through digital predistortion (DPD), enables a reduction of EVM. For example, Table 3.1 illustrates the relationship between the channel bandwidth and the associated number of RBs.

3.4.4.4 Subcarrier Modulation

The different signal constellations that may be used to modulate the subcarriers in a composite OFDM signal are QPSK, 16QAM, and 64QAM. Each of the signal constellations allows specific number of bits per symbol. For the high-order signal constellations, since the signal points in the constellation are closer, relative to the lower-order signal constellations, the transmit power is relatively higher. The addition of multiprotocol encapsulation–forward error correction (MPE-FEC) to the outer layer of FEC coding allows for a containment of the transmit power requirements for the higher-order modulations. For example, the use of MPE-FEC allows the use of 16-QAM at a transmit power, which is equivalent to that of a lower-order modulation (QPSK) without MPE-FEC.

3.4.4.5 Differential Modulation

In a differential modulation scheme, the transported data bits induce a shift in the phase angle of the subcarrier, which occurs relative to the phase angle transmitted on the same subcarrier in the preceding OFDM symbol. This type of modulation scheme utilizes noncoherent detection, which is different from coherent detection used in LTE, where reference symbols are required for channel estimation. While noncoherent detection avoids the overhead of reference symbols, there is a trade-off between overhead reduction and the SNR performance impairments associated with noncoherent detection. For example, the data bits 01 would shift the subcarrier's phase by 90°. The benefit of the differential modulation scheme is that it allows the receiver to only determine the phase difference between the current and the previous phase angles of the subcarrier. The benefit is that the overhead of reference symbols for channel estimation are avoided in this noncoherent detection mode, leading to higher spectral efficiencies. On the other hand, there is performance degradation relative to the coherent detection mode, which renders differential modulation less than adequate for mobile connectivity.[*]

3.4.4.6 Coherent Modulation

In this type of modulation, the data bits are mapped to a specific constellation point. In this type of modulation, the data bits are mapped to a specific constellation point, for coherent detection and processing. For example, the data bits 00, in the case of QPSK modulation, are always mapped

---

[*] Markus, H., Michael, S., Stefan, S., and Markus, R., "Performance evaluation of differential modulation in LTE-downlink," in *20th International Conference on Systems, Signals and Image Processing (IWSSIP)*, Bucharest, Romania, July 2013.

to a phase angle of 45°, while the data bits 01 and 10 are respectively mapped to the phase angles of 135° and 225°. The benefit of the coherent modulation scheme is that unlike the differential modulation scheme, it does not incur a 3 dB SNR overhead.

Enhanced spectral efficiencies and higher bit rates are afforded by the use of 16-QAM and 64-QAM signal constellations, within the coherent modulation scheme.

3.4.4.7 Spatial Diversity

The nature of signal propagation is such that multiple signal paths occur as a result of reflections. These multiple signal paths are leveraged, using the MIMO technique—using multiple spatially separated antennas—for throughput improvement opportunities. The spatially separated antennas allow the recognition of signals appearing along different paths. For example, the antenna distributions may be configured as 2 * 2, 4 * 2, 4 * 4, and 8 * 8 combinations. The physical form factor demands of these antenna distribution configurations may be accommodated relatively easily in the case of base stations. It is quite a different story—in the case of a mobile device, with the inherent geometry constraints—since MIMO requires the antennas to be spaced at least half a wavelength apart.

The information transfer rates—base station to mobile device—hinge on the modulation order and the antenna configuration. For example, the application of higher modulation orders and antenna configurations to the nominal information transfer rates for the baseline (initial) release of LTE increases the peak data rates, as follows:

- Peak downlink speeds with 64QAM and different antenna configurations
  - 100 Mbps (SISO); 172 Mbps (2 * 2 MIMO); 326 Mbps (4 * 4 MIMO)
- Peak uplink speeds with different orders of modulation and antenna configurations
  - 50 Mbps (QPSK); 57 Mbps (16QAM); 86 Mbps (64QAM)

The benefits of multiple antenna configurations together with high orders of modulation are also beneficial in terms of radio-link robustness and capacity for the following baseline ingredients of LTE-enabled mobile connectivity:

- Duplex, bidirectional information transfer, modes include both FDD and TDD.
- Mobility speeds supported by radio access include low speed (e.g., 10 miles/h) and high speed (e.g., 75 miles/h).
- Transport channel transition latencies—idle to active—in the vicinity of 100 ms; transfer of small packets (e.g., voice) of the order of 10 ms.
- Modulation type for downlink and uplink: QPSK, 16QAM, 64QAM.

These nominal targets and expectations are intended to set the stage for enhanced radio access—allowing a variety of IP multimedia information transport demands, in terms of appropriate levels of QoS (e.g., conversational voice, interactive applications such as gaming, and peer-to-peer applications.)

The benefits of multiple antenna transmission techniques include benefits to system performance and service experience. Within LTE access technology, MIMO capabilities may be categorized as transmission diversity and multiple information stream transmission, together with beam forming as an auxiliary feature. Deep signal fading conditions observed at a single antenna are mitigated, where the multiple antennas serve as portals—transmission diversity—for averaging the signals received via multiple antennas. Transmission diversity within LTE is based on space-frequency block coding (SFBC)* in conjunction with frequency switched transmit diversity (FSTD),[†] for example, in the case of multiple antenna ports. Typically, diversity in transmission is utilized over common downlink radio-frequency channels, where channel-dependent scheduling (system load and channel conditions) cannot be leveraged. In the case of latency-sensitive services, such as voice over IP (VoIP), where the preservation of user-experience is vital, transmission diversity is attractive since channel-dependent scheduling would add overhead, relative to the small size of the voice packets. These aspects of transmission diversity imply the potential for improvements in both system capacity and increased base station coverage footprints. The transmission and reception of multiple streams of information, through the use of multiple antennas at the base station and at the mobile device, provides a model for parallel bidirectional information transactions over the same radio-frequency channel. For instance, with four antennas configured at the base station and at the mobile device, a fourfold increase in the information transfer rates is feasible. This potential for an enhancement in the information transfer rates hinges on the size of the coverage area and the serving base station load. In the case of small cells—small base station coverage footprint—and light load conditions, the use of multiple antennas promote parallel bidirectional information streams and high information transfer rates are realizable. In other words, this also implies improved radio resource utilization for providing mobile connectivity. On the other hand, in the case of large coverage footprints and heavy load conditions, the availability of the basic quality of the radio channel is constrained and limits the loading of multiple information streams to preserve the basic quality of the radio channel. In such scenarios, the multiple antenna transmissions are used for beam-forming a single stream of information, to improve the information quality, over the radio channel.

Both the peak and the average information transport speeds—over the radio-access highways—are subject to augmentation through the use of higher-orders of MIMO and beam-forming. Augmentations—in these highways of information transport—include base station edge performance and resilience to downlink/uplink information transfer asymmetry, intercell/intracell interference, and multipath reflection delay spread. Beam-forming is an attractive technique—in concert with MIMO—for base station edge performance augmentation, including spectrum utilization, under adverse radio-signal conditions.

MIMO schemes are widely used, allowing enhanced robustness through the use of the diversity of the received signal or, alternatively, higher throughput (using spatial multiplexing techniques). The transfer of information copies over multiple paths has beneficial implications such as enhanced information integrity. Elements of MIMO techniques consist of the following types of functions: spatial multiplexing, precoding, and diversity coding. Each of these function provides the capabilities to enhance spectrum utilization, signal integrity, and throughput—all catalysts in promoting information transport efficiencies for radio access.

Multiple path transport has two adverse phenomena—ISI and selective channel fading. In the case of ISI, OFDM symbols may overlap in the time domain with neighboring OFDM symbols. The concept of a CP—a cyclic extension of the OFDM symbol—introduced earlier serves to minimize the potential for ISI. Selective channel fading results in different channel characteristics—amplitude and phase—for each subcarrier. The channel characteristic variations are subject to compensation techniques at the receiver. Measurement of channel conditions and the detection of neighboring base stations and channel estimation are other procedures at the receiver. Procedures

---

* 3GPP TS 36.101.
† Torabi, H., Jemmali, A., Conan, J., "Analysis of the Performance for SFBC-OFDM and FSTD-OFDM Schemes in LTE Systems over MIMO Fading Channels", *International Journal on Advances in Networks and Services*, vol. 7, no. 1 & 2, 2014, http://www.iariajournals.org/networks_and_services/

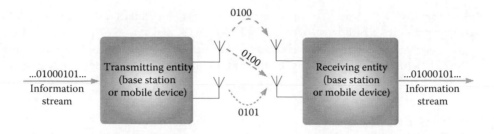

Figure 3.12   Multiple information copies over diverse wireless paths of signal propagation.

at the receiver for properly recovering the information also include link adaptation (LA), channel estimation, frequency synchronization, and symbol timing accuracy.

The nature of wireless signal propagation allows the transfer of a multiplicity of information copies over multiple paths—over the radio interface—between a source and a destination, as depicted in Figure 3.12.

The MIMO approach exploits spatial diversity and spatial multiplexing. This enables different information streams to be transmitted and received simultaneously, over a multiplicity of transmit and receive antennas. The resultant simultaneity of information stream transactions yields enhanced information presentation experience for the mobile device user, through accelerated information transfer rates. From a system perspective, the network capacity—base station capacity—is enhanced through improved efficiencies in information transactions between the mobile network and the mobile device. Knowledge of the radio-frequency channel conditions, through feedback information from the radio receiver to the radio transmitter, facilitates the MIMO technique to leverage beam-forming to further optimize the information transfer rate and the efficiency of radio resources and spectrum utilization.

The performance requirements for LTE radio access hinge on the use of MIMO techniques for the transmission and reception of information. Higher-order antenna configurations for MIMO are allowed that involve multiple transmit and receive antenna configurations to promote spatial information transfer diversity—for improvements in system capacity and cell-edge information throughput. Closed-loop MIMO with codebook-oriented linear precoding can be applied on the downlink. This technique enables spatial multiplexing with dual-code word transmission, together with a fast rank adaptation. Further, an Alamouti-type transmit-diversity technique referred to as SBFC is available on the downlink. On the uplink, a multiuser (virtual) MIMO technique is included to promote capacity enhancement. The multiuser MIMO (MU-MIMO) enables pairs of spatially near-orthogonal users to transmit concurrently on the same PRBs, resulting in uplink capacity improvement.

The evolution of LTE radio access marks a milestone in the ceaseless demand for human beings to communicate and come together—a nomadic pursuit in a virtual landscape—to enrich life's experiences, in the evolving, changing seas of information. Beyond the inception of the foundational LTE technology framework specifications, commonly referred to as Release 8,* the innovation, the design, and deployment initiatives continue unabated toward new horizons—a treasure trove of new possibilities, driven by unbounded thought and imagination, and an intersection weaved in creative thinking and analytical endeavors. The implementation and deployment

* 3GPP—Third-Generation Partnership Project specification landmark: *A global standardization initiative.*

scenarios will serve as a realization of the ideas and concepts embedded in the mobile broadband radio-access capabilities that characterize the LTE system. A realization of benefits offered span performance, operational simplicity, and economies of scale. These benefits render a compelling value proposition for investments. It is an incubator—attracting innovation—ensconced in a vibrant mobile connectivity ecosystem.

3.4.4.8 Error Correction

The FEC coding scheme is a significant function over the radio link in addition to the modulation function. The benefits of the enhanced error correction techniques include higher information transfer (data rates) rates, lower transmission power (translating to improved battery life for the mobile device), and signal integrity at the radio receiver. Figure 3.13 illustrates the processes in a model for FEC. Between the transmitter and the receiver, the encoding of data provides enhanced levels of robustness against signal corruption, as a result of radio-frequency signal propagation behaviors, such as fading, scattering, reflections, interference, delay spread, etc.

The use of FEC codes allows error correction at the radio receiver. An additional level of error correction redundancy promotes better error correction capabilities at the radio receiver. A significant parameter of the FEC code is Rc (code rate) expressed as k/n, where k is the number of input symbols and n is the number of output symbols. For example, a convolution coder with a code rate of 2/3 has 2 input symbols, 3 output symbols (output from the encoder), and a parity bit of one. The redundant symbol added is (3 – 2) = 1. The fractional redundancy added is 1/3.

In Figure 3.14, the benefits of FEC are illustrated in terms of BER and SNR improvements. The coding gain exemplifies a preservation of the signal BER for reduced SNR conditions, during the propagation of the signal from a transmitter to a receiver. The BER improves as redundancy is increased in the encoding stage of data transmission.

The encoding process results in the inclusion of redundant bits to the bit stream that contains the original information. The encoded bit stream, which is referred to as a code word, is decoded in the reception process to extract an estimate of the original data. A larger redundancy introduced in the encoding process reduces the susceptibility to corruption impacts, with the trade-off that the actual data throughput is reduced together with increased latencies.

For real-time application data, this trade-off between latency and integrity must be balanced for an acceptable performance, in conjunction with the application layer enhancements to optimize the user experience.

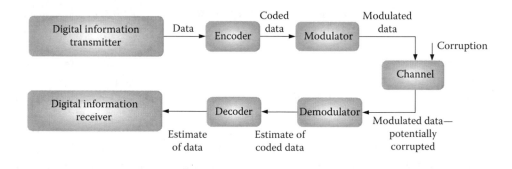

Figure 3.13   FEC model.

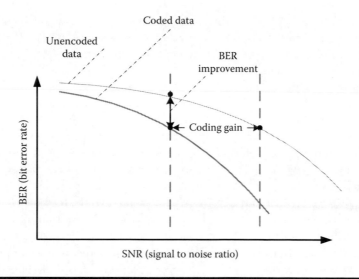

Figure 3.14   FEC benefits—BER and SNR perspectives.

3.4.4.9 Time and Frequency Interleaving

The information transport, over the radio link, is susceptible to signal fading causing errors in the transmission of information. To minimize this inherent susceptibility (nature of the radio link), the information is reorganized over time and/or frequency, by rearranging the associated OFDM symbols for transmission over the radio link. The rearrangement—interleaving process—spreads the impact of signal fading induced errors over the radio link over noncontiguous segments of the information stream. The receiver performs a reverse interleaving process to extract the intended information stream. A spreading of the impact of errors in the information stream maximizes the error-correction potential of the FEC decoder.

3.4.4.10 HARQ

To mitigate the impacts of channel fading conditions, HARQ is utilized. This is an optimized combination of FEC with automatic repeat request (ARQ). HARQ is widely applied for communication reliability, over nosy radio links. HARQ provides a mechanism to compensate for LA errors and facilitates better granularity of coding rate, for an enhanced throughput performance. The different types of HARQ schemes are HARQ type 1, HARQ type 2, and HARQ type 3. In a basic HARQ-1 scheme, when an erroneous packet is detected via a cyclic redundancy check (CRC), then a packet retransmission request is sent to the transmitter and the erroneous packet is discarded at the receiver. The erroneous packet is retransmitted, up to a specified maximum number of packets, until the packet is properly decoded at the receiver. A variation of the HARQ type 1 scheme is one where the erroneous packets are buffered and the corresponding bit level values are combined according to the received signal SNR weights. This is known as packet combining or Chase combining. The advantage of this approach is that with the knowledge of a previously received erroneous packet, it is more likely that the retransmitted packet will be decoded correctly.

The HARQ type 1 method for information recovery at the receiver always includes the error correction bits together with the information transmission, adversely impacting the information throughput as result of the FEC encoder at a high SNR, where the signal integrity is likely to be sufficient for information transport. In the HARQ type 2 approach, the FEC encoder is rate adapted to the integrity of signal transmission.

In the HARQ type 2 approach—referred to as full incremental redundancy technique—the coding rate is gradually decreased with each retransmission by including additional redundancy bits. The latest received information is compared with the same piece of previously transmitted information, stored in the buffer, to form stronger error correction codes—a better estimate of the transmitted information—which in turn improves the decoding process at the receiver. In this approach, the error correction bits are only transmitted in case errors have been detected at the receiver. The disadvantage is that in the HARQ type 2 approach, the incremental error correction bits are not self-decodable—in other words, it requires the knowledge from the previous transmission of the same information to properly decode the information. However, if the initial or the previous transmission has extensive errors, then the decoding process is likely to fail. Therefore, a self-decoding capability for the incremental error correction bits is desirable. This capability is rendered in the form of complementary punctured codes (CPC) for a new type of HARQ, namely, HARQ type 3.

A partial incremental redundancy variant of HARQ is referred to as HARQ type 3. In this technique, the coding rate decreases with each transmission, as a result of the inclusion of additional redundancy bits, similar to the HARQ type 2. The difference in the case of HARQ type 3 is that the ability of self-decoding is preserved for each retransmission. The diversity gain can be enhanced via Chase combining with previously received packets of the same information. The self-decoding capability of HARQ type 3 decoder utilizes two passes of decoding—the first uses the latest received packet, and the second use the combination of all the received packets of the same information. The use of CPC codes are easily decodable and offer coding gains without excessive bandwidth overhead. The HARQ type 3 offers an enhanced performance, relative to the HARQ type 2, under lower SNR conditions, while HARQ type 2 performs better under higher SNR conditions, where the initially transmitted information packet does not contain any error correction bits.

The LTE radio interface leverages the HARQ technique to combat the impacts of noise, interference, and signal fading, which may translate into information transmission errors, while preserving high-performance levels. The low-overhead feedback mechanism of the lightweight HARQ protocol can be complemented with a highly reliable, more processing intensive, selective ARQ protocol, if needed to balance both robustness and latency.

3.4.4.11 SFNs

Information bearing signals from multiple base stations may be combined at the mobile device—similar to the reception of multipath signal components from a single base station. An instantiation of information bearing signals from multiple base stations, single-frequency network (SFN), is one where several spatially distributed transmitters in base stations transport the same information stream over the same frequency channel. The mobile device receiver is subjected to multipath signal propagation—several copies or echoes of the same signal creating self-interference conditions for signal fading. The time spread signal copies may also produce ISI. Since the OFDM signals are inherently composed of lower-rate subcarriers, the use of a guard interval allows for the mitigation of ISI. Further, the use of subcarriers within the OFDM signal hinges on narrowband components and is therefore less susceptible to the impacts of signal fading, which is frequency selective over a single wideband channel. FEC could be selectively at the receiver, as a countermeasure for narrowband components of the OFDM signal that are exposed to significant fading conditions.

The guard interval—referred to as the CP—determines the maximum distance between the transmitters in the SFN, including the size of the SFN. The delayed multipath signals arrive at the mobile receiver, displaced in time, from the different transmitters. The delays must be shorter than the guard interval to allow for the proper reception of the OFDM signal at the mobile device receiver. The operating bandwidth for the SFN is narrower to accommodate a guard interval for managing multipath interference. Spectrally, SFNs have a spectral efficiency advantage over multifrequency networks (MFNs) since the information transport occurs over the same frequency channel.

3.4.5 Convergent Directions

How much information can the highway transport simultaneously? How many points can the highways connect? How long does it take? A timeless quest since anything has been in motion— slow or fast. Over ages people have moved in search of food and shelter and in pursuit of comfort— measured and molded by prevalent and emergent expectations.

Flexibility and adaptability to service demands are primary motivators among the emerging directions for connectivity enabling capabilities in an untethered world. The demand for data capacity is triggered by the widespread appeal of IP multimedia services, and the availability of heterogeneous RATs, terrestrial and satellite categories, requires flexible radio-access capabilities and mobile devices. The current and forward-looking incarnations of mobile devices—smartphones, tablets, wearable gadgets, and the like with yet-to-be-unveiled sensory interfaces—yield products shaped in imagination.

The shifting sands of imagination and technology evolution—catalysts shaping and being shaped by the endless possibilities—are draped in usage scenarios, incubated in the rich, unbounded nature of the information age. Pushing the limits of flexibility and power consumption efficiency, software-oriented radio architectures amenable to programmability reside behind the scenes, which are attractive toward customizable paradigms for mobile devices to suit individual usage nuances. These trends engender a need for multimode devices, providing highways of connectivity over different RATs.

The need for multimode devices—heterogeneous radio-link interfaces—is an artifact of evolution and a catalyst enabling flexible migration across a proliferation of evolutionary possibilities. The term "heterogeneous" in this context implies both a variety of wireless coverage area sizes as well as a variety of radio-interface technologies. A realization of this generalized notion of heterogeneity in connectivity is anticipated in the evolution toward Internet of Everything (IoE)—a confluence of connectivity among humans and machines. Markets and business models are enabled to leverage this flexible potential to explore new service opportunities in the information age. Choices fostering creativity are the primary driver for the production and consumption of end-user experience enhancing services. Low power consumption, high throughput, and multimode chipsets are pivotal elements in the implementations toward programmable software-oriented radio architectures. A virtualization of LTE-propelled wide-area coverage, together with heterogeneous radio-link interfaces, is aligned with satisfying the demands for flexibility and efficiency, shaping new horizons for untethered connectivity.

Choices for access, coupled with low-power utilization, demand flexibility orchestrated through the artifact of programmability rendered in cost-effective software-defined radio (SDR) technologies. Flexibility—pivotal in the experience of seamless mobility—is elemental in a sea of mobile broadband access possibilities.

3.4.5.1 LTE-Advanced

The expectations for connectivity and service are reasonably well defined in the wired segments of the Internet. Consistency and ubiquity of the access experience to a universe of information—in a constant state of creation and change—are pivotal. The elements of this experience expectation are mired in the ingredients of speed—information transfer rates—and the QoS. The performance of wide-area wireless access technologies, such as LTE-A (4G and beyond), is required to provide a high rate of information transfer. They require an enhanced SNR operation, especially for preserving the user experience at the cell edges. These ingredients are elemental in the absorption of the wireless access segments into the Internet universe. The mobile experience—in this profound journey of human existence—disappears into the creation and consumption of information-centric conveniences. The ease of configuration, reduction in the cost per bit of information transport, and compatibility to coexist with the current system, while encouraging a migration, are some of the significant considerations in the enhancements for 4G radio access.

To minimize the overhead, the system information (base station information) acquisition by the mobile device for gaining access is transferred over channels that are different from the information transport channels. The system information utilizes containers, such as the master information block (MIB) and the system information block (SIB). This strategy provides for flexibility and efficiency in the transport of information, through an awareness of the capabilities of the base station and the mobile device, via an exchange of system information between them.

Flexibility and agility, in addition to interference resilience, are inherent characteristics of OFDM in terms of scalable bandwidth. This is a particularly attractive asset, where uneven spectrum distribution constraints in the mobile broadband electromagnetic highways are detrimental to a uniform quality of the radio link, across the different lanes of access. The available frequency bandwidth—width of the transport lanes—determines the number of component carriers in the OFDM signal, which has an impact on the symbol length, among other aspects. The larger the available frequency bandwidth (wider transport highway, with an aggregation of multiple lanes), the larger the capacity of information transport.

The aggregation of both contiguous and noncontiguous spectrum for information transport provides opportunities for both cost reduction per bit of information transport and wider bandwidths for higher information transfer rates. The mobile device capabilities for such enhancements require specification and design innovation, for the minimization of cost and power consumption demands resulting from the complexities of sophisticated signal processing. These capabilities are realized through enhanced radio-frequency filter designs that provide improvements in interference mitigation and isolation while allowing an aggregation of radio-frequency carriers distributed across a given frequency spectrum. In the LTE-A radio-access parlance, the widening of the information highways is referred to as carrier aggregation (CA). Multiple, potentially disparate—in terms of heterogeneous radio-access technologies—narrower pathways consisting of radio-frequency transport are aggregated as contiguous lanes to get wider highways of rapid information transport.

The narrower the information transport lanes, the larger the proportion of the signaling overhead—the smaller the useful proportion of the lane width for information transport. The peak information transfer rates are lower for each lane (channel) as the width of the lanes gets narrower. If each aggregated lane is narrower, the handover overhead is higher. Performance enhancements are achieved by allowing the scheduling and transfer of information across different available lanes (fast scheduling that enables a frequency hopping diversity) although the resource management gains diminish as the width of the lanes gets narrower, akin to managing traffic over different lanes of a highway. The frequency hopping is applicable within a subframe (intrasubframe) or across subframes (intersubframes). For example, in the FDD-type frame structure, each subframe is 1 ms, with 14 OFDM symbols. There are two time-domain slots defined for each subframe. An RB corresponds to one slot in the time domain and 12 subcarriers in the frequency domain.

Frequency hopping techniques facilitate information transfer performance gains, by exploiting frequency diversity and interference averaging. The benefits of these radio-access enhancements translate to improvements in the user experience for multimedia services real time (e.g., VoIP) and non–real time (e.g., file download/upload).

The use of asymmetric bandwidths for information transport—higher in the downlink and lower in the uplink—are typical. This is valid both from a practical view—information consumption by most users will exceed that produced. The search for information, understanding, knowledge advancement, and related access are likely to sustain this asymmetry. This asymmetry is favorable to the power consumption and cost constraints implicit for a mobile device.

The objective of the LTE-A initiative is to evolve beyond the capabilities of the initial release of the LTE crafted radio access. The significant enablers in the LTE-A suite of radio-access capabilities include distributed and diverse topologies—heterogeneous networks, macro systems, femto systems, pico systems, and relays. The movement of evolution in this direction is to adapt to the inexorable demand for access capacity, in the age of information. The approach provides an umbrella of topologies—ubiquitous coverage, capacity handling, and user choice—where the edges of the network are moving closer to the user. The LTE-A radio access allows a sharing of frequency bands with the initial release of the LTE radio-access capabilities essential to serve as a bridge in nascent and evolutionary markets steeped in change. The aggregation of lanes, frequency carriers of information toward wider highways, promotes increased channel bandwidths of up to 100 MHz and ushers the potential for high speeds of information transport.

Much like the rapid distribution of the scribe's work, enabled by the invention of movable type, wider highways provide both flexibility and quality for information transport, encapsulated in the generic language of data. A scalable expansion of these lanes, for wider highways (wider effective bandwidth for information transport) is a significant distinguishing feature of LTE radio access in the directions of a viable mobile broadband ecosystem.

Among the other enhancements associated with the radio interface are cognitive radios, interference management/suppression, inter–base station coordination, hybrid access OFDMA*, and SC-FDMA in the uplink. Dual transmit antenna solutions for single-user MIMO (SU-MIMO) and diversity MIMO, precoding, and FEC are also among the subjects for advancements. Distribution of connectivity is promoted through heterogeneity in local-area and wide-area coverage profiles

---

* Balakrishnan, R., Canberk, B., "Traffic-Aware QoS Provisioning and Admission Control in OFDMA Hybrid Small Cells," *IEEE Transactions on Vehicular Technology*, vol. 63, pp. 802–810, 2014.

and autonomic and automatic operation, characterized by self-optimization, self-configuration, and self-healing capabilities.*

The aggregation of different carrier frequencies to get wider highways of transport is one of the main highlights of the LTE-A (4G) RATs. The baseline† operational bandwidths in LTE scale are 1.4, 3, 5, 10, 15, and 20 MHz, in paired and unpaired frequency bands. The LTE-A capabilities enable the scaling of bandwidth beyond 20 MHz, through the aggregation of the building-block bandwidths of 1.4, 3, 5, 10, 15, and 20 MHz. Arrangement of different bandwidths to allow for an aggregate of up to 100 MHz is envisioned. In addition to the wider highways enabled, via the aggregation of component frequencies, flexibility is provided to stitch together different bandwidths to suit a variety of spectrum holdings for radio access. The different bandwidths are represented by different carrier frequencies referred to as component carriers (CCs). The flexible width lanes of transport connect the base station (eNodeB—eNB) and the mobile device (user equipment—UE).

3.4.5.2 CA

In his novel *The Time Machine* circa 1895, H.G. Wells mused, "There is no difference between time and any of the three dimensions of space, except that our consciousness moves along it." This is suggestive of a communication experience across time and space, over a wireless communication channel. The transfer of information across time and space is foundational in mobile connectivity, where the capacity of the information communication channel affects the information transfer rate and its quality. In turn, this influences the mobile service experienced by the user. Beyond the benefits that accrue from coding and modulation efficiencies, the use of wider bandwidths allows for capacity enhancements.

For higher peak information transfer rates and capacity, wider communication channel bandwidths are required. Wider channel bandwidth allocations, up to 100 MHz, are accomplished in the LTE-A radio interface, through the use of the CA feature, which combines the various baseline operational bandwidths to establish a variety of wider channel bandwidths. Since the constituent bandwidths are aligned with the baseline bandwidths, backward compatibility with non-CA-aware mobile devices is preserved. The aggregation of frequency carriers that are contiguous within a single frequency band (intraband) have a center frequency spacing that is a multiple of 300 kHz, which is a multiple of the 100 kHz radio channel raster in the baseline version of LTE. Since 300 kHz is also a multiple of 15 kHz, the orthogonality of the subcarriers is preserved by definition. The channel raster is a frequency sweeping step size for establishing the center frequency and the bandwidth based on energy detection. This allows backward compatibility with the baseline version of LTE. The mobile device configuration for the CA feature, together with the applicable frequency carriers for aggregation, is provided by the radio-access network, via the system information signaling procedures. Some of the challenges include intermodulation and cross-modulation impacts across transceivers.

CA provides a mechanism to circumvent the limited availability of contiguous wide-spectrum bandwidths, while also providing an avenue for redundancy in mobile connectivity, since more than one frequency carrier is utilized. In conjunction with interference management, CA allows cross-carrier scheduling for enhancing spectrum utilization and heterogeneous network deployment

---

* 3GPP (Third-Generation Partnership Project), TR39.912.
† 3GPP (Third-Generation Partnership Projects) TS 36.101 E-UTRA: User Equipment (UE) radio transmission and reception.

scenarios. This then translates into system capacity improvements. The notion of cross-carrier scheduling enables the configuration of a physical downlink control channel (PDCCH) on a frequency CC to schedule a physical downlink shared channel (PDSCH) and physical uplink shared channel (PUSCH) on another frequency carrier for aggregation purposes, by using a carrier indicator field (CIF) that is inserted within the PDCCH message information. Cross-carrier scheduling promotes intercell interference coordination (ICIC) in a heterogeneous network environment containing a combination of high-power macro cells and low power small cells, by appropriately scheduling information transfers on a secondary CC via the PDCCH on an initial or primary CC. This implies that the CA capability can minimize or eliminate intercell interference on PDCCH. The use of CA promotes a leveraging of the different propagation characteristics of the component frequency bands, being aggregated, for improving the integrity of mobile connectivity.

The constituent information streams that are transported over the aggregated constituent frequency carriers are aggregated above the medium access control (MAC)* layer. The use of hybrid ARQ transmission procedures is independent of the constituent frequency carriers. This allows the independent configuration of each constituent frequency carrier, with respect to the choice of transmission parameters, such as the coding rate and the modulation scheme. The autonomy of configuration of each of the constituent frequency carriers is especially beneficial from a CA perspective, since the radio-channel characteristics of the constituent frequency carriers in different frequency bands are likely to be different and therefore require different handling for optimal operation. At the baseband level, there is no difference in the processing based on whether the aggregated frequency carriers are within the same frequency band or in different frequency bands.

The CA capability, for example, could concatenate five CCs, each with a bandwidth of 20 MHz to achieve a cumulative bandwidth of 100 MHz. Compatibility of the CA feature with the initial capabilities of the LTE radio access is maintained since the CC bandwidth definitions are preserved (reused). Figure 3.15, illustrates the two categories of CA, where in one category the aggregation of bandwidths occurs within the same frequency band, while in the other category the aggregation of bandwidth occurs across different frequency bands.

In Figure 3.15, frequency carriers and bandwidths are associated flexibly to meet the demands of deployment scenarios, which are driven by the corresponding frequency spectrum holdings. The nominal bandwidths are 1.4, 3, 5, 10, 15, 20 MHz, while the frequency carriers, utilized in various CA combinations, are within the same frequency band or across different frequency bands

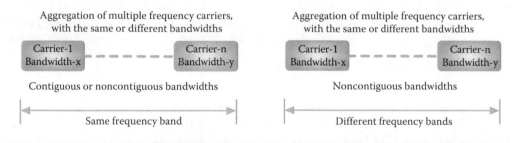

Figure 3.15 CA types—intraband (same frequency band) and interband (different frequency bands).

---

* Same as Layer 2, in the radio-access protocol stack.

as specified in the standardized frequency band arrangement plan.* CA techniques allow wider aggregate bandwidths, leading to wider wireless information highways, through innovation and advancement of techniques related interference management, coexistence with other frequency bands, and mobile device radio-frequency front-end designs.

The CA feature is applicable in different categories of spectrum availability scenarios. These categories are foundational to bolster a market-driven reality—fragmentation in the spectrum landscape, where both bandwidth availability and an uneven distribution of interferers contribute to impairments in the performance of the highways for information transport. The categories are as follows:

- Interband noncontiguous frequency carriers
  - *Aggregation of lanes across disparate highways*
    In this scenario, CCs, separated in frequency across different frequency bands, are combined to provide information transport enhancements by exploiting different radio resource management (RRM) opportunities, across the disparate frequency bands.
- Intraband noncontiguous frequency carriers
  - *Aggregation of disparate lanes across the same highway*
    In this scenario, CCs, separated in frequency across the same frequency band, are combined to provide information transport enhancements by exploiting expanded RRM opportunities similar to the previous scenario. This objective of this scenario is similar to the previous one and is governed by the nature of the spectrum holdings, where segments of frequency ranges within the same frequency band are separated from each other. Wider information transport lanes are allocated for the users being served by aggregating different ranges within the same frequency band.
- Intraband contiguous frequency carriers
  - *Aggregation of adjacent lanes for wider lanes*
    In this scenario, CCs within the same frequency band where the bandwidths are adjacent to each other are aggregated, providing another flavor of RRM opportunities for wider information transport highways.

Wider highways, through the aggregation of component lanes, provides for information throughput (trucking) enhancements, while a relatively small overhead is traded off in spectral efficiency for the flexibility in transport resource allocation. Flexible lane configuration capabilities provide for asymmetric downlink and uplink transfer of information. The highways may consist of a single frequency carrier or multiple frequency carriers where a multiplicity of carriers contributes to configuration flexibility for aggregation. A single frequency carrier is limited to 110 RBs, where each RB corresponds to 180 kHz of bandwidth.

Complexities in the implementation of the CA capability are particularly relevant from a mobile device perspective as a result of the limited geometry and power. Among the complexities are error vector magnitude (EVM), a measure of signal quality, intermodulation product impacts, receiver selectivity, receiver sensitivity, spurious emissions, and output power stability. Each of these parameters adversely impacts the integrity of the channel (lane) for the transport of information. The further the separation of the frequency carriers between the downlink and the

---

* 3GPP (Third Generation Partnership Project), "Evolved Universal Terrestrial Radio Access (E-UTRA); User Equipment (UE) radio transmission and reception," TS 36.101.

uplink, the greater the difference in the path loss between the downlink and uplink. Consequently, enhancements are required in the estimation techniques, prevalent in the non-CA radio access, for the channel characteristics, such as ICIC/reference signal received power (RSRP) for handover and uplink power control, etc.

3.4.5.3 CoMP Access

Imagine a group of horses working in unison to transport a carriage—both the load and the speed of transport are enhanced. Coordination is the essence here, for performance improvements. Coordination among a multitude of base stations, interconnected via a high-speed backhaul for the transport of information between the mobile device and the base stations, provides system performance gains in the presence of interference conditions. These ideas are particularly useful for evolutionary directions in mobile communications, since they are inherently aligned with the demand for large information traffic volumes, while also enabling low-latency operation for real-time services, such as VoIP, gaming, videoconferencing, instant messaging, etc. To meet the diverse demands of mobile multimedia services, performance measures at the edge of wireless mobility (radio-access network) in terms of the average base station,[*] information throughput and the information throughput at the edge of the base station coverage affect the mobile service quality.

While OFDM promotes deployment flexibility, and signal integrity, the characteristics of OFDM are such that there is a susceptibility to frequency offsets and creates narrowband interference. Inter–base station interference is a significant source of noise in the system, which adversely impacts system capacity. Noise reduction—accruing from a coordination of transmission and reception—contributes toward an augmentation of system capacity. This coordination[†] enhances the signal strength of the intended signal at the mobile device. In turn, this improvement reduces the transmission interference emanating from the mobile device, with respect to other mobile devices in the vicinity of a mobile device transmission. The concepts of cooperation and coordination among base stations to enhance performance and capacity are embodied within the coordinated multipoint (CoMP) transmission and reception framework. This framework consists of a suite of enabling capabilities for information transmission and reception, while reducing interference impacts and increasing the system capacity and the quality of mobile multimedia services.

The CoMP framework provides a suite of enabling capabilities for effective coordination of information streams between the radio-access network and the mobile device. The downlink multicell MIMO capability for information transfer is attractive for QoS and capacity improvements, in the delivery of mobile multimedia services. The two categories[‡] of multicell MIMO, within the downlink CoMP framework, include coordinated scheduling and joint transmission. In the case of coordinated scheduling, the spatial diversity benefit of MIMO together with the joint selection of a MIMO precoding vector is leveraged across cooperating base stations. The selection considers the radio-channel conditions between the serving base station and the served mobile device, together with a minimization of the interference to mobile devices at the base station or cell coverage edges. In the case of joint transmission, as indicated earlier, the benefits of average

---

[*] Also referred to as "cell."

[†] Karakayali, M.K., Foschini, G.J., and Valenzuela, R.A., "Network coordination for spectrally efficient communications in cellular systems," *IEEE Wireless Communication*, vol. 13, no. 4, pp. 56–61, August 2006.

[‡] 3GPP TR36.814, "Evolved universal terrestrial radio access (E-UTRA); further advancements for E-UTRA physical layer aspects," Technical Report, v9.0.0, March 2010.

information throughput and the base station coverage edge throughput accrue at the expense of radio resource consumption in the uplink for providing the CSI feedback estimates from the mobile device to the radio-access network.

In the downlink direction, joint transmission by cooperating base stations serving a mobile device enhances the signal quality at the mobile device, thereby reducing the impact of interference. The performance enhancements also appear in the form of reduced interference from this joint transmission approach on other mobile devices in the vicinity. The trade-off is that there is a higher consumption of radio resources—reduction in spectral efficiency—when joint transmission is utilized by cooperating base stations to serve a single mobile device. This can be mitigated by the use of MIMO* techniques, where multiple mobile devices are served via joint transmissions performed by cooperating base stations. For coordinated or joint transmission of information from multiple base stations to a group of served mobile devices, the radio-channel state information of these mobile devices is required at the base stations. This increases the uplink information load relative to conventional information transactions between a base station and a mobile device, where joint processing or coordination is absent. Cooperation among the serving base stations, in terms of radio resource allocation and the use of MIMO precoding, while only one of the base stations transmits information streams to the mobile device being served, allows a reduction of the uplink multi–radio channel condition feedback information load. Besides the reduction in the uplink signaling overhead, since only one among the cooperating base stations† is transmitting to the mobile device, interference to the other mobile devices in the vicinity is avoided, when the neighboring base station transmission times and the frequency resources overlap. The MIMO precoding scheme utilizes beam-forming transmission concepts. Transmission diversity and spatial multiplexing are the elemental attributes of MIMO precoding within the LTE specifications.‡ The proper recovery of information streams via multiple receiver antennas, in a MIMO system, hinges on the radio-channel conditions. This recovery is impaired, if the SINR of the radio channel is too low. MIMO precoding enables improved information integrity at the receiver antennas, through the use of preconfigured codebooks that are created to suit different MIMO antenna arrangements (2 * 2, 4 * 4, etc.). The proper selection of a MIMO precoder requires knowledge of the channel conditions at the base station transmitter, through closed-loop feedback from a mobile device receiver. The radio-channel measurements at the mobile device provide hints for an estimate of the precoding matrix indicator (PMI), channel quality indicator (CQI), and/or rank index (RI). These estimates are based on the dynamic radio-channel conditions perceived by the mobile device receiver and the associated configuration of the mobile device, with respect to the PMI and the CQI. This feedback information to the base station triggers a suitable modification of the MIMO precoder codebook selection for an enhancement of the system performance. There is a trade-off between the rate of change of the radio-channel conditions and the latency of the feedback information. A reduction in the information feedback would reduce the signaling overhead at the expense of fewer MIMO precoder codebook selections, thereby limiting the effectiveness

---

* Gesbert, D., Kountouris, M., Heath, R.W. Jr., Chae, C.-B., and Salzer, T., "Shifting the MIMO paradigm: From single user to multiuser communications," *IEEE Signal Processing Magazine*, vol. 24, no. 5, pp. 36–46, October 2007.

† Liu, L., Zhang, J., Yu, J.-C., and Lee, J., "Inter-cell interference coordination through limited feedback," *International Journal of Digital Multimedia Broadcasting*, vol. 2010, Article ID 134919, 2010.

‡ 3GPP TS 36.21, "Evolved universal terrestrial radio access (E-UTRA); physical channels and modulation," Technical Specification, v11.2.0, March 2013.

of MIMO precoding for system performance enhancements. Insights into the optimization of trade-offs across system performance, preservation of information quality including at the edges of wireless coverage, in terms of feedback information signaling overhead, MIMO precoding effectiveness, are revealed through flexibility in the estimation of radio-channel conditions. The latter conditions are influenced by interference, noise, and the correlation between the antenna and the radio channel, where the level of correlation determines the proper recovery of the code words bearing information streams.

In the uplink direction, cooperating base stations serve as a spatially distributed array of receive antennas, resulting in a spatial diversity of signal reception for improvements in signal detection from the mobile device. For noncooperating base stations, in a traditional radio-access network, signals from a mobile device not being served appear as interference. The logical interface between base stations, referred to as X2, serves as a means for cooperation and coordination of information among a cluster of interconnected base stations. The information shared across the base stations, via the X2 interface, is utilized in the coordination of uplink signals for improved mobile device signal detection and service. This is in contrast with the CoMP operation in the downlink, where feedback signaling from the mobile device is necessary for multicell radio-channel state estimation. Additionally, for effective joint transmission across base stations, MIMO precoding schemes are required, together with a signaling of the precoders to the mobile devices being served.

The challenges related to CoMP, which has the potential for system performance and capacity enhancements, include low-latency feedback and radio-channel state estimation efficiency for base station cooperation and coordination. Since the interface is logical, a direct physical link between cooperating base stations may not exist, which implies that site-specific configuration of interconnections would be required to meet the demands of low-latency operation, for joint processing and scheduling in the downlink.

Optimizing the network utilization (backhaul connectivity) and processing latencies among cooperating base stations, such as by reducing the amount of information to be coordinated among the base station to achieve the performance enhancements, are noteworthy considerations within the CoMP framework.

3.4.5.4 Broadcast Access

A broadcast type of access is one where information is transported from one point to multiple points (multiple mobile devices). The multicast type of access is a specific type of broadcast access where only a group of mobile devices, subscribed for multicast access, are enabled to acquire the multicast. Broadcast access is a push type of information transport service. The mobile broadcast multicast service (MBMS) is a rendition of point-to-multipoint access, with both broadcast and multicast modes of operation. With increasing demand for capacity driven by a relentless growth of information proliferation, broadcast provides an approach for the transfer of information of broad interest to mobile devices. The demand profile includes linear TV, video, and a variety of nonlinear multimedia content. Broadcast access provides opportunities that ameliorate the impact on available capacity and chooses a more wireless resource-efficient point-to-multipoint arrangement of information traffic lanes, instead of a point-to-point arrangement, for the distribution of any given broadcast information service—concerts, sporting events, conference, alerts, social groups, advertising, etc.

The enhanced MBMS (eMBMS) broadcast access provides both TDD and FDD variants for sharing a single-frequency carrier, with a bandwidth of up to 20 kHz between the unicast and the eMBMS broadcast access. An instantiation of eMBMS broadcast access,* with a dedicated broadcast carrier, is integrated mobile broadcast (IMB). The notion of SFN is a capability within the eMBMS framework. The factors that influence the attractiveness of broadcast contents across different broadcast service channel, akin to moving across television channels, is dependent not only on the throughput but also on the user-perceived service channel switching time. The broadcast service experience requires a preservation of the user experience in different environments—urban or suburban—including at the edges of coverage, while sustaining spectral efficiency targets. In the use of SFN, the mobile device is served with broadcast content from a group of synchronized base stations. The broadcast content transmissions that arrive at the mobile device, from the different synchronized groups of base stations, appear to come from a single base station, provided that they arrive with delays that are smaller than the CP, at the start of each symbol, to avoid any ISI. The use of the SFN approach in IMB provides the flexibility for interference mitigation, allowing optimizations such as antenna tilt, positioning, and power. The benefits of IMB include congestion avoidance for information transport, low-cost content delivery, and high-bandwidth availability for information broadcast.

Content distribution and its attractiveness, given the potential diversity of possibilities such as alerts, social content, programmed content, etc., are subject to the nuances and the benefits of the long tail.† It is a paradigm where the consumption and demand for content is not restricted to a few widely popular items but to a potentially unlimited filed of niche possibilities. Convenience— choice and flexibility in the long tail—is promoting the creation and consumption of content. IMB offers the possibility of using TDD for broadcast access to reduce the capacity demand on FDD channels.

3.4.5.5 Heterogeneous Access and Handover

In a Web of Things—sensors, smartphones, tablets, etc.—connected over a variety of radio-interface technologies, the notion of always best connected (ABC) is significant in the effective rendering of multimedia services. The selection of an available access technology provides choices to optimize the QoS requirements of a service. This provides the potential for an enhanced utilization of access resources, which then promotes improvements in the utilization and availability of access capacity. The heterogeneity of available access technologies allows a diversity of access resources capabilities, such as wireless coverage area, bandwidth, delay, security, power consumption, access cost, mobility speed, and user preferences. The distribution of access resources, via heterogeneous access technologies in a geographic region, facilitates opportunities to optimize the balancing of connectivity demands.

Coverage is of paramount significance in the pastures of a nomadic world, enabling the colorful experiences for both the individual and the collective, in a changing and evolving social fabric. A resonance with the innate heritage and destiny of humankind vis-à-vis experiences embedded in the ethos of communications. In these directions of technology evolution, the enabling artifacts of mobile access technologies blend indistinguishably into the intended experiences in the human journey. It is in this mode that technology reveals its potential as an effective, nonintrusive catalyst wielding an irresistible appeal.

---

* 3GPP (Third Generation Partnership Project), "Multimedia broadcast/multicast service (MBMS); architecture and functional description," 3GPP TS 23.246 v11.1.0, 2012.
† Anderson, C., *The Long Tail: Why the Future of Business Is Selling Less of More.* New York: Hyperion, 2008.

Dynamically changing mobility and service requirements affect the changes of the connectivity demand load. Topological distribution and decentralization of access technologies serve as a catalyst to support variations in connectivity demand load conditions. Autonomous decision making and allocation of resources for connectivity is an attractive facet of distributed heterogeneous access technologies to meet the demands of dynamically changing connectivity, through appropriate network selection and handovers. Coordination and control signaling are necessary to orchestrate handovers to an appropriate access network, based on mobility, QoS, and connectivity load conditions. The intelligence for the coordination and control function may reside in the mobile device or in the access network, while cooperative signaling occurs between the mobile device and the access network. This intelligence is embodied in a selection function that processes a variety of criteria—mobility, QoS, and connectivity load conditions—to perform coordination and control, for the discovery and selection of an appropriate access technology. The selection of a specific access technology, within an available landscape of heterogeneous connectivity, requires an architectural model that includes the discovery and selection function.

A logical model that illustrates the interaction between the mobile device and the network, with respect to the coordination and control signaling, is shown in Figure 3.16. This signaling orchestrates the information flow, over a mobile wireless transport segment, as the attachment point is prone to change dynamically as a function of mobility.

The awareness of the coordination signaling is required both in the mobile device and in configured domains of heterogeneous networks that provide connectivity over geographic regions. The discovery and selection function selects a RAT interface based on the availability and suitability of a discovered RAT that is suitable for connectivity between the mobile device and the network. The suitability of a selected RAT is governed by mobility, QoS demands, and load conditions, which are dynamic considerations. Network configuration, policy, and pricing are among the relatively static attributes that affect suitability, in the selection of a RAT.

A logical model for the positioning of the discovery and selection function is depicted in Figure 3.17. The discovery and selection function, for the establishment of a suitable RAT for connectivity and handover management, could be resident either in the mobile device or in the network. Coordination capabilities for coordination between the mobile device and the network are required in the mobile device and in the network. Coordination allows the mobile device and the network to converge to a suitable RAT, identified by the discovery and selection function,

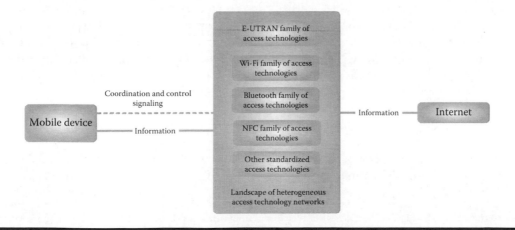

Figure 3.16   Distributed heterogeneous access technology oriented connectivity.

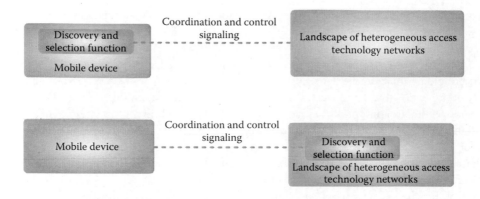

Figure 3.17   Selection and discovery function—logical model.

based on autonomous conditions known exclusively by either mobile device or network entity, as a result of their distinctive function and usage roles, and the corresponding architectural tenet of information hiding. The coexistence of diverse power levels across small cells and macro cells to suit different coverage areas is a critical consideration in the realization of a heterogeneous mobile connectivity environment. The base station types in this environment may be characterized in terms of small cells, macro cells, and RAT types such as E-UTRAN, Wi-Fi, etc. The mechanisms to dynamically manage the signal transmission power levels across different types of base stations are vital to avoid conditions known as the tragedy of the commons,* where two different types of base stations progressively ratchet their transmission power levels, based on perceived interfering signals, rendering both transmissions unusable.

Enhancements in capacity, coverage, and the quality of experience (QoE) rely on a supportive harnessing of the distinctive benefits of small cells[†] and macro cells. In this heterogeneous mobile connectivity fabric, an effective management of interdependent considerations across the cells is vital for a realization of these enhancements. While macro cells manage wide-area mobility, small cells offer traffic offload opportunities for targeted high-demand, pedestrian mobility, and local-area footprints. Coordination of connectivity across these modalities of connectivity promotes intelligent traffic offload operation, governed by user mobility patterns. On the other hand, for a realization of these cooperative benefits, effective radio-frequency interference management techniques, such as CoMP transmission and ICIC, are pivotal. CoMP enables the transmission and reception of information between cooperating cells (macro cells and small cells) and a single mobile device to improve the cell edge mobile device rendered QoE. Along complementary directions, ICIC enables a scheme for muting macro cell high-power transmissions through the use of almost blank subframes (ABSs), which only contains a relatively small amount of coordinating signaling information—lower interference energy relative to payload information—to increase the probability of traffic offload to small cells, within a heterogeneous environment to realize gains in capacity and system capacity utilization.

The positioning of the mobile device, relative to the cell center, affects the arrival times at the serving cell of the uplink transmissions from different mobile devices. This is a result

---

* Garrett, H., "Extensions of 'The Tragedy of the Commons'," *Science*, vol. 280, no. 5364, pp. 682–683, May 1, 1998.
† The term "cell" and "base station" are used synonymously.

of the relative position-driven differences in the signal propagation delays. These signals from the different mobile devices are processed collectively at the cell or base station for a time domain to frequency domain conversion followed by demodulation for information recovery. Consequently, an alignment of the signal arrival times at the base station from the differently positioned mobile devices within the coverage area is vital for the information recovery process. Cooperation between the base station and the mobile device, via feedback from the base station to the mobile device, is applied to adjust the uplink signal transmission timing to suit the information recovery process cycles at the base station. These cooperative behavior is extensible to a heterogeneous connectivity fabric as well that includes variable coverage footprints, realized in a distribution of macro cells and small cells. In such a scenario, CA of component frequency carriers—one allocated to a macro cell and the other allocated to a small cell—provides system capacity enhancements through traffic offload to the small cell. This implies large differences in signal propagation delays between mobile device transmissions over the different component frequency carriers, as a result of different coverage footprints of macro cells and small cells. The grouping of the necessary time alignment adjustments, based on a component frequency carrier for mobile device transmissions, ensures a proper signal recovery operation at a corresponding macro cell or a small cell. The characteristics of the heterogeneous connectivity fabric are harnessed in an interdependent and cooperative manner to realize the proper recovery of mobile device signal transmissions. This is reflective of functional collaboration across a variety of coverage topologies, where interdependence is leveraged to resolve the challenges, and to optimize the trade-offs, associated with connectivity for diverse mobility patterns. Interdependent behaviors serve as a foundation for the operation of significant capabilities such as CA and traffic offload for system capacity and coverage enhancements. In turn, the mobile user experience is bolstered.

3.4.5.6 Relay

Radio access—rendered in many technologies, including LTE—includes the relaying of wireless links to fill in the geographic areas where the coverage for wireless access for communications and services is insufficient. The relaying principle improves coverage by retransmitting information over radio-access links to spread the availability of information. Relays may simply retransmit the information to targeted geographic areas or may also decode the information before repackaging for transport over new radio-access links. The decoding of information before retransmission allows for a selective forwarding of traffic to users located in a targeted geographic area. This approach minimizes the interference to other radio transport lanes, which is very desirable in an ever-increasing mobile information consumption and creation world.

The LTE radio-access advancements utilize both in-band and out-of-band relaying of information for mobile access coverage enhancements. Figure 3.18 illustrates the architectural model of the LTE radio-access relay principle. There are two interfaces—one between the base station and the relay entity and the other between the mobile device and the relay entity. From a mobile device perspective, the relay entity appears as a unique base station with a specific identity and synchronization channels together with reference symbols for the proper synchronization of information transfer.

The relay entity reconstructs the signaling and the user information channels—radio transport lanes—for selective coverage enhancements propagating profusion in user experience. The partitioning of radio transport resources is accomplished using time-division multiplexing, where each user is served one at a time. The move toward getting the network attachment points closer to

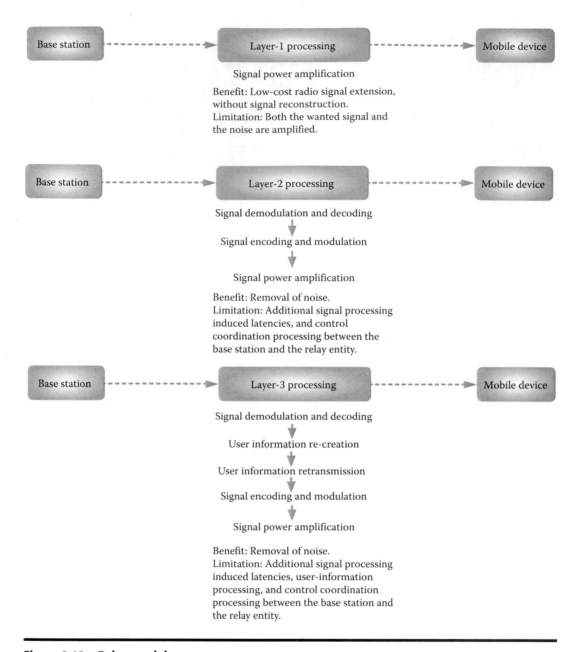

Figure 3.18 Relay model nuances.

nomadic users is vital in the rendering of high information transfer rates. Relay entities are a step in this direction by reducing the separation between the network and the mobile device. The base station serving one or more relay entities is referred to as a donor cell.

The donor cell, in addition to serving relay entities, may also serve mobile devices in the conventional manner. Relay entities may simply be a signal amplification intermediary, which forwards analog signals to a targeted radio coverage area. Such intermediaries may perform this amplified forwarding either passively (nonintrusively), where its presence is not detected by the donor cell or the mobile devices, or actively, where the operation of the intermediary is controlled in terms of when to forward and alter the amplification power of the forwarded analog signal. The decision logic, such as resource allocation, scheduling, and mobility management, is a function of the donor cell. A passive mode of operation is depicted as a layer 1 relay model, as shown in Figure 3.18.

In an active operational mode, the relay extracts and repackages information between the donor cell and the mobile device. An active mode of operation includes both layer 2 and layer 3 aspects of the relay model, as shown in Figure 3.18. This intermediate processing incurs additional information transfer delays—longer than the subframe interval of 1 ms for LTE radio access—but avoids the replication of noise, as a result of the information parsing and retransmission. The impact of delays can be mitigated using LA techniques for an appropriate selection of coding schemes to manage information transfer rates, based on the channel conditions. In an operational mode where the relay performs the functions of a base station, albeit with reduced power demands and form factor such as mobility, scheduling, and HARQ-oriented retransmission, transparency for different vintages of mobile devices is preserved.

Since the transmission and the reception of information are likely to overlap between the donor cell or relay intermediary and the mobile, interference mitigation via gaps in the transmissions between the relay intermediary and the mobile device is desirable. An alternative would be to utilize network transceiver (donor cell or relay) reference signals in each LTE radio-access subframe. An information broadcast technique—referred to as multicast broadcast single-frequency network (MBSFN) in LTE radio access—utilizes subframe signaling to indicate to the mobile device that other than the reference OFDM symbol, downlink information transmission to the mobile device is absent for the remaining part of the subframe. In this absence or gap, the transmissions between the donor cell and the relay intermediary occur, thereby diminishing interference impacts at the mobile device.

With a wide variety of coverage footprints and assorted information transfer rates demanded by multimedia services and machine-type communications, the relay entity serves as a supportive capability to meet these demands while optimizing investment costs. In the arena of small cells, the relay entity has the potential to move the wireless attachment points closer to the mobile user. This translates to lower battery consumption and improvements in information throughput, which naturally implies improvements in the user-experience for multimedia services.

3.4.5.7 SDR

How can flexibility be accomplished by leveraging common characteristics across different standardized next-generation access technologies? SDR provides opportunities for evolution through programmability—an ingredient for flexibility.

OFDM principles are an intrinsic attribute of next-generation technologies embodied within the LTE system, with interoperable functions and interfaces, allowing flexibility, and innovation in implementations and deployment scenarios. Functions are partitioned into distinct layers that cooperate to render end-to-end features and capabilities. The distinctions are identifiable in terms of layer 1, layer 2, and layer 3 of the radio-access system. In the case of layer 1, the distinctions are with respect to physical channel descriptions, frame structures, subcarrier spacing, FEC schemes, and reference signal characteristics. Variations manifest at layer 2 and layer 3, in terms of procedures for base station selection and handover.

Within a given RAT, flexibility allows a relative ease of migration since the evolving capabilities can potentially be designed for the corresponding upgrades. This ability translates to operational efficiency enhancements. Experiential information—deployed system behaviors—can be leveraged for the application of algorithmic techniques, which are inherently dynamic for enhancements or for corrections to an observed problem.

The interdependence of capabilities, such as multiple modes of radio access, MIMO, etc., together with the high information transfer rate targets in next-generation technologies, imposes much higher processing demands relative to 3G technologies. Traditional digital signaling processing techniques—widely applied in 3G and preceding generations—are relatively constrained in terms of minimizing the associated power consumption. A combination of desirable features—minimization of power consumption and an enhancement of operational efficiencies through convenient upgrades for new features and bug fixes—elevates the significance of SDR for promoting flexibility in deployments. Implicitly, SDR concepts harness flexibility through an adaptation of common features in an evolving ecosystem.

The partitioning of the modem components between configurable hardware and SDR is pivotal for next-generation multimode transceivers for enhanced power consumption efficiencies. The main processing blocks for next-generation transceivers include

- Digital front end (DFE) for data sampling and filtering of adjacent channel interference
- Resampling of data for sampling frequency error correction
- Estimation of radio-frequency impairments and correction
- Use of FFT for a conversion of time-domain data into the frequency domain

Since the DFE block functions are relatively common across different renditions of OFDM implementations, it is akin to the FFT core. However, it demands high processing power requirements as a result of a high data sampling rate in its operation. The DFE therefore lends itself to a static realization as configurable hardware. In the receiver, a core function is frequency-domain processing. The elements of this function include the following:

- Interference measurement
- LA
- Compensation for frequency errors, gain, and timing
- Channel estimation
- Performance of decoding schemes

The characteristics of OFDM demand the processing of vector-oriented operations. Frequency-domain processing demands both high power consumption and programmability. SDR-driven flexibility lends itself naturally to the demands of vector-oriented operations within the frequency-domain processing function. The FEC function—characterized by standard coding techniques, for example, HARQ, convolutional codes, or turbo-convolutional coding techniques—is an attractive candidate for harvesting the benefits of flexibility afforded by SDR, or software configurable hardware.

The replication ease of common functions across different RATs using OFDM modulation principles raises the attractiveness of SDR-oriented implementations. Examples include LTE radio access in the realm of multimode 4G and beyond RATs. The prominent features of SDR-oriented renditions include

- Scalability through flexible processing elements
- High computational power

- Optimization of processing and power consumption*
- High utilization of computational capacity, with power consumption efficiencies approaching traditional application specific integrated circuit (ASIC) realizations
- Intelligent partitioning of task for an efficient utilization of computation resources

The assortment of OFDM-oriented RAT families imposes challenges in terms of balancing the demands of technology evolution with the risks associated with an adoption of rigid, hardware-oriented processing platforms. A scalable and programmable embedded processing fabric transcends the evolution of technologies and the associated enabling standards. The architectural model of SDR enables relative ease in the processing and in the delivery of information. The model is reflective of a synergistic, configurable, scalable combination of high-performance processing and switching-interconnect entities, which are attractive attributes in an evolving next-generation mobile ecosystem.

3.5 Decentralization and Shift toward All-IP Systems

The Internet is now an established sanctuary for a multimedia information experience. In this haven, IP is firmly embedded as the de facto transport protocol. The convergence of the information media types—voice, data, video—has brought the appeal of the Internet endorsed all-IP vision in the evolution of mobile broadband. Information media type convergence and the economics to realize this convergence trend are the twin motivators for the shift toward all IP. Attractive trends for architectural considerations in the EPS are embodied in the mobile broadband framework envisioned in the system architecture evolution (SAE)—a pivotal theme in the fabric of the LTE initiative. The all IP–inspired ingredients of the SAE, coupled with the wireless RAT performance enhancements, are the signature characteristics of the LTE system. The all-IP design principles invite flexible choices in terms of mobile broadband services and operations.

The ingredients of the EPS to enable wireless mobility may be classified as attachment, movement, and scalability. The all-IP nature of the EPS promotes each of these ingredients to orchestrate a personalized connectivity experience. The attachment between the mobile device and the EPC network within the EPS occurs at the IP layer—also known as the network layer in the open systems interconnect (OSI)[†] model. The movement of the mobile device across network points of attachment is managed at the network layer, with IP as the protocol for mobility management. The mixed media types—inherent in multimedia communications—impose diverse QoS demands. The management of these demands is implicit in the nature of IP—routing and treatment of information flows transported as discrete packets. The packets of information are associated with mobility-driven attachment points to a serving network and transferred between a source and a destination. The packets belonging to different information flows are treated according to the related QoS demands, using specific information embedded in the packet headers.

The EPS embodies principles that accommodate a flat IP architecture—a reduction of intermediate entities in the connectivity link between the mobile device and a remote entity. This reduction, in turn reduces the connection establishment states, which implies a lower overhead for information

---

* Meeus, W., Aa, T.V., Raghavan, P., Stroobandt, D., "Hard versus Soft Software Defined Radio," *2014 27th International Conference on Embedded Systems*, 2014, pp. 276–281.
† Open systems interconnect, a layered model of protocols for communications.

transfer. These principles applied in the EPS architecture enable a reduction in latencies for the transfer of information over the connectivity link. The enhancements in the performance of the connectivity link foster improvements in the user experience of IP multimedia services.

The wireless bandwidth constraints, in typical 3G systems, encouraged an intelligent conservation of the wireless segment of the connectivity link. These implied a release of the physical resources of the wireless segment, when no information transfers to or from the mobile device were taking place for a predefined time interval. The wireless connection segment state was then deemed to be in an idle state—dormant state for the mobile device, implying no information transfer. A resumption of information transfer would require a change in the connectivity link state to an active mode—triggered either via a user request for a service launch, or an automatic process in the mobile device, or in the network that required the transport of information.

For example, the state transition delays are of the order of 1 s—idle state to active state—in the case of HSPA, for the wireless connectivity segment. In the EPS architecture, the idle to active state transition delays are much shorter (of the order of 0.1 s) for the wireless connectivity segment. In addition to the flat IP model embraced by the EPS architecture, the wireless connectivity segment is decentralized—removal of the radio network controller (RNC), prevalent in 3G systems. The absence of the RNC in the EPS architecture provides base stations with a greater autonomy in connection establishment, which allows for a distribution of signaling and processing overhead—enabling performance improvements for the connectivity link. The reduced idle to active state transition delays usher the arrival of an experience of "always-on" in mobile broadband communications.

The innate capabilities of IP allow for a sharing of the transport paths between a source and a destination, much like the wagons in a train, which may be partitioned according to a QoS, for example, class grades. Shared paths and information persistence (caching) enable opportunities for resource utilization enhancements and a reduction in transport latencies.

3.5.1 Assorted Mobile Broadband Connectivity: Ubiquity of IP

The ubiquity of IP is harnessed in the next-generation network (NGN) model for mobile connectivity—a global endeavor endorsed by the ITU. The NGN model encompasses a packet-oriented architecture, with IP as a unifying communication protocol at the network layer in an ocean of protocol renditions for connectivity and service at the other layer of the OSI model. Connectivity is enabled through the use of assorted-broadband, QoS-embedded, transport technologies. Service is enabled, such that it is decoupled from the nuances of the transport technologies. The NGN paradigm offers unencumbered choice for users to connect and enjoy access to virtually endless service possibilities. NGN is a stepping stone in the connectivity fabric with ubiquitous service potential. It is a model along the path of evolutionary ideas toward a widespread virtualization of the mobile connectivity fabric.

The attractiveness of the ubiquity-inspired NGN model is manifested in a variety of forms including interoperability across heterogeneous connectivity network technologies and an unfettered information sharing potential—a promising departure from the disparate islands of connectivity bridged by complex solutions. These solutions were motivated by a circuit-switched technology heritage, originating in a voice-centric era. The potential of IP architectures, hinging on the NGN model, promotes a blending of connectivity technologies—wireless fixed, wireless mobile, or wired.

The architecture in Figure 3.19 depicts the generalized potential of IP as a fabric of information transport. A vision—blending a multitude of connectivity pathways—is realized through an umbrella of terrestrial and satellite systems. An orchestration of information sharing in diverse

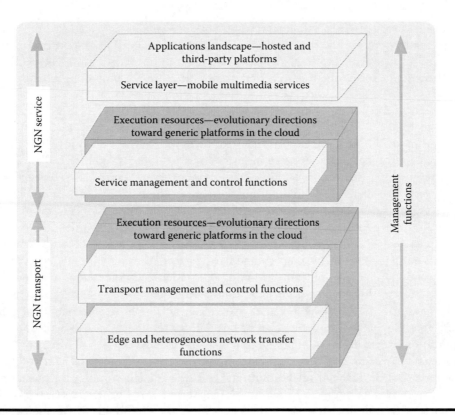

Figure 3.19 NGN architectural model.

technologies is manifested in innovative compositions of connectivity pathways. Indeed, veritable scaffolding for embedding the vital ingredients of mobility, interoperability, open interfaces, handover, QoS, seamless service, extensibility, scalability, and security across a web of connectivity pathways. The ubiquitous IP scaffolding enables both vertical handover (across connectivity technologies) and horizontal handover (across different domains of the same connectivity technology). The two main categories of functions within the NGN architectural model, depicted in Figure 3.19, are identified as "NGN Transport" and "NGN Service." The NGN Transport category of functions harnesses the ubiquity and universality of IP as the protocol of choice for transport of information encoded within different types of access technologies—wired or wireless.

Both the UE and the network equipment in the NGN architectural model utilize IP for the transport of information. The NGN Transport functions coordinate the allocation of the required resources to satisfy the multimedia service related QoS demand across the access and the core network segments of connectivity. The NGN Service category of functions provides both session-oriented and non-session-oriented services. The nonsession-oriented services include messaging schemes for instant messaging and a subscribe/notify scheme for managing presence information. The session-oriented control functions include the registration, authentication, and authorization of services requested by a service user. These control functions also control the allocation of the media resources associated with the user requested services.

The NGN Service category of functions also includes the coordination of service user profiles and related control information within cooperating databases. The support for open application programming interfaces (APIs) is an integral aspect of the NGN Service category functions. Open APIs allow and encourage the participation of third-party application and service providers to innovate and influence an evolving market of mobile multimedia services. The NGN Service category functions provide the resources and capabilities through processing platforms (servers and gateways) in the NGN Service layer of the NGN architectural model.

The management functions span the NGN Transport and the NGN Service categories of functions, to enable a coordination of the capabilities within an implementation of the NGN architectural model. The coordination is required to ensure that the functions within the NGN architecture deployment are operating to meet the intended QoS, reliability, availability, billing, charging, accounting, and security requirements while enabling experiential excellence in mobile service delivery. The charging and billing functions are configurable for either offline (postpaid) or online (prepaid) near real-time metering of fees for applications, services, and connectivity resource utilization, to suit a variety of NGN-oriented connectivity networks. The NGN architectural model generalizes the connectivity interfaces to heterogeneous access technologies and assorted UE interfaces that include mobile and fixed devices. The NGN architectural model provides an evolutionary vision while not restricting innovation and potentially unbounded variety in implementation. This includes a virtualization of functions in the NGN Service and NGN Transport categories of functions, together with a generalization of processing platforms. The vision inherently opens an enormous field of possibilities for widely distributed topologies and decentralization for the deployment of mobile connectivity networks.

This separate, yet loosely coupled model is a significant architectural construct, since it allows a variety of deployment orientations and business models through universal and well-defined, open interfaces. It enables a landscape of assorted wired or wireless broadband access technologies, with resource allocations shaped to meet the QoS demands of multimedia services. The generalizations embodied in the NGN architectural model sets the stage for a long-term evolution of functional organization and distribution to cater to the multimedia service demand at the edge of mobile connectivity for people and things. A generalized mobility and service paradigm—high reliability and high availability service delivery—allows virtualization, scaling, and flexibility for connectivity and service providers. User-driven choices are allowed in the NGN paradigm to promote innovative vistas for connectivity and service creation, orchestration, and delivery. The NGN architectural model reflects technology mediation in a contextually oriented world of experience.

The prominent characteristics of the ITU-inspired NGN architecture—shifting horizons toward distributed and mobile connectivity—are shaped by the following:

- Packet-oriented connectivity
- Separation of control functions across bearer capabilities, call/session, and application/service
- Decoupling of network/open interface provisioning from service provisioning
- Service enablers for a diverse array of services
- End-to-end QoS
- Open interfaces for a wide variety of connectivity interworking options
- Mobility management
- User-driven open-access capabilities for connectivity to broadband services
- Domain names services for interdomain routing of connectivity, through a resolution of domain names to IP addresses

- Seamless service experience for the user
- Service convergence—mobile and fixed
- Independence of the connectivity layer from the service layer

These malleable ingredients engender implicit cooperation interdependence among the inherently independent and diverse ingredient. It is this independence, within interdependence, that promotes a rich connectivity fabric fostering ubiquitous, broadband communications, with mobility as the crown jewel. It is mobility—with connectivity anytime, anywhere for anyone—that promotes individuality in the collective. Akin to the human experience, it is a fundamental note in innovation.

Unlike the traditional telecom networks where the stored program control exchangers, such as a mobile switching center, utilized circuit-switching techniques, the NGN-oriented telecom networks leverage packet switching. The guiding principles of the NGN architecture allow connectivity via terrestrial or satellite systems at the edges of the connectivity ecosystem—flavored with the dominant traits of the Internet interconnectivity. This dominance has shaped the evolution direction of the telecom network, where IP is the protocol of choice—common glue across the disparate islands of connectivity technologies—often driven by historical and market trends specific to different regions around the globe. This approach enables cooperation and management across the heterogeneous connectivity technologies, through an alignment and control of network resources, states, and behaviors. The IP-oriented transport accommodates a plethora of payload types such as voice, data, image, and video, independent of the payload types and any related processing or signaling—a powerful artifact of the generalization inherent in the NGN model.

The evolution of the Internet architecture—foundational in the evolution of all-IP mobile broadband communications—has been shaped by global collaboration in a dominant theme of openness, rich in diverse perspectives that have preserved its sustainability and aligned with incessant change in the information age. The separation of the network layer (IP) and the transport layer (Transmission Control Protocol—TCP) is profound in its implications, specifically with respect to the support for ubiquitous mobility, where IP provides an elegant, extensible, and evolutionary potential. The notions of subnets, autonomy within networks (inherently distributive in nature), and domain names provide enormous possibilities in terms of topologies and policies. These notions have a profound impact on extensibility of the all-IP framework, in terms of scale and adaptability to suit a variety of RATs. These capabilities—inherent to the nature of the IP paradigm—enable features such as mobile IP (MIP), QoS, network address translation (NAT), packet routing, and handovers (vertical and horizontal) across assorted RATs. The corresponding architectural tenets that leverage the amorphous nature of the IP paradigm provide both scale and adaptability—essential ingredients in an assorted RAT ecosystem. Distribution is both from an end-to-end perspective and from a radio-access perspective. Compatible ideas are essential since mobility inherently implies a distribution. It is recognition of the unchanging principles—distribution, autonomy, ubiquity, adaptability, extensibility, and scale—that have led to the adoption of IP everywhere. This is manifested through the global presence of the Internet, where mobile systems are emerging as a centerpiece of this compelling communication web. The implications of these trends are enormous and unprecedented, shaping a variety of socioeconomic and behavioral trends in the human journey.

The arrival of 3G mobile technologies ushered in the era of IP-oriented mobile access networks—Universal Terrestrial Radio Access (UTRA) and 1xEV-DO—for wide-area packet-oriented mobility-enabled access. The latter systems were a precursor toward 4G wide-area mobile technologies classified as Evolved UTRA (E-UTRA). The E-UTRA embodies a global convergence of a wide-area RAT for 4G and beyond in the long term. The primary motivational tenets are

- Separation of the signaling and transport procedures
- Optimization of operation and maintenance

These foundational tenets are a basis for the emerging architectural frameworks in the evolution of the mobile access networks. The building blocks in the transformation of mobile communication networks toward the distributive paradigm of an all-IP network are

- Definition and separation of the control plane (signaling) and the user plane (transport)
- IP Multimedia Subsystem (IMS)
- Definition of IP technology extensions to suit different RATs

The proliferation of hotspots (Wi-Fi radio access) satisfies the connectivity demand for stationary to pedestrian mobility speeds, where connectivity is enabled across smaller coverage areas or local areas of coverage. These islands of connectivity complement the wide areas of coverage enabled mobile connectivity, via E-UTRA and satellite technologies. The integration of Wi-Fi radio access, together with 3G and 4G types of RATs, is pivotal in the evolution toward seamless IP multimedia services, satisfying the universal drivers of ubiquitous access and a rich end-user experience. The entities in the radio-access segment of mobile connectivity include the base station, a transceiver that handles the radio-access link, and the base station controller (BSC), a transceiver that handles the RRM and handover. The access network subsystem, which serves the radio-access segment, utilizes the IP blueprint for a common connectivity framework—wireless or wired. This is the seed for evolutionary perspectives in the next-generation horizons, in the quest for a user-centric confluence of mobile connectivity and service.

In combination, the radio-access segment and the IP-oriented network connectivity segment provide the elements of generalized next-generation mobile connectivity architecture. The radio-access segment serves to process the message transactions, between the mobile device (an endpoint of connectivity) and an intermediary IP-oriented network (a multimedia gateway or an edge router). The IP-oriented connectivity segment has the intelligence to manage the connectivity across the same RAT/domain of coverage area or a different one, depending on the identifiers embedded in the IP transport, associated with a remote destination. This completes the end-to-connectivity for communications. The types of intervening links may be homogeneous or heterogeneous, therein lies the power and flexibility of IP-oriented connectivity within the NGN architecture, which is intrinsically adaptable, binding different types of RATs while allowing the tools for operational efficiencies.

3.5.2 Assorted Coverage Footprints

MIP multimedia services—data-centric, distributed, and dynamic in nature—demand diverse connectivity choices. The availability of assorted RATs and topologies provides innovative opportunities for both the access to and the creation of MIP multimedia services. The interplay of business models, RATs, and usage scenarios allows a landscape of possibilities to optimize the service experience.

The LTE access technology architecture, interfaces, and signaling provide a framework to orchestrate a variety of coverage topologies—macro cells, small cells, etc.—for flexible connectivity optimization opportunities. These opportunities provide benefits for both the user and the access provider—the former, in terms of service access choices, and the latter, in terms of resource utilization enhancements. Assortment of radio-access choices together with interworking across the corresponding coverage footprints is foundational in the directions toward seamless mobility. The elemental components enabling seamless mobility consist of a systematic integration and interoperability of a variety of access technologies. The access technologies may be delineated in terms of coverage, shaped by a variety of parameters, such as capacity, QoS, bandwidth, load conditions, etc. Regardless of whether or not a single RAT, such as LTE, is adopted for all types of coverage footprints, the need for different coverage footprints is expected to be driven by unique and capacity-driven market and usage scenario requirements. The dynamic and evolutionary nature of data services intrinsically allows attractive choices for access—a user-centric paradigm.

In this paradigm, IP serves as the protocol of choice (universal glue) across disparate islands of RAT coverage. It is a pivotal enabler separating the service logic from the transport logic. The IP-oriented packetized transport provides a generic mechanism for mobile access to IP multimedia services, where the service intelligence is at the network edges. This model allows IP to serve as the backbone of a packet core network that facilitates the binding of disparate islands of coverage footprints, served by a homogeneous or a heterogeneous RAT environment.

From a user-centric perspective, the availability of individual preference-driven services and applications is paramount, not the features of an underlying RAT, which serve as enablers behind the scenes. Availability of access to services and applications—anytime, anywhere—defines the usage and adoption potential. This reflects a shift from the closed hierarchical mobile access systems of yesteryear, where mobile services were confined to voice and related circuit-switched derivatives, such as paging and text messaging. In contrast, IP transport networks, allowing flat, decentralized, autonomous islands of RATs, forge an environment for a common core network. This serves as a foundational fabric in directions toward distribution and virtualization of the functions resident in the core network.

The transition toward a common core network enables innovative revenue generation opportunities rendered via constituent horizontal capabilities—flexible policy, charging, and management frameworks. Commonality at the IP layer—resident in the core network—is a bridge across current access capabilities and evolutionary access trends. The various coverage footprints serve as access pathways for both services and peer-to-peer human communications—unicast, multicast, or broadcast IP multimedia content.

Given the nature of the access heritage, human communications—largely voice and multimedia communications—is the central theme in mobility scenarios within and across diverse coverage footprints. In this regard, the access context, bandwidth, and latency are significant parameters for a preservation of the service experience. This in turn implies smooth handovers across disparate coverage island served by different serving access gateways. The confluence of a mass market notion and user-level flexibility is an imperative in the information age—an interdependent landscape. Flexibility of access—both network and user perspectives—is for a personalization capabilities limited only by resource availability and imagination. The use of multimode—enabled with heterogeneous RATs—mobile devices provides elegant seamless mobility capabilities across disparate access technology islands. This is in contrast to more complex network-oriented handover schemes—complexities resulting from the fact that access networks serve as connectivity intermediaries between communicating endpoints. The latter is fraught with inherent latencies resulting from signaling, context, and resource management processing requirements.

The primary motivating consideration for leveraging disparate RATs is a market reality—there is no "one-size-fits-all" panacea that can satisfy the vibrant nature of user-centric preferences. An axiomatic truth—a high level of optimization for any given measure, which is a part of a multitude of elements that characterizes the whole—inherently limits flexibility. The limitations of narrow optimizations are manifested not only in terms of technology but also in terms of business model and usage scenario preferences, which are both innately diverse. It is this diversity of opportunities that creates value and drives innovation.

3.5.3 Hybrid Access

Assorted access technologies, with different coverage footprints for information creation and consumption, are vital ingredients in the milieu of diverse user demands—mobility scenarios, social patterns, and the related usage behavior changes in the age of information. Mobile broadband coverage and capacity enhancement potential for an experience-rich service environment are the twin pillars of motivation for the rollout of a hybrid network. There are several categories of beneficiaries. Examples of categories of mobile broadband users, for a variety of multimedia services, include the following:

Nomadic users—People on the move (e.g., personal, business, or other savvy Internet users) with relatively variable bandwidth demands and interactive usage scenarios. Broadcast/multicast/point-to-point services (e.g., public safety, emergency, traffic, informational, telemedicine, weather, news), where the individuals are primarily receiving services from an authorized management entity, with limited or no interaction

Communities—Broadcast/multicast/point-to-point services, where data traffic is aggregated (e.g., rural areas, building or other construction sites, resorts, trains, planes, ships)

Institutions—Companies or organizations with diverse geospatial coverage requirements for access to multimedia services (e.g., governmental, law enforcement, news media), with high bandwidth, security, and reliability requirements

Standardized, open connectivity network capabilities provide opportunities for widespread interoperability, fostering a broad participation, and opportunities for business partnerships, collaboration, and competition. This is a trajectory toward a cooperative system of connectivity and services—networks, devices, and applications—to orchestrate a compelling user experience. The proliferation of mobile data (information) traffic over the Internet fabric is an affirmation of the viral nature of user-oriented services. Applications, in conjunction with assorted access highways, provide a framework for innovative services, fostering an unlimited potential for service expansion over the Internet landscape. Personal mobile devices (e.g., smartphones, netbooks) provide users with "anytime, anywhere" connectivity to services. Cooperative systems of mobile connectivity foster a transformational paradigm, in terms of yet-to-be-envisioned usage patterns, behaviors, lifestyles, and cultures. These transformational trends enhance the opportunities for service growth measured by user experience and market vibrancy. An ensemble of terrestrial and satellite access networks and technologies is poised to play a significant role in these transformational and evolutionary trends. Viral and organic, access diversity is a powerful paradigm, revealing new horizons for sustainability and market expansion.

With the worldwide variation in spectrum availability including variations in regional regulatory regimes, the historical and traditional silo-oriented architectural and business models for enabling flexible mobile broadband connectivity frameworks are inherently fraught with limitations. These legacy models worked well for low-traffic volumes associated with voice-centric services. Such models are by design susceptible to severe limitations associated with the efficiency of resource utilization in terms of coverage, capacity, extensibility, scalability, and cost. Sustaining such models, or marginally modifying them, would be deficient in establishing a direction to leverage the enormous revenue generating potential, implicit in the nature of mobile multimedia services. Innovation, global standardization, and openness to leverage distributed architectural models—heterogeneous networks, self-organizing networks (SONs), spectrum sharing, etc.— are some of the ingredients necessary to meet the diverse and challenging coverage and capacity demands to support user-centric IP multimedia services.

The multifaceted benefits of advanced wireless connectivity technologies include enhancements in terms of battery life, application responsiveness, system resource utilization, and service experience.

With the enhanced spectral and operational efficiencies (cost of connectivity per bit) of the 4G and beyond broadband connectivity technologies, the potential for earnings is high, relative to the investments in the terrestrial segment (ATC*-enabled base stations) in the hybrid mobile terrestrial satellite system (HMTSS) framework. Further, the hybrid nature of HMTSS offers the potential for innovative capacity management techniques between the terrestrial and the space connectivity segment to optimize the user experience, coverage, and capacity. These capabilities foster more opportunities for widespread collaboration across a variety of connectivity and content providers for mass-market-oriented services—residential and enterprise.

An open approach with standardized HMTSS interfaces promotes widespread interoperability and a multivendor participation in products and services. The inclusion of terrestrial usage of the mobile satellite service (MSS)† frequency band promotes a realization of mission critical— public safety, emergency communications, etc.—and user-oriented mobile broadband services. Examples of the MSS frequency bands are the L-band‡ and the S-band.§ An adaptation of the E-UTRAN terrestrial radio interfaces and the packet core network provides heterogeneous connectivity that includes ATC-enabled E-UTRAN base stations.

The challenges associated with terrestrial mobile broadband and HMTSS are mainly in the realm of managing spectrum availability, radio-interference management, and the form factor of a hybrid mobile device—a dual terrestrial satellite device (DTSD) type. In concert with technological advances (DTSD types and network distribution), business model innovation and industry-wide partnerships, bolstered by new revenue sharing opportunities, are enabling considerations to meet the challenges associated with a realization of ubiquitous (rural or metropolitan) mobile broadband connectivity. The enabling of multifaceted access enriches the mobile connectivity environment, by expanding choices for customizing user experience, coverage, capacity, and cost. A heterogeneous RAT approach is foundational to meet the dynamic and geospatial coverage requirements, for an attractive rendition of IP multimedia service experience and availability. Intelligent detection and selection of a terrestrial or satellite mode of connectivity transcends the limitations associated with any single RAT. In roaming partnerships, where terrestrial connectivity providers, satellite providers, and hybrid connectivity providers cooperate, the availability of capacity and coverage is promoted together with an attractive and expansive mobile broadband ecosystem.

---

* Ancillary Terrestrial Component.
† Mobile Satellite Service.
‡ L-band: downlink (1525–1559 MHz), uplink (1626.5–1660.5 MHz).
§ S-band: downlink (2190–2200 MHz), uplink (2000–2010 MHz).

3.6 Ideas and Concepts in Connectivity Patterns

An ensemble of elemental concepts curates the lens of evolution and innovation. It is the spirit of concepts that sheds light on the nature of complexities and softens boundaries to reveal interdependence in the information age, analogous to the natural world around us.

Connectivity through wireless links bridges the time and space separation with the freedom of mobility. This capability of wireless links comes with its innate probabilistic nature, which sets it apart from wired links that are relatively deterministic conduits of information. The probabilistic nature of wireless links is prone to corruption between a point of attachment to a wired network edge and a mobile device, as a result of scattering, dispersion, and interference of electromagnetic waves that form the wireless link. Additionally, mobility adds time–space variations that undermine the robustness of the wireless link. Wireless links permit a variety of choices, in terms of both device distance from attachment points and the device mobility across attachment points. This has widespread ramifications related to types of applications and services that are rendered over these conduits of information. It promotes a variety of connectivity choices in an intrinsically social and machine-oriented communication fabric. In turn, these choices serve as foundations for innovation in both the wireless link and the possibilities associated with applications and services.

The wireless links themselves are subject to designs with different characteristics for enabling different ranges of separation (coverage area) between an attachment point in a network and the device. These differences also translate into the amount of information (capacity) that is transportable over a wireless link for acceptable performance measures. The different wireless link designs span different types of technologies—heterogeneous connectivity—to accommodate different coverage and capacity capabilities. The available coverage and capacity metrics for a given wireless link technology are a function of frequency spectrum ranges and cost trade-offs. The trade-offs play out in terms of operational efficiencies, networks, devices, and frequency spectrum utilization.

The possibilities for connectivity that arise out of the choices in coverage and capacity provide the ingredients for crafting both user-centric and machine-centric services. In both cases, the wireless link is a foundational building block of mobile connectivity. This is the essence that replicates the nature of communications, where connectivity is pervasive in a ubiquitous rendering of applications and services. It serves as untethered bridges of information and expression in a mobile world. The stage of connectivity vanishes in the experiential story and in the creation and consumption of information.

A long-term evolution trajectory for wireless mobile connectivity has adopted the following considerations as being essential:

- Simplification of the system-level architecture that embraces the notion of open interfaces
- Reduction in the power consumption of devices
- Flexible service provisioning for enhancements in user experience
- Reduction in the cost per bit of information transport
- Flexibility in the introduction and use of new frequency spectrum
- Leveraging of terrestrial and satellite connectivity for widespread coverage
- Coexistence and coordination across variations in coverage footprints and technologies

A flexible multiplexing of a variety of radio-access systems and services are among the ingredients that unveil new vistas for agile and optimal mobile connectivity and service. The mobile connectivity deploy-and-serve cycle, fueled by technology evolution and innovation, is a catalyst in the acquisition process for more radio-access spectrum, as the landscape of new services continues to expand. Along these directions, prominent capabilities include dynamic spectrum sharing and a service-oriented RAN evolution, with augmented trunking efficiencies over the radio interface and service awareness. In the latter case, akin to a virtualization* of functions in the core network, a virtualization of radio-frequency carriers and computing resources opens avenues for a simplified management of radio-layer resources and services. These capabilities are a well spring for enabling the availability of bandwidth on-demand and for spectrum-sharing choices across mobile connectivity providers such as mobile network operator (MNO) or mobile virtual network operator (MVNO). Furthermore, the capabilities enable enhanced efficiencies for resource sharing between services associated with unicast (point-to-point) information traffic or MBMS traffic.

The partitioning of the system-level architecture into an access segment (wireless entities) and a core segment (wired entities) is elemental in separating the nuances of each, while at the same time allowing the necessary coordination and collaboration between the two segments to optimize the exchange of information necessary to provide a rich service rendering and experience over a wireless mobile connection. These directions embody a shift toward opportunities for decentralized arrangements for connectivity. Decentralization is a centerpiece for flexible adaptations to localized connectivity management and market demands. In these changing directions, the challenges shift from a centralized control to coordination across decentralized islands of connectivity.

3.7 Coordination in Connectivity

The access and the core segments of connectivity collaborate over open and interoperable interfaces to orchestrate the establishment of an end-to-end mobile information conduit. This conduit is tuned to enable a user-centric experience, where the device and the network play supporting roles to embellish the potential for innovation in the creation and delivery of services. Traditional business models and strategies for mobile connectivity then evolve into a service centric paradigm, where new horizons for highly imaginative services emerge as the central theme in human communications. With this backdrop, the demands on mobile connectivity for providing the appropriate levels of QoS become increasingly complex and demand awareness across the access and core segments through nonhierarchical cross-layer collaboration and coordination.

The LTE system exemplifies these directions toward the creation and delivery of user-oriented services. The directions embody shifting paradigms in the orchestration of mobile communications, where both the type of media and the modalities can be arranged in virtually unbounded combinations to cater to customizable service experiences. Much higher information rates, akin to those available over the fixed Internet, are foreseeable in the mobile Internet. These advances are complemented with advances in reduced latencies, always connected potential, scalable bandwidths, adaptable service quality, and a graceful coexistence with other types wireless mobile access technologies. These advances are afforded in large part through the flexible nature of the LTE radio interface and the associated packet-oriented system architecture that allows configurable choices for dynamically and incrementally trading-off capacity and coverage while optimizing a mobile service experience.

---

* Network functions virtualization (NFV): An industry specification group (ISG) endeavor for virtualization in ETSI.

Figure 3.20 Logical model of entities in the LTE connectivity system.

The notion of coexistence and interdependence plays out across the protocol layers and across peer layers, for an optimization of mobile wireless link integrity vital for mobile service quality management. Radio-layer requirements, driven by the nature of wireless links, are coordinated with link integrity mechanisms such as TCP for end-to-end link integrity over the wired segments of an end-to-end connection. The end-to-end performance of a connection that includes both wireless and wired links is largely characterized by throughput, packet loss, and delay.

The LTE system utilizes wider and flexible bandwidths that allow an assortment of choices and features to manage the link integrity under variable signal and mobility conditions. This facilitates enhanced coordination capabilities across the radio and network layers to satisfy the connectivity demands associated with diverse applications. A logical model of entities that serve as a context for the LTE mobile connectivity system is depicted in Figure 3.20.

The LTE radio link is characterized by OFDMA. This foundational technology allows higher spectral efficiencies, lower operational costs, lower latencies, and higher information transfer rates. The forward-looking building blocks foster a coordination and collaboration across the radio and network layers for adaptation to the diverse demands of mobile multimedia services. Topological flexibility is afforded through architectural flatness and decentralization. These attributes are pivotal in the directions toward a localization of service experience while effectively enabling a management of capacity and coverage. The combination of the radio-network E-UTRAN and the EPC is referred to as the EPS. The EPS embodies the advances described earlier with sustainable forward-looking direction, at the radio-layer and at the network layer for mobile connectivity. The related specifications are crafted within the 3GPP standards. Colloquially and loosely, these new directions are often referred to as LTE and its evolution as LTE-A. Enhanced throughput and information transfer rates over the radio interface were realized through enhanced capabilities at the radio layer to meet the ITU qualification requirements for 4G, which corresponds to the LTE-A suite of specifications.

3.8 Architectural Framework Horizons

The move toward packet-oriented connectivity, fueled by an increasing demand for multimedia content and resource utilization efficiencies, realized through IP is the cornerstone of the mobile connectivity fabric. Implicit in this significant direction is the migration of the traditional circuit-oriented telephony service transport, such as voice, messaging, etc., toward architectural tenets that are essential and compatible with the notions of a unified multimedia service transport, where voice and messaging are simply different modalities among potentially unbounded combinations of media types.

The EPC framework hinges on the guiding principles of the NGN* vision. The vision embodies the enablement of anytime, anywhere and personalized mobile communications while allowing flexible information transfer rates, multimedia content, contextual services, and variable wireless

---

* ITU-T Recommendation Y.2012 (2006), "Functional requirements and architecture of the NGN release 1."

coverage footprints, indoors and outdoors. In other words, the EPC architectural theme facilitates the NGN vision and enables the attributes of ubiquitous computing. These attributes include a matrix of devices—smartphones, sensors, transducers, etc.—within the Internet of Everything (IoE), where they function nonintrusively and seamlessly to promote automation and to enhance the human experience.

The centerpiece of the NGN framework is service convergence. This is exemplified by its defining characteristic, which is the ability to enable the delivery of heterogeneous services, consisting of assorted data content such as voice, video, audio, picture, text, etc. The types of services include both session-oriented and interactive or transaction-oriented modes. Service delivery is enabled via unicast and broadcast modalities of mobile wireless or fixed packet data transport. The NGN framework astutely separates the service and the transport functions in a loosely coupled fashion to allow a variety of physical instantiations of the archetypal architectural model within the NGN framework. The definition of the various and assorted types of content combinations are left to design imagination and innovation, while the focus of the NGN framework is to provide interoperable and ubiquitous transport pathways (wireless and wired) where content shapes and is shaped by the human experience. The NGN framework is crafted to preserve service consistency over wireless or wired transport and devices.

With the facet of mobility, the mobile device is a natural interface for a continuous exploration of experiential enhancement in the consumption and creation of assorted content provided over packet-oriented architectures aligned with the NGN framework. Figure 3.21 shows the entities in the NGN model. The IP serves as a de facto glue in the NGN framework, as a result of its logical nature, in a packet-oriented information landscape. In the NGN model, the distinct segments of functionality consist of the transport stratum and the service stratum, together with the management, end-user, and other network interoperability functions. The transport stratum provides a

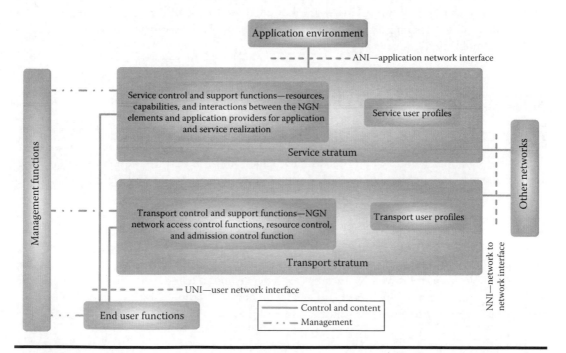

Figure 3.21   NGN model.

generalized connectivity suite—users, service platforms, heterogeneous access technologies, and networks that are geographically distributed with topological variations to suit diverse business models. The service stratum provides a generalized service environment—geographically distributed servers associated with different types of services such as voice, video, web, data files, etc., that require the invocation of a set of applications associated with a service profile or a service classification, real time or non–real time. Each stratum—transport and service—has its own constituent planes consisting of the user/data plane, control plane, and management plane. Further, each stratum is decoupled from an architectural perspective, such that the administrative roles,* provisioning, and functions are distinct. The elements for satisfying end-to-end QoS demands, within the NGN framework, consist of coordination and cooperation across the transport stratum (access network and core network segments) and the service stratum (application resource requirements).

The notion of resource sharing is paramount, for enabling multimedia content and services, over physical channels—wired or wireless. The use of packet switching central to the theme of the EPC framework to facilitate the management of multiple information transfer sessions over a physical channel for single or multiple mobile device users. This implies an improvement in the utilization of the available physical channel capacity, as a result of intelligent sharing, relative to a dedicated resource utilization approach embodied in a circuit-switched architecture. The packet-switched scheme consists of distinct, service, control, and transport functions to enable physical resource sharing across real-time and non-real-time services. The packet-switched model allows flexibility, with respect to the introduction of new features, as a result of commonality and ubiquity inherent in the use of IP. Distribution and diversity of RATs represent a valuable facet of the EPC architectural profile. From a business and deployment perspective, the use of a common access framework for information transfer (voice and multimedia data) is a compelling direction for optimizing both capital and operational costs, while scaling efficiently. From a user-experience perspective, transparency pervades, where the underlying technologies are invisibly subsumed within a virtually unbounded landscape of services. These avenues are a cornerstone in the shaping of architectural directions toward ubiquity—universal interoperability and mobility.

3.9 Embodiment of Paradigm Shifts

In the transition toward an all-IP mobile connectivity network, the dominant guiding principles for the LTE EPS are based on the NGN framework. These guiding principles consist of a separation of the signaling (control) and user (traffic) planes, and the optimization of the user plane. These principles are vital to address both dynamic mobility conditions and scalability of proliferating information flow volumes—Big Data—characterized by multimedia content. The loose coupling of the signaling plane and the user planes allows variability of scaling in terms of information flow types—control or traffic. Control-oriented information volumes are driven by the number of users, while traffic-oriented information volumes are driven by the number of services. The traffic-oriented information volumes are virtually unbounded in potential, since mobile services and application are driven by attractive experience and demand. This drives the expansion of Big Data, since the mobility aspect is aligned with personalization, which in turn implies large and increasing multimedia content volumes.

---

* ITU-T Recommendation Y.110 (1998), "Global Information Infrastructure principles and framework architecture."

The optimization of the traffic plane, in terms of reduced hierarchical connectivity segments, under mobility conditions, then becomes a vital architectural consideration. Direction toward flatness in the traffic traversal path of connectivity is a dominant facet in the paradigm shift toward an all-IP mobile connectivity network. The shift away from a hierarchically oriented mobile connectivity architectural model toward flatness in the arrangement of network entities is depicted in Figure 3.22. These entities naturally have relatively more functional autonomy to cooperatively engage in the orchestration of mobile connectivity. These autonomous functions process and enable the flow of information transactions. The EPS model reflects autonomous directions and is an illustration of a reduction of the number of network entities required to orchestrate mobile connectivity. This direction sets the stage for implementations that scale smoothly to support coverage and capacity expansion. The information transactions that are characterized by multimedia packet data traffic are enabled over heterogeneous mobile access technologies through open and interoperable interfaces—a prominent facet together with the flatness of the EPS architectural model. Logical oneness is promoted via IP at the network layer, which serves as a conduit for information flows over heterogeneous RATs. The separation of the signaling information flows (signaling or control plane) from the user information flows (user or traffic plane) is a guiding principle within the EPS architecture that facilitates the design tenet of loose coupling, where the two different types of information flows are governed by distinct requirements to establish and maintain mobile connectivity.

The signaling information flow scale is associated with the number of users and connectivity devices, while the traffic information flow scale is associated with the volume of multimedia content creation and consumption. These distinct requirements serve as the dominant themes to optimize the utilization of resources, associated with the signaling and the user plane, in the forward-looking logical architecture that characterizes the EPS. The availability of choices for decentralization or centralization that are guided by the distinct nature of the plane of functionality (signaling or traffic plane) is an essential facet of a next-generation mobile connectivity architectural model. The

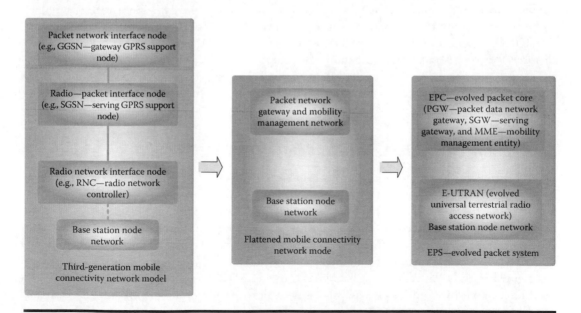

Figure 3.22 Toward flatness—next-generation architectural model.

distinct nature of the signaling plane relative to that of the traffic plane functionality is characterized by their different objectives. The signaling plane provides a coordinating function, while the traffic plane provides pathways for the flow of user- or service-related information.

The information flows could be among humans or machines in the context of IoT. The signaling plane, in its role as a coordinating function, allows distributed or autonomic systems to behave in a coherent fashion. This is akin to the physiological model of the starfish, which has a distributed nervous system, where each of its five limbs is autonomous. Signaling across the distributed nervous system allows the starfish as a whole to move or act in an integrated manner. The signaling plane functionality is typically centralized via the corresponding network entities since its role is to coordinate the flow of information. In a virtualized rendition, the control functions are open to a range of possibilities—centralized or distributed—where the available physical resources are shared. This is in contrast to the traffic plane functionality, which allows decentralization and heterogeneity of access technologies and topologies since no topological coordination of information is required. The distribution of network entities, engaged in the transport of information, allows both operational and performance efficiency enhancements. The enhancement in operational efficiency accrues through flexible deployment choices and scaling of these entities to suit capacity and coverage demands. The enhancement in performance efficiency accrues through an optimization of information flow paths, based on the proximity between the source and destination of the information flows. Protocols that enable mobile connectivity at the network layer are logical signaling abstractions that provide mobile connectivity support, based on extensions to IP, such as MIP, proxy mobile IP (PMIP), and GPRS Tunneling Protocol (GTP). These guiding principles allow implementation innovation and technology heterogeneity while promoting interoperability across mobile connectivity network and devices. The EPS architectural model embraces these guiding tenets, and mobility enabling signaling, to enable flexibility in deployment preferences that are shaped by business models.

The EPS is intrinsically a packet-oriented architecture, which leverages the IP all the way to the edges of the network to foster rich choices in the rendering of mobile connectivity. The edge of the network is where the maximum uncertainty exists as a result of the dynamic nature of mobility and its interplay with the probabilistic nature of the radio link. The edges of the network, realized through the bases stations (eNBs—eNodeBs), collaborate not only with the EPC but also among themselves to orchestrate the coordination of mobile connectivity in an effective manner over distributed IP interfaces, identified in the EPS model as X2. Using the context for LTE connectivity depicted in Figure 3.20, the specific logical entities and their connection arrangement within the EPS are shown in Figure 3.23.

3.9.1 Connectivity for Service Mobility: EPS

The directions toward network decentralization and distributed have widespread ramifications in terms of evolutionary considerations motivated by new perspectives and insights, for providing connectivity for service mobility. The power of these hidden insights is derived from a holistic view of the architectural tenets and the larger context of human communications and experience within which these interdependent ideas operate. The architectural and technology directions toward mobile broadband, including the radio and the network layer, are collectively represented in the EPS framework. In this text, the term EPS is used while referring to the forward-looking visions and insights associated with next-generation mobile broadband systems. The EPS architectural model, which includes the radio-layer and network-layer components, embodies open interfaces that permits architectural flatness, decentralization, and distribution.

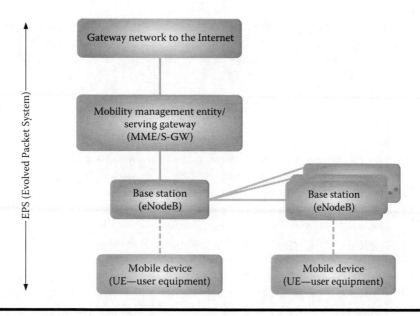

Figure 3.23 EPS model.

In Figure 3.23, the packet data network gateway (PGW) entity, within the gateway network to the Internet functional block, serves as a gateway for connectivity to the Internet. The PGW is an anchor in the EPC for end-to-end IP multimedia services (real-time and non-real-time services), with diverse QoS demands. The EPS transport bearers for providing service connectivity are shaped to support service-specific QoS demands. An EPS transport bearer consists of an IP packet flow with a specific QoS allocation between the mobile device and the PGW entity. The different QoS demands are satisfied through the use of differently shaped EPS bearers, according to the QoS demands of a multimedia service that utilizes the EPS bearer. Different EPS bearers can also be utilized for connectivity to different packet data networks (PDNs), via different PGW entities. For instance, an EPS bearer for a real-time service such as VoIP would be shaped with a different QoS profile (low latency, small packets) relative to a non-real-time service such as File Transfer Protocol (FTP) or web browsing. Security and privacy are intrinsic system aspects for a preservation of user experience and access network protection. The QoS allocations are end to end, within the scope of the EPS, with the mobile device at one end and a PDN environment at the other. The radio-layer and the EPC interfaces collaborate to enable device mobility, while the QoS allocations are established and preserved according to the configured and available EPS resource conditions.

3.9.2 EPS Framework Characteristics

The EPS framework has characteristics that are associated with a flat and distributed all-IP network. The entities in this framework are categorized in terms of whether they are radio specific or core network specific. The functions associated with these entities span the handling of control and user plane information. The eNB (radio-specific entity) interacts with the entities in the EPC for the handling of the control and user plane information for the connection and mobility management of the mobile device or UE. The mobility management entity (MME) within the EPC handles the control plane information required to manage the mobility of the mobile devices while enabling virtually seamless user plane connectivity. Within the EPC, generic routing entities handle the flow of user plane traffic between the mobile device and a PDN. The use of generic routing entities, with well-defined and specified protocol interfaces, is a step toward lowering capital and operational costs. The control, traffic/bearer, policy, and database functional entities and their interfaces, which are the building blocks of the EPS framework, are shown in Figure 3.24.

The prominent characteristics of the EPS framework are reflected in architectural flatness and in the separation of the control plane and the user plane (synonymous with bearer or traffic plane) information. This separation is realized via the split of information flows to and from the base station (eNB), with the MME in the path of control information and the serving gateway (SGW) in the path of the traffic or bearer information. MME provides critical control plane functions in the LTE EPS architectural model. The MME is the primary entity for the processing of mobility control and management information. Consequently, it terminates the non–access stratum (NAS) signaling protocols, associated with the mobile device, and maintains the state information context associated with the mobile device. The functions of the MME consist of the mobile device authentication, management of mobile device states (idle, connected, etc.), supervision of handovers across base stations, location of the mobile device, and establishment of transport bearers of voice and data, as needed for connectivity to the PDN, within the EPC and the Internet, based on the mobile device context. The radio resources are allocated and managed by the base station (eNB), since these resources are local to the base station, for wireless connectivity to the mobile device. The MME and the user plane traffic/bearer gateway entities (PGW/SGW) are intrinsically unaware of the radio resource conditions—a classic rendition of functional role-oriented information hiding and loose coupling—for architectural and design elegance. This separation of the core network entities from the base station stems from the inherently probabilistic nature of radio link, which is in stark contrast with the relatively deterministic conduits of transports in the core network. Consequently, the MME scope of control does not include resource allocation associated with the base station.

Within the EPC, the functional role of the MME is extensive, and its prominent interfaces are shown in Figure 3.25. The PGW performs generic information packet processing and routing

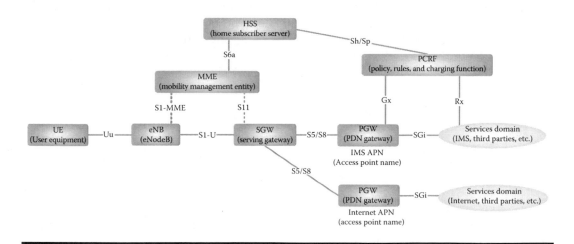

Figure 3.24 Arrangement of logical EPS entities and interfaces.

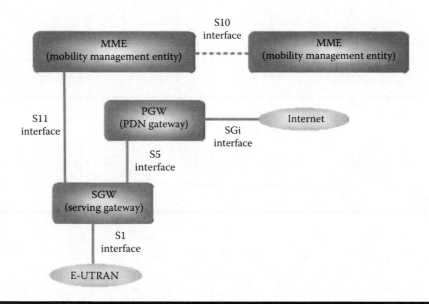

Figure 3.25 Control and bearer plane entities and interfaces.

functions, for information flows that are either sourced from the mobile device or destined for the mobile device. The PGW serves as an information anchor entity and is a gateway for a PDN within the EPC.

The arrangement of the logical pathways of information in the EPS framework is depicted in Figure 3.26. The architectural characteristic of this arrangement is reflective of hierarchical or vertical flatness in terms of both the control plane and the user plane. The benefits that accrue from a model of architectural flatness can be delineated in terms of higher levels of fault tolerance enabled through a distribution of EPS network entities and reduced latencies for the traversal of information packets between a source and a destination.

The challenge naturally translates to effective coordination among relatively autonomous network entities, through an efficient handling of coordination, via the control plane. The

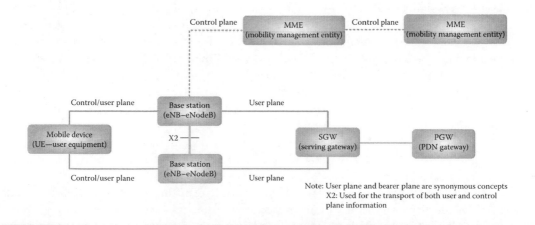

Figure 3.26 Logical pathways of information—control and user plane.

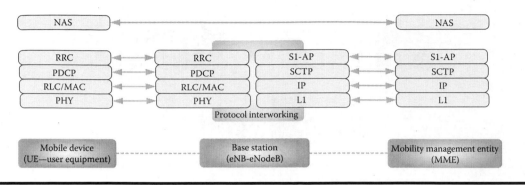

Figure 3.27 Control plane information—protocol-oriented flows and EPS entities.

logical organization of the control plane between the MME and the mobile device is shown in Figure 3.27. The separation of the control plane EPS functions from the EPS bearer functions promotes an independent scaling of the number of sessions and the traffic or the volume of user-oriented information rendered through assorted multimedia services. The MME is the primary control plane entity in the EPS framework that provides the EPS control plane function, which coordinates mobile connectivity in the bearer plane for user-oriented information. The SGW is the gateway at the edge of mobility, which serves as the entry point in the bearer plane for connectivity to the PDN in the EPC. The MME function scaling is driven by the mobile connectivity demand, measured in terms of the number of sessions of connectivity. The SGW function scaling is characterized by a measure of the traffic volume demand across the bearer plane.

The separation of the control plane and the bearer plane functions allows flexibility in the topological placement of the corresponding physical entities to optimize the performance of the EPS, to suit deployment scenarios. For example, the MME and SGW functional distribution and the corresponding one-to-many or many-to-many relationships could be configured based on mobile connectivity load conditions. Further, if these functions are virtualized, the configuration could be dynamic, based on the concepts of SONs.

In addition to the pivotal concept of separating and distributing the control and user planes for flexible choices and independent scaling characteristics, the use of generalized mobility protocols, such as MIP and PMIP, based on the interface affected by mobility, is among the tenets embedded in the EPS framework. The bearer plane information flows are mediated by the associated protocol flows that include the radio link control (RLC)/MAC and the Packet Data Convergence Protocol (PDCP) layers of processing. The PDCP layer is resident in both the mobile device and the base station. It operates in conjunction with the RLC/MAC layer. Toward the EPC, the PDCP layer transfers information to the GTP layer, which serves as a tunnel between the PDCP layer and the EPC. Information transfer in the reverse direction from the EPC toward the base station occurs via the GTP tunnel and the PDCP layer to the RLC/MAC layer. The latter in turn processes the information over the radio interface to the mobile device. The PDCP layer is a bearer plane protocol layer, which is configured by the RRC layer—control plane protocol.

For the establishment of an RRC layer connection between the base station and the mobile device, the mobile device transmits a random access preamble message, as a part of the random-access channel (RACH) procedure. The base station processes this random access preamble message to estimate the transmission timing associated with the mobile device and provides a preamble response to the mobile device containing an appropriate timing advance indication. The mobile device utilizes the timing advance indication to align its transmission timing with the base station. The alignment of transmission timing between the mobile device and the base station is critical as a result of the variability in the signal propagation delays between the mobile device and the base station. The propagation delay variability is a function of the mobility-driven changes in the distances of the mobile device from the base station, which adversely impacts signal transmissions, in terms of timing misalignment or collisions. The RACH procedure for an RRC layer connection and timing alignment, between the mobile device and the base station, is applicable in the following four scenarios:

1. Initial access and connection establishment, from the RRC_IDLE state or a radio-access failure state
2. Handover operation, where RRC connection establishment is required
3. Information to send or receive at the mobile device, in the RRC_CONNECTED state, after the mobile device physical layer (PHY) connection has lost timing synchronization with the base station, as a result of a variety of access conditions, such as a power-saving mode
4. Information to send from the mobile device to schedule the allocation of uplink buffer resources at the base station, via a scheduling request in the physical uplink control channel (PUCCH)

The main objective of the RACH procedure is to avoid contention or a disruption of the information transfers between the mobile device and the base station.* The PHY protocol is the air-interface or the radio-interface protocol layer. This layer transports all the information to and from the MAC transport channels. The information includes the base station search for an initial or handover synchronization, for mobile connection establishment, power control, and measurements related to radio resource control (RRC), within the EPS and with other heterogeneous systems for interworking.

The MAC transport channel is logical and multiplexes and demultiplexes the RLC information with respect to the PHY transport channels. The MAC layer manages the prioritization of the logical channels for the same mobile device and performs dynamic scheduling of logical channels between mobile devices and HARQ. The RLC layer provides a logical transport channel for the PDCP encapsulated information (protocol data units—PDUs). The RLC layer is capable of operating in different modes, characterized by different levels of reliability. Corresponding to a level of reliability for information transport, the RLC layer provides functionality such as ARQ error correction, duplicate information detection, ARQ error correction, and reordering of information packets for in-sequence delivery.

The PDCP layer interfaces with the RRC layer (control plane) and the IP layer (bearer plane), where information is exchanged in terms of service data units (SDUs). With respect to the RRC layer, the PDCP layer provides integrity protection and information transport with ciphering. With respect to the IP layer, the PDCP layer provides the transport IP encapsulated information, with capabilities for header compression, such as ROHC, and ciphering. Depending on the RLC mode of operation, the IP layer is supported with capabilities such as duplicate information detection, in-sequence information delivery, and the retransmission of its own SDUs (encapsulated information) to support mobility-triggered handovers. The RRC layer handles the broadcast of

---

* 3GPP specifications (36.133 and 36.321).

system information, which is associated with bearer plane, and the control plane functions, such as paging, handover, QoS, mobile device measurements pertaining to intersystem mobility across different RATs, management of security keys, and the establishment and release of RRC connections, between the mobile device and the network. The distribution of RRC and management function across the base stations, through the X2 interface, promotes lower latencies for information transfers during handovers. The termination of the RLC and PDCP layers in the base station partitions the radio-related connectivity aspects from the core network–related connectivity aspects.

The RRC layer manages the configuration of the PDCP, RLC/MAC, and PHY layers with the parameters that are required for their runtime functionality. The radio bearers exist between the RRC and the PDCP layers. These radio bearers are mapped to the logical channels between the RLC and the MAC layers. The mapping of logical channels, transport channels, and the physical channels is orchestrated to establish mobile connectivity and to move information between the connectivity network (core and radio network segments) and the mobile device. The logical organization of the bearer plane between the bearer gateways and the mobile device is shown in Figure 3.28.

The distributed nature of the EPS architectural model promotes the avoidance of latency and performance bottlenecks associated with a centralized processing model. The management of radio resources distributed across base stations allows a minimization of the propagation of radio-access measurements report information from the mobile device through the network in a hierarchically oriented centralized processing model. In a distributed RRM environment, the base stations are enabled to cooperate and coordinate their loading information with neighboring base stations. This loading information can then be proactively utilized as a look-ahead cue for the management of handovers across base stations to facilitate low latency and reliable mobile connectivity to a candidate target serving base station.

Context transfer at the PDCP layer mitigates the need for a transfer of the detailed PDCP SDUs information, encapsulated as RLC PDUs over the radio interface, thereby improving the utilization of the radio-interface spectrum resources. The transfer of PDCP layer context to a target base station allows the avoidance of retransmission PDCP information from the target base station to the mobile device. The SDU information is preserved as it traverses each next lower layer of the protocol stack, through the process of encapsulation, concatenation, segmentation, or padding in PDU information. In the case of encapsulation, the PDU information encapsulates the SDU information, typically with header information. In the case of concatenation, the SDU information is chained together within PDU information, while in the case of segmentation, the SDU information is split across multiple pieces of PDU information. In the case of padding, the SDU information is filled with unused bits to suit the format of the PDU information. The arrangement

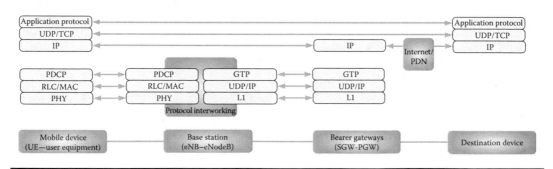

Figure 3.28　Bearer plane—protocol-oriented information flows and EPS entities.

of SDUs and PDUs facilitates a generic structure for the exchange, sorting, and processing of information to meet the requirements of functions and capabilities in the EPS framework that facilitate mobile connectivity. The use of generic processing modules for core network and the radio-access network functions allows for opportunities for functional virtualization that allows reductions in both capital and operational costs associated with the EPS.

3.9.3 Radio Resource Management

RRM is applicable to a variety of functions across the physical (layer 1), MAC (layer 2), and signaling layer (layer 3) of the radio-access segment of mobile connectivity. The component functions of RRM span these three layers. The layers include both the bearer plane and the control plane of the radio-access segment. At layer 3, the component functions are associated with admission control, QoS profiling, semipersistent scheduling, etc., which are relatively static, since they are executed either during system configuration or during the establishment of a new session. The RRM functions at layer 1 and layer 2 are dynamic with a scheduling granularity in the time domain of one transmission time interval (TTI), which is 1ms and contains 14 OFDM symbols, where 3 of these OFDM symbols is used for control signaling. In the frequency domain, the scheduling granularity is one PRB, which consists of 12 consecutive OFDM subcarriers. The allocation and scheduling of PRBs for information transfer to a mobile device occurs at the granularity of one TTI. The requirement for this allocation and scheduling of resources for a given mobile device is that the same modulation scheme is used across the PRBs for each information transmission. Higher levels of scheduling granularity are to be expected at the lower layers relative to the upper layers. Sensitivity to changes as a result of mobility or service is naturally higher at the lower layers, which are the initial information processing layers that detect and process changes before the upper layers are impacted.

The scheduler in the base station, together with the LA and HARQ functions, forms the RRM framework, which performs an allocation of system resources across the mobile devices being served. The allocation of system resources to the mobile devices hinges on the signaling information feedback received from the mobile devices, via ACK/NACK handshakes, and channel quality information (CQI) reports, for each TTI and for each PRB. The spatial domain information, such as whether single or multiple transmission streams are demanded across one or more mobile devices, affects the allocation of system resources. In the case of SU-MIMO, there are two information streams to a mobile device. In the case of MU-MIMO, there are two information streams for each mobile device among multiple devices. The RRM framework in the base station distributes the system resources across the mobile devices being served, based on mobile device resource demand priorities that are derived from signaling information feedback, CQI, and spatial domain information.

The RRM functions over a downlink shared data channel can be augmented through information diversity enabled by MIMO capabilities. The MIMO rank reflects the number of spatial information streams, while the MIMO code word is utilized to jointly encode up to two spatial streams. The feedback to the base station from the mobile station CQI measurement indicates an average SINR. The CQI is quantized, formatted, and reported to the base station. The reporting can be either periodic or aperiodic, and the resources utilized for CQI reporting purposes is governed by the base station. The LA function, in the presence of MIMO, is further enhanced, when the CQI feedback to the base station of the CQI measurements is made for each of the spatial streams that are separately encoded. As a trade-off to minimize performance impairments resulting from the signaling overhead of the CQI feedback process, the spatial streams are grouped, so that the grouped CQI measures are sufficient for MIMO-oriented information transmission. The channel adaptation at the mobile device receiver is achieved through the use of three types of feedback information contained in the channel state information (CSI),* which is fed back to the serving base station, based on the channel conditions perceived by the mobile device.

The purpose of this feedback is to optimize the transmission of information to suit the channel conditions at the mobile device receiver. Knowledge of the radio-channel conditions, characterized by the CSI feedback, allows the base station to alleviate the interference generated from the transmission of multiple spatially distributed information streams. The first type of feedback information is referred to as the rank indicator (RI), which indicates the number of streams of information preferred by the mobile device. For example, an RI of one would indicate that the same information stream is being transmitted across transmissions from base stations, which would increase the coverage. On the other hand, an RI that is greater than one would indicate that different information streams are being transmitted simultaneously, from base stations, which would increase the system capacity. Since the same modulation scheme and coding scheme is utilized for all the allocated PRBs, it can be assumed that the same RI can be used for these allocated PRBs.

The second type of feedback information is referred to as CQI. This parameter is used by the mobile device to indicate a suitable information transmission rate, or in other words a modulation and coding scheme (MCS). The mobile device receiver perceives the signal-to-interference-plus-noise ratio (SINR) associated with the information channel conditions. The mobile devices characteristics, such as the number of antennas, supported modulation schemes, and the type of receiver based on the number of inputs and outputs, affect the CQI estimation process. These characteristics influence the selection of an MCS at the base station that is compatible with the mobile device, for the same SINR value. The CQI feedback value is used by the base station for channel adaptation and in the scheduling of information streams to the mobile device.

The third type of feedback information is referred to as the PMI, which is utilized to assist the transmitter† to select the best codebook for precoding information streams for information integrity through error correction. The PMI feedback provides the hint necessary to the base station in the selection of a suitable precoder to apply on spatially multiplexed transmissions of information. The system performance is enhanced with the PMI feedback from the mobile device since the base station has a better estimate of the channel conditions at the mobile device prior to the transmission of information. This is a closed-loop feedback process, where the base station iteratively adaptively to the PMI determined by the mobile device, based on the channel conditions perceived by the receiver. The PMI feedback delay increases with the mobility speed of the device, which naturally has an adverse impact on system performance, in terms of BER, relative to stationary conditions. The PMI feedback–oriented channel adaptation provides a mitigation strategy for performance degradation, for the device under stationary or mobile conditions.

Under high mobility conditions, the CSI information may only consist of the RI and CQI information (open loop or limited feedback mode) since the closed-loop feedback mode operation to leverage PMI iteratively is likely to be limited in effectiveness. The feedback overhead incurred may not effectively mitigate performance impairments, resulting from rapidly changing channel conditions at the mobile device receiver.

---

* 3GPP TS 36.213, "Technical specification group radio access network; evolved universal terrestrial radio access (E-UTRA); physical layer procedures (Release 11)," v11.4.0, September 2013.
† Love, D.J. and Heath, R.W., "Limited feedback pre-coding for spatial multiplexing systems," in *IEEE Global Telecommunications Conference*, San Francisco, CA, 2003.

The PUCCH is utilized as the feedback channel for CQI, MIMO feedback information consisting of RI, PMI, and the requests for scheduling uplink information transfer. The PUCCH consumes two PRBs for the uplink transfer of the feedback signaling information. Either BPSK or QPSK modulation is utilized for this information transfer. These forms of modulation are simpler and require fewer bits per symbol of information (one bit per symbol for BPSK and two bits per symbol for QPSK) and function well under low SINR conditions, to provide robustness for lower bit rate (smaller packets) signaling information transfer. This signaling channel uses radio resources at the edge of the system bandwidth consisting of one PRB for each transmission, followed by a transmission at the other end of the system bandwidth, in the next slot, for leveraging frequency diversity.

The CSI feedback scheme includes both periodic feedback and aperiodic modes of operation. In the periodic CSI mode of operation, the mobile device periodically transmits the CSI to a base station, while in the aperiodic CSI mode of operation, the base station solicits the CSI from a mobile device via a signaling trigger. The mobile device responds with a CSI that corresponds to the radio-channel condition measurements. In the periodic feedback mode, the CSI is delivered to the base station via PUCCH. The CSI is delivered to the base station via PUSCH in the aperiodic feedback mode. If the CSI feedback requires a high level of granularity or accuracy, which is of the order of a subband, then the CSI is delivered over the PUSCH, where the nature of the feedback operation is periodic or aperiodic. For example, periodic CQI reports, which are part of the CSI feedback information with a period of 5 TTI, which corresponds to 5 ms, are considered as typical.[*,†] For an optimal adaptation to the radio-channel conditions, separate CQI measurements are performed for each spatial information stream that is separately encoded. CQI feedback overhead is minimized by grouping together the spatially distinct information streams into groups, with a maximum of two streams per group.[‡] Each group is encoded separately and corresponds to one code word. This approach avoids significant performance degradation while adapting MIMO transmissions to radio-channel conditions that are susceptible to change dynamically.

The use of CSI feedback in a MIMO configuration from the mobile devices being served by a base station is utilized by the base station for radio resource allocation and scheduling for each of the mobile devices. The rate at which the radio-channel adaptation occurs to suit a MIMO configuration has an impact on the overall system performance. For low mobility scenarios, the rate of channel adaptation of the order of 100 ms results in only about 5% degradation in system performance relative to a fast rate of channel adaptation,[§,¶] where the MIMO rank selection is performed, during each scheduling period.

---

[*] Pokhariyal, A. et al., "HARQ aware frequency domain packet scheduler with different degrees of fairness for the UTRAN long term evolution," in *IEEE Proceedings of Vehicular Technology Conf*erence, Dublin, Ireland, May 2007.

[†] Pedersen, K.I. et al., "Frequency domain scheduling for OFDMA with limited and noisy channel feedback," in *IEEE Proceedings of Vehicular Technology Conf*erence, Baltimore, MD, October 2007.

[‡] 3rd Generation Partnership Project; Technical Specification Group Radio Access Network, "Evolved universal terrestrial radio access (E-UTRA); physical channels and modulation," 3GPP TS36.211, v8.1.0, Tech. Spec., November 2007.

[§] Tuomaala, E. et al., "Performance of spatial multiplexing in high speed downlink packet access system," in *Fifth International Conference on Information, Communication and Signal Processing*, Bangkok, Thailand, December 2005.

[¶] Wei, N. et al., "Efficiency of closed-loop transmit diversity with limited feedback for UTRA long term evolution," in *Personal, Indoor and Mobile Radio Communications*, Helsinki, Finland, September 2006.

3.9.4 *Quality and Policy Coordination*

The allocation and coordination of resources along the radio network access and the core network access segments are essential functions for mobile connectivity. Within the EPS, the GTP is utilized between the SGW and the PGW (S5/S8 interface) for the transfer of information, over this bearer plane segment. The EPS bearer plane is terminated at the PGW, which controls the management of resources to meet the QoS demands. Signaling within the bearer plane on-path model is utilized to allocate the required resources to suit the QoS associated with the requirements of the information being transferred. The policy and charging rules function (PCRF) operates as a decision point entity for policies associated with the mobile connectivity. The enforcement of the corresponding policies for QoS demand-oriented resource allocation is enforced by the policy and charging enforcement functions (PCEFs) entity within the PGW. The PCRF transfers the packet filters and the QoS parameters, to satisfy the QoS demand-oriented resource allocation, to the PCEF, over the Gx, Gxa, Gxb, and Gxc interfaces. The PCEF performs the filtering of the information being transferred, in terms of bit rate and charging for service based on the policy and charging control (PCC) rules. The filtering performed by the PCEF is also applicable to information arriving from external networks—Internet or enterprise virtual private networks (VPNs).

The information transfer across the mobile connectivity segments occurs as service data flows (SDFs) that are created to suit the QoS demands related to the information being transferred. The information pertains to the user-associated traffic, such as web browsing, file transfer, VoIP, email, messaging, etc. The SDFs are bound with the transport bearers, in a manner that is governed by the relevant PCC rules. The binding between the SDFs and the transport bearers is established at the PGW and the mobile device (UE), using traffic flow templates (TFTs). The TFTs contain packet filters (SDF filters) that are used to identify and map the user-associated traffic to specific transport bearers. A TFT is associated with a transport bearer. These packet filters are configurable based on the PCC rules and contain at least the following information, typically referred to as the 5-tuple:

1. Source IP address
2. Destination IP address
3. Source port number
4. Destination port number
5. Protocol identifier (i.e., TCP or UDP)

The PCRF is a centerpiece—coordinating entity for policy and resource allocation—within the EPS that enables opportunities to optimize the user experience for mobile connectivity while optimizing network resource utilization and new horizons for multimedia service innovation. The distribution, coordination, and virtualization are aspects that are amenable in the realization of a PCRF function, such as the MME, SGW, PGW, and PCEF, together with the IMS framework components. QoS-driven resource allocation for the transport of information flows is orchestrated by establishing PDN sessions between the mobile device and the EPC. A PDN session consists of a relational binding between a mobile device and PGW, with user-specific attributes and an allocated PDN IP address. One or more information transport bearers, between a mobile device and the PGW, may be handled within a single PDN session. These bearers—logical in nature—are categorized as either default bearers or dedicated bearers. A default bearer is established during the initial attachment process of a mobile device with the PGW. The default bearer

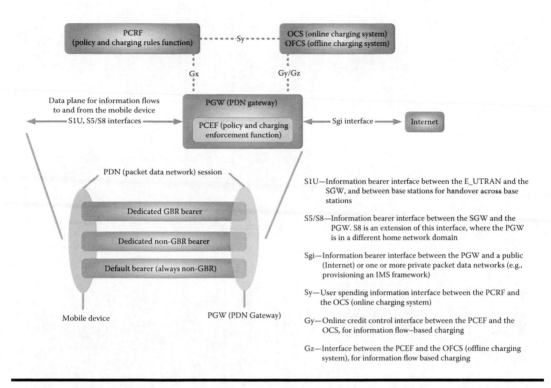

Figure 3.29 PCRF architectural model.

is maintained as long as the PDN session exists. An illustration of the PCRF model is depicted in Figure 3.29. A default bearer is best-effort transport for information flows and corresponds to a non–guaranteed bit rate (non-GBR) resource allocation, where there is no dedicated bandwidth allocation. Dedicated bearers, on the other hand, have a variety of QoS allocation levels for the transport of information flows. These types of bearers are dedicated for a guaranteed minimum bandwidth between the mobile device and the core network. Resource reservation techniques are applied to reserve a minimum bandwidth, based on the QoS demands of the information flows, while the best-effort information flows are transported over a default bearer, which increases the availability of resources for QoS-constrained information flows.

These types of additional dedicated bearers—distinct from a default bearer—have a non-GBR or a GBR QoS allocation level. In the case of a GBR QoS allocation level, there is a certain minimum reserved information flow rate or a bit rate. The setup of a dedicated bearer occurs to allocate the necessary resources to meet the specific QoS demands of an information flow. TFTs are utilized by dedicated bearers for specific treatments of the transport for certain information flows that have specific QoS requirements, such as voice over LTE (VoLTE). Dedicated bearers are always linked to a default bearer and therefore require no new IP address allocation since the IP address is the same as that of the default bearer, which is established when the mobile device initially attaches to the network. Every default bearer has a unique IP address. A group of information flows is also referred to as an SDF, which is an aggregate of the group of information flows that match a specific set of SDF filters to satisfy the QoS requirements of the associated information flows. Each SDF is associated with a single set of SDF filters to realize a single treatment for all the related information flows within the SDF. Each TFT can have multiple SDF filters corresponding

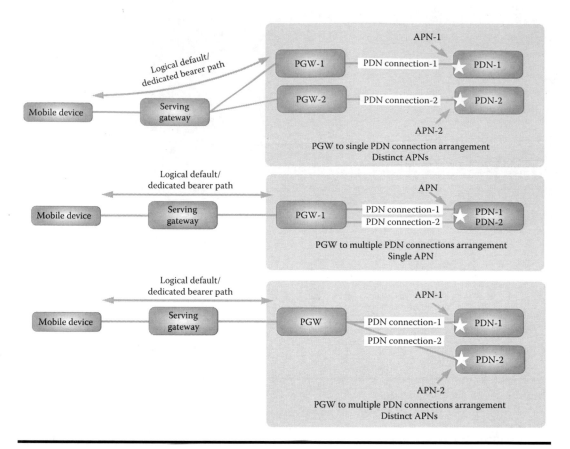

Figure 3.30 Model for PDN connections and APNs.

to multiple SDFs. The SDF filters are applicable for uplink, downlink, or both uplink and downlink information flows. These filters use the 5-tuple of IP address (source and destination), port number, (source and destination), and the protocol ID (UDP, HTTP, etc.) for matching the information flows. The information flows are associated with user-specific services, such as file transfer, email, web browsing, VoLTE, multimedia streaming, etc. To delineate separate PDNs, APNs are utilized, and each APN has a separate default bearer. This is useful to distinguish PDNs associated with the IMS framework from those associated with the Internet. The conceptual model of APNs, PDN connections, and bearers is depicted in Figure 3.30.

Each SDF can only be associated with a single QoS class identifier (QCI), which determines the treatments such as packet forwarding, scheduling, configuration of link-layer protocols, queue management thresholds, etc., for the associated information flows. The QCI values are represented within a one-byte data size and can range from 1 to 9, where all the other values (10–255) are specific to a mobile connectivity provider.

Every bearer—default or dedicated—has a QCI marking. A mobile device can establish up to eleven separate bearers, consisting at one default bearer, it is based on the data size of the EPS bearer identity (EBI)* field (4 bits), where the first four values are reserved. In addition to QCI, the

---

* 3GPP TS23.401, "GPRS enhancements for E-UTRAN access," v12.3.0, 2013.

bearers also have an associated access retention priority (ARP), which is used for admission control and overload control treatment of a bearer by the control plane. The ARP value is used by the admission control function to determine whether the establishment of the bearer or its subsequent modification should be allowed or disallowed. The PCRF utilizes the bearer-related attributes for the rendering of QoS policies for the information flows.

The attributes that describe a dedicated bearer include the maximum bit rate (MBR), which is a maximum limit for a sustained information flow rate, and the aggregate MBR (AMBR), which is the total information flow bit rate of non-GBR bearers. The use of the AMBR attribute allows a differentiation of information flow rate limits across users, based on subscription priority levels for mobile connectivity. The provisioning of an AMBR value is either on a mobile device basis or on an APN basis. In the case of an APN-oriented AMBR, the AMBR value is stored as part of the user subscription profile in the home subscriber server (HSS). For a mobile device–oriented AMBR, it is provisioned as a maximum information flow rate limit for an aggregate of all the non-GBR information flows (SDFs).

As the wireless conduits of information become commoditized lanes of transport, differentiation, business sustainability, and opportunities based solely on mobile connectivity are limited. On other hand, harnessing virtually unbounded potential of multimedia service delivery over the wireless conduits of mobile connectivity unveils new horizons for business endeavors and user experience. Over-the-top (OTT) services and mobile connectivity provider rendered services are candidates to promote widespread revenue sharing with third-party application providers and to mine a variety of unique value-added features for service differentiation across the mobile connectivity landscape. The PCRF entity allows the coordination of policies for resource allocation that allows the customization and configuration of dynamic service delivery choices, where the service sources may be native or third party. The generalization of service sources, coordinated by the PCRF, allows specific choices in terms of collaboration, coordination, and customization to suit business models and user-demographic preferences. The shaping of these user-oriented preferences can be dynamic, where learning systems adapt and suggest, based on usage scenarios and choices. These types of information can be used to dynamically evolve policies and coordination across participating OTT providers to optimize the user experience in a generalized fashion, as well as being gracefully adaptable to user-driven preferences and choices. Along these architectural, design, and deployment directions, complexity is handled behind the scenes, while simplicity and intuitive behaviors are exposed both at the network operational level and at the user level.

The functional building blocks of the PCRF entity are derived from the conceptual attributes that primarily consist of resource prioritization, bandwidth reservation, information rate limiting, and access blocking. The decision-making and enforcement process utilizes these conceptual attributes based on static or dynamically interpreted policies bases on subscription-specific or service-specific profiles. Static policies are configured initially, while dynamic policies could be updates to static baseline policies that are adapted to suit network and load conditions. For example, a service-specific profile could be the basis for policies pertaining to VoLTE service and a subscribe-specific profile could be the basis for policies pertaining to access to VoLTE service. The policies dictate the allocation of services that require a guaranteed bandwidth, where an overall maximum bandwidth is allocated for certain types of traffic or to dynamically adapt to network load conditions in near real time. In other words, the policies perform the function of admission control. Value-added services, which have the potential to provide a niche experience, together with VoLTE, would be the beneficiaries of dynamically adaptable policies. The ability to provide and enforce dynamically adaptable policies to suit services with a variety of QoS and latency demands, over multiple dedicated information transport bearers, within the EPS framework is especially attractive for agility in service delivery. The use of shared resources for the delivery VoLTE, with the required QoS attributes, allows the harnessing of a common infrastructure that simultaneously permits a policy-driven customization of multimedia information transport. Roaming is another facet of harnessing shared resources across different administrative domains through the use of policies, which is particularly useful for managing a consistent QoE across different attachment points that appear based on mobility patterns. Shared resources may also appear in the form of a local breakout of the mobile connectivity segment—bypassing a home network—and connects to the Internet segment, to minimize latencies for delay-sensitive services such as VoLTE.

OTT service–related information flows are transported over default bearers, which are devoid of any specific treatment for quality or bit rate guarantees. All the related packets of information are subjected to a best-effort quality and are liable to degradation in terms of latency and jitter, under network congestion conditions. On the other hand, for the information flows associated with managed or prioritized services, the scheduler within the core network together with PCRF policies steers the information flows over dedicated bearers for a QoS-aware rendition of service experience. With this approach latency-sensitive information flows, such as for a VoLTE service, are prioritized over non-delay sensitive information flows, such as for file transfer, email services, etc. The ability to render a QoS-aware service experience for services managed by a mobile connectivity provider enables a significant differentiation potential relative to OTT service providers.

The crafting and application of policies that are aligned with customizable choices that augment service experience sets the stage for innovative differentiation and brand evolution for a mobile connectivity provider. The retention and growth of subscribers are among the benefits that accrue as a result of adaptable and attractive policies. These aspects can be harnessed, utilizing the interoperable functions that shape the capabilities around the PCRF framework that utilizes an ensemble of default and dedicated bearers to realize multimedia service-demanded QoS. Unique combinations of connectivity capabilities, QoS awareness, pricing, brand crafting/recognition, and business models provide opportunities for molding a lifestyle-enhancing service experience. These complex objectives are realizable through sophisticated implementations of the PCRF framework of building blocks that promote compelling differentiators, relative to third-party OTT services that are unaware of the conditions of the wireless segment of mobile connectivity, at the edges of the network. For instance, different types of bearers—default or dedicated—could be established to meet the complex demands of service-oriented QoS, charging models, subscription profiles, and dynamically changing network load conditions. Some services may be delivered to a user based on simple charging quotas or limits based on the volume or rate of information consumption. Other types of services of a real-time nature such as VoLTE or on-demand video may have a relatively complex charging model, as a result of intricate routing and traffic shaping/optimization of information flows. The evolution of multimedia service complexity is a natural motivator for innovative realizations of the PCRF framework, where intelligent and near-real-time policy decisions are demanded to suit service requirements and to maximize the quality of user experience. Dynamic policy behaviors are a direction toward SON ideas, where self-healing concepts—alternative connection possibilities based on heterogeneous network awareness—are realized based on events, such as base station congestion conditions or fault conditions. Other avenues toward dynamic policy and coordination include event triggers that are available via authorized and innovative inspection schemes pertaining to service-related information flows. These triggers may be effectively leveraged to promote similar or enhanced service capabilities that offer new vistas for experiential improvements and revenue potential. Event triggers coupled with location coordinates serve to augment the potential for new service exploration as a result of spatial information that could reveal opportunities for improving the experience, based on user-specific interests in shopping, entertainment, local maps, transportation, health and well-being centers, social events, etc. The PCRF framework unleashes a suite of capabilities for the delivery of customized services that render enriching experiences for the user, together with network optimizations and revenue potential for the mobile connectivity provider.

3.9.5 Mobility and Handover Management

Mobility has influenced a global shift toward customized and individualized vista of communications and services. This has spelled a departure away from a geographically dependent (fixed place-to-place) and isolated modalities of information exchange. This shift is profound in terms of social, personal, and professional behavior patterns. Information exchange, embellished with rich multimedia content, bolstered with mobility—ranging from high-speed to pedestrian or stationary with wireless connectivity—allows a transcending of established boundaries delineated by location and lifestyles.* Mobile device connectivity to a core network is sustainable in two modes of connectivity—idle mode and connected mode. In the idle mode, the mobile device does not have a user information bearer link with the network. In this mode, the mobile device listens to the system information broadcast over the radio interface for session state maintenance or clearing actions, between the mobile device and the network. In the connected mode, the mobile device has a user information bearer link with the network, for voice, video, or data service–related information transport.

The EPS framework hinges on a simplified arrangement of resources, along the path of mobile connectivity—radio interface, base station, wired interface, and the EPC. The wired interfaces logically map into the S1 interface (between the base station and the EPC) and the X2[†] interface (between base stations). The X2 interface provides efficient and reduced latency information transfer across base stations, since the overhead of requiring an SGW in the EPC to mediate the information transfer across the base stations is avoided. The information transfer over the X2 interface facilitates a low-latency handover for mobile connectivity as the mobile device moves across base stations.

For interworking of the EPS with other access technologies, MIP provides a generic mobility management capability. As an alternative to GTP, an MIP protocol–oriented rendition of mobile connectivity (PMIPv6) is applicable between the SGW and the PGW, over the S5/S8 interface, for information transfer over the bearer plane segment associated with E-UTRAN access. For non-E-UTRAN or other types of radio technology access, the application of the PMIPV6 is applicable over the S2a/S2b/S2c interfaces for information transfer between the SGW and the PGW. The PGW in this case is unaware of the bearer information related QoS requirements and only handles the MIP signaling with the mobile device. Consequently, the PCRF-provided QoS rules for enforcement in the PCEF within the PGW are insufficient. Such rules are provided by the PCRF for the SGW, where a segment of the bearer information is terminated in the case of non-E-UTRAN access. The bearer binding and event reporting function (BBERF) within the SGW configures the QoS rules received from the PCRF. The BBERF performs bearer binding to ensure that the appropriate resources are allocated for the QoS demanded by the information being transferred. The bearer binding process may result in the establishment of a new bearer path

---

* Wellman, B., Boase, J., and Chen, W., "Networked nature of community: Online and offline," *IT & Society*, vol. 1, no. 1, pp. 151–161, June 2002.
† 3GPP TS36.420, "Evolved universal terrestrial radio access network (E-UTRAN); X2 general aspects and principles," v11.0.0 September 2012.

or in the modification of an existing bearer path to suit the QoS demands. If dual stack mobile IPv6 (DSMIPv6) is the mobility protocol of choice, then the information being transferred (user traffic) is filtered by the BBERF. The user traffic flows in this model of mobile connectivity are tunneled between the PGW and the mobile device. The outer header IP address of this tunnel is signaled from the PGW to the BBERF, via the PCRF. The revelation of the outer header IP address allows the BBERF to have access to the user traffic flows. The BBERF is thereby enabled to apply the appropriate packet filters to the user traffic flows that are contained within the tunnel.

The concept of tracking area (TA) consists of a group of base stations that is tracked by the MME within the EPS framework. The location of a mobile device in the idle mode is known with a resolution that corresponds to a TA. Each time a TA boundary is traversed, the mobile device performs a tracking area update (TAU) procedure to update the MME about its location. This enables the MME to locate the mobile device in the appropriate TA, where it is currently located. A priori knowledge of the current location of the mobile device enables the network to rapidly page the mobile device, in case there is information such as that related to a VoLTE service or other multimedia services that are destined for the mobile device in the idle mode. The handover of a mobile device as it traverses the boundaries of base stations is handled using different techniques. One of the techniques is a predictive type of handover, where a source serving base station utilizes radio-link measurement reports to select a potential candidate target base station for handover. The source base station queries the candidate target base station to determine whether sufficient resources are available to serve an incoming mobile device. If sufficient resources are available, the target base station arranges the necessary resources to serve the incoming mobile device, when the source base station initiates the process of a handover of the mobile device.

In the EPS framework, a handover is always a hard handover, which implies a break before make of the radio-link connection between the mobile device and the network. Typically the "detach time"—interval when the mobile device has no radio-link connection to the network—is about 50–80 ms during the handover process across base stations. The interruption in the information flow associated with a real-time service, such as VoLTE or audio/video streaming, should be below perceptible thresholds* in terms of delays and jitter. The typical thresholds for delay are of the order of less than 100 ms for conversational voice and less than about 20 ms for jitter. Application design and implementation account for these typical thresholds, with an appropriately sized jitter buffer for a preservation of the perceived quality. Higher delay tolerances are acceptable for non-real-time services, although a minimization of delays and quicker responses are aligned with experiential improvements. The significant components of end-to-end delay consist of processing delay, propagation delay, queuing delay, and serialization delay. Processing delays are incurred to set up the information flows for appropriate forwarding and are typically small (of the order of microseconds). Queuing delays are a function of queue size, which implies that real-time information flows require prioritization for a scheduled entry into low-latency queues, which have relatively lower delays for departure to a destination over an associated interface. Propagation delays are a function of the physical medium of transport—wired or wireless—and have a degree of variability over the wireless segment in terms of radio-link conditions. The delays associated with serialization and deserialization, where information is arranged in serialized bits for transport and reassembled at the receiving interface, are relatively small over high-speed links.

To minimize information loss during a handover process for information destined for the mobile device, either the bicasting of information or buffer forwarding over the X2 interface

---

* G.1010, "Series G: Transmission systems and media, digital systems and networks—Quality of service and performance," ITU-T, 11/2001.

between base stations is utilized over the downlink. In the case of buffer forwarding, the source base station forwards information destined for the mobile device to the target base station, when the handover process is initiated. On the other hand, in the information bicasting approach, the SGW bicasts/multicasts the information destined for the mobile device to a set of candidate target base stations, together with the source base station. Relative to the buffer forwarding approach, bicasting demands a significantly higher backhaul capacity and may still not be able to eliminate information loss during the handover process. The challenge with the bicasting approach is that an early triggering of the bicasting/multicasting process consumes more backhaul capacity, resulting in an inefficient utilization of backhaul capacity, while a late start of the process raises the probability of information loss during the handover process. With these trade-offs, the buffer forwarding approach provides an optimized solution for redirecting information destined for the mobile device within the EPS framework. The nature of the information (real time or non–real time) destined for the mobile device determines the priority of buffer forwarding. For instance, information associated with VoLTE requires a higher-priority forwarding to avoid adversely impacting the maximum tolerable latencies.

The information packets in the source base station buffer, during the handover interval, consist of unacknowledged RLC layer packets, packets in the buffer being arranged for forwarding, and packets arriving from the associated SGW. These information packets are forwarded to the target base station during the handover interval. After the mobile device has signaled a handover complete message to the target base station, the forwarded packets are delivered over the radio downlink to the mobile device. The target base station signals the MME to switch the S1 interface path between the SGW and the base station from the source base station to the target base station. The MME then updates the SGW with the target base station address for a new downlink path for the bearer plane and confirms a relocation of the S1 interface path via a path acknowledgment response to the target base station. During the handover interval, as a result of the S1 interface path relocation to the target base station, information packets destined for the mobile device could arrive at the target base station from the SGW and from the source base station at different times, resulting in out-of-order information packet delivery to the mobile device leading to an impairment of TCP behavior. The SGW initiates a release of the old S1 interface path and bearer resources associated with the source base station by sending an "end marker" signaling message to the source base station. When the target base station receives this message, it starts the delivery of the information arriving from the SGW to the mobile device over the radio downlink.

The handover interval when the mobile detaches from the source base station and attaches with the target base station is referred to as the "detach time" and has the potential to adversely affect the QoE, based on the latency sensitivity of the information being transported. During this transition time interval, the mobile device is not connected to the network, resulting in delays in the arrival of the packets destined for the mobile device. The delays may result in information packet retransmissions, duplication, or out-of-order arrivals. The objective of the handover process across base stations is to preserve the originally negotiated QoS that was in effect before the commencement of the handover process, where the existing applications or services are not perceptibly impacted. The information bearer plane header compression techniques, such as robust header compression (ROHC)* techniques, are employed to minimize the information transport overhead over the radio-interface resource and improve resource utilization. Header compression is not applied over the control plane since signaling information is limited in size and requires rapid processing to optimize the system behavior and performance. On the other hand, over the bearer

---

* RFC 5795, "The RObust header compression (ROHC) framework," in *IETF*, March 2010.

plane, for every 32 bytes of voice information, 40 bytes of protocol header overhead is incurred (RTP, UDP, and IP*) when IPv4 is utilized. In the case of IPv6 transport over the radio interface, the overhead in terms of protocol headers increases to 60 bytes, which implies a significant wastage of radio-link resources for a given size of actual voice content. The application of ROHC diminishes the protocol header overhead significantly, resulting in an improvement of the radio-link resource utilization. This is particularly useful for information transport where the ratio of protocol header size to the actual information size is large, such as in the case of voice packets. Header compression techniques may also be applied to other types of information transport, where this ratio is large to realize radio resource utilization gains.

The use of buffer forwarding in the "backward handover" approach can be enhanced to minimize the potential for adverse impacts to the QoE. The enhancement consists of a forwarding of packets from a source base station to a target base station, where the association of the service requirements of each information flow that belongs to a group of flows with an SDF, corresponding to a radio bearer instance, is preserved. This classification of information flows at the source base station, based on the related service requirements, is propagated using the existing forwarding policies for a forwarding of the information packets to the target base station. The detection of duplicate packets and any reordering of information packets during the handover process can utilize the PDCP sequence numbers.[†] The use of the PDCP instance at the target base station allows an identification of the information packets arriving from the source base station. Information flows with the same PDCP instance belong to the same SDF, although each of the constituent information flows may have different service requirements, such as in the case of information flows that represent mixed media content.

3.9.6 Quality of Experience

The management of system resources is a pivotal function in a heterogeneous mobile connectivity environment, to suit the bandwidth demands of innovative multimedia services with diverse QoS requirements. The heterogeneity of radio access is shaped in terms of cooperating and coexisting technologies that provide access over a variety of radio coverage footprints—macro cells and small cells. An optimization of operational costs and an efficient use of system resources for mobile connectivity require the coordination of radio resources management functions, associated with the disparate access technologies, via a network management overlay. Fault detection, fault diagnosis, and fault recovery are among the dominant themes of capabilities within network management. These attributes embody the characteristics of self-organization that are well suited to the highly distributed nature of heterogeneous, macro and small cell, mobile connectivity network environments. A rapid release and redistribution of allocated resources associated with fault scenarios promote resource utilization efficiencies while minimizing QoE impairments. A minimization of fault recovery times provides directions toward a sustainable QoE, implying a sustainable revenue strategy for connectivity providers. The automation of the fault recovery process—self-healing—is a localized and relatively autonomous process. This process includes RRM parameter adjustments based on diagnostics within the RRM framework. The key performance indicators (KPIs) of a serving base station and its neighbors are utilized to trigger any corresponding RRM parameter

---

* RTP, Real time protocol; UDP, user datagram protocol; IP, Internet protocol.
[†] TS 36.323, "Evolved universal terrestrial radio access (E-UTRA); packet data convergence protocol (PDCP) specification," 3GPP—Third-Generation Partnership Project.

adjustments, as required. The requirements* of the KPIs† for connectivity are described in terms of (1) accessibility, (2) retainability, (3) integrity, (4) availability, and (5) mobility.

The dynamic nature of mobility patterns and wireless link conditions requires a softening of the rigid delineations of the protocol layers. These delineations operate effectively under deterministic conditions, in contrast with dynamic conditions, which demand cross-layer collaboration. Collaboration across the application, transport, network, and link layers allows for an adaptation of the system to dynamic mobility and wireless link conditions. Adaptation implies an adjustment of behaviors of each of the layers to realize a specific service delivery profile based on related QoS requirements over a mobile connection. For instance, in an application such as VoLTE where the packet error resilience is less significant relative to packet delay, the signaling overhead associated with HARQ is reduced together with an increased bundling of packets for adaptability to both radio-link conditions and the demanded QoS requirements. Adaptability to radio-link conditions and the QoS requirements (cross-layer collaboration) translate into desirable experiential qualities. The logical nature of the IP layer (network layer) allows an intrinsic universal adaptation of diverse applications, which implies that it promotes application independence. The information derived from a collection of QoS requirements is objective in nature and specifies a set of attributes that must be satisfied for the tolerable performance of a given application. Resource allocation, management of signaling overhead to suit either error resiliency or delay sensitivity, and load conditions represent the main aspects of adaptation across the protocol layers to realize a specified set of QoS requirements associated with an application. Coordination across the different protocol layers is required to fulfill or exceed the QoS requirements and to translate them into a QoE, which is fundamentally subjective in nature. An astute optimization of cross-layer adaptation to satisfy an application-specified set of QoS requirements provides novel avenues to translate objective QoS attributes into the subjective realm of QoE.

The adaptation of the radio link consists of adjusting the transmission to suit the radio-link conditions. Prominent radio link adaptation techniques consist of AMC and HARQ. Adjustments in the type of modulation and the format of coding are accomplished by AMC, to suit changes in the radio-link conditions. Feedback from a mobile device receiver allows the base station to estimate the radio-link conditions. At favorable locations closer to the base station, higher-order modulation schemes, with higher coding rates such as 64-QAM and ¾ turbo codes, are suitable to optimize the information transfer rates. On the other hand, along the base station coverage boundaries (unfavorable radio-link conditions), lower-order modulation schemes, with lower coding rates such as QPSK and ½ turbo codes, are likely to suit the radio-link conditions. Adaptation to suit radio-link conditions via coarse adjustments of information rates is performed by AMC, while finer information rate adjustments are performed by HARQ. The trade-off with just a coarse scale adjustment is that excessively lower information transmission rates to suit radio-link condition could potentially lead to radio-link underutilization, while adjustments to excessively increase information transfer rates could imply an increase in retransmission overhead. Cross-layer coordination between AMC and HARQ schemes, an optimization of radio-link adaptation, is enabled. Optimization of radio-link adaptation in collaboration with application-specific QoS profiles enables a service delivery–aware mobile connectivity.

---

* 3GPP TS 32.451, "Key performance indicators (KPI) for evolved universal terrestrial radio access network (E-UTRAN); requirements."
† 3GPP TS 32.450, "Key performance indicators (KPI) for evolved universal terrestrial radio access network (E-UTRAN): definitions."

The delivery of multimedia services with the appropriate allocation of mobile connectivity resources in terms of QoS characterized by latency, jitter, error rates, and packet loss bounds shapes the QoE. The mobile connectivity rendered by the EPS facilitates identification and treatment of multimedia service flows to adapt to the related QoS characteristics. A guarantee of the QoS characteristics desired by diverse multimedia service flows, especially those pertaining to real-time service flows, are the most demanding with respect to efficient and timely connectivity (radio and core network) resources. For example, voice rendered over the EPS (VoLTE), including mixed-media QoS profile–oriented service flows, is an important objective measure of the mobile connectivity system performance, which naturally affects the QoE (a subjective measure).

3.9.7 Interdependence in the Communication Landscape

Fragility is detectable and can be used to trigger decisions intelligently to promote intended behavioral patterns within a communication system. On the other hand, the Black Swan* effect is one where the risks of rare events cannot be calculated or predicted. Heterogeneous communication systems for mobile connectivity, with virtually unbounded arrangements of coverage topologies, vertical overlays, and usage scenarios, are complex in terms of sustaining experientially attractive behavioral patterns, at both the user level and the system level as a whole. Rare events, such as wireless link instability in a homogeneous, well-engineered macro system, are likely to increase as a result of complexity and interdependence among the assorted forms of mobile connectivity. Dynamic, cooperative, and feedback schemes are vital to alleviate the behavioral uncertainties in the quality of connectivity over the wireless segment in a heterogeneous connectivity environment. The quest for attractive experiences in on-demand, and service-oriented mobile connectivity is pivotal.

The spatial reuse of frequencies for information transport is among the arrangements to harness topological distributions of mobile connectivity for enhanced radio-frequency spectrum utilization. Small cells interspersed with macro cells are a rendition of flexible and distributed access to a global web of services. The range of coverage of small cells, with coverage radii of ten to hundreds of meters, contrasts with the much larger coverage areas of macro cells, with coverage radii of tens of kilometers. The other types of wireless attachment points in this varietal web of mobile connectivity include remote radio heads (RRHs) and relay nodes. Coexistence considerations are central in this theme of distributed heterogeneous connectivity. User-driven choices are enabled through the mobile device to selectively utilize different radio-interface technologies, in addition to the heterogeneous network-driven access availability to services. Relative to the homogeneous access technology, such as the E-UTRA, which is the technology of choice in a distributed, heterogeneous connectivity landscape, a multi-RAT includes Wi-Fi access. The use of E-UTRA technology across heterogeneous access networks simplifies the management of vertical handovers and the preservation of service quality, in the presence of shifting attachment points resulting from mobility, load balancing, or fault scenarios. Offloading of information flows via alternative, vertically arranged topological areas of coverage provides attractive pathways of connectivity to adapt to network conditions, mobility speeds, and service transport demands. Cognitive capabilities within small cells facilitate an agile allocation of radio resources to mitigate adverse interference impacts at the mobile device to orchestrate a seamless service experience within coverage areas and across handovers. Dynamic allocation of radio resources are performed through a sequence of detection and decision phases to establish potential spectral connectivity opportunities for a desired integrity of the transport channel conditions between the mobile device and the network.

---

* Taleb, N., *Antifragility: Things That Gain from Disorder*. New York: Random House, 2012.

Cognition of network conditions provides the intelligence for mobile devices to discover and select overlapping coverage footprints of assorted footprints—small cells and macro cells—together with non-LTE RATs, such as Wi-Fi. The utilization of radio and network resources has the potential to be optimized, together with a selection of an appropriate selection of network to suit the demands of a service, which is aligned with improvements in the QoE. Cooperation across connectivity domains through distributed policy framework provides avenues for self-organization and increased availability and quality of mobile connectivity. The broadcast of information about the conditions across cooperating networks provides a level elasticity for satisfying QoS and connectivity demands that shift dynamically both temporally and spatially, based on mobility usage scenarios such as social events, commuters, entertainment, etc. The benefit of cooperation in an arrangement of distributed access networks is that the variations in the service-related QoS tolerances, such as latencies or throughput demands, are harnessed to optimize both network resource utilization and user experience. Cooperation between the network and the mobile device provides opportunities for hybrid decision-making strategies,* where network assistance provides information to the mobile device about network conditions. The availability of broadcast network condition information for decision making at the mobile device for access network selection reflects the concept of leveraging common or global information for local application and individualization, for mobile connectivity at the edges of a heterogeneous network landscape.

The hybrid decision-making strategies provide directions toward a distributed cognition of the availability and the effective utilization of radio layer and the network layer resources. With the smaller footprints associated with heterogeneous networks, there is a high likelihood of variations of resource demands across the neighboring islands of coverage, governed by user mobility patterns. From a system resource (radio and network layer) perspective, users are likely to maximize their utilization of the available resources to suit their service usage profiles. This behavior corresponds to a Nash equilibrium relevant in a noncooperative game, where the demands suit each user, independent of those of another. The downside of such a noncooperative game in the context of leveraging the mobile system resources is that eventually it could render the system unstable and unusable, for even the user that may have temporarily benefitted, from accessing resources without restraint. The transformation of system resource utilization from a noncooperative equilibrium, with a Nash equilibrium as the ideal, to a Stackelberg equilibrium, where there is a level of cooperation and collaboration between the network and the mobile device to optimize and distribute the availability of system resources to support the delivery of multimedia services in the presence of mobility. This hybrid approach to promote equilibrium condition utilizes a global awareness of resources conditions, which are available to the network in its role, relative to that of the mobile device, where its awareness is limited to the resource conditions in the vicinity of the attachment point and its own status. The hybrid approach harnesses the simplicity and elegance of the Nash equilibrium, with a minimum level of hierarchically oriented control plane information to optimize the utilization of system resources and stability, under a variety of system load conditions. In other words, intelligence is flexibly distributed between the network and the mobile devices to promote self-organization, to maximize the quality of user experience on an individual user basis, while optimizing network resource utilization and operational efficiency. With a global awareness of distributed resource availability and load conditions, the network in its role is positioned to signal the status of these conditions to mobile devices. This status information allows the network to

---

* Elayoubi, S.E., Altman, E., Haddad, M., and Altman, Z., "A hybrid decision approach for the association problem in heterogeneous networks," in *IEEE Conference on Computer Communications (INFOCOM)*, San Diego, CA, March 2010.

maximize its resource utilization while guiding the mobile device behaviors to maximize access to resource demands, within the guidance thresholds indicated by the dynamic resource conditions in the network. This strategy embodies the hybrid decision approach to establish equilibrium in the system while gracefully maximizing resource utilization, collectively and individually. In other words, the decision-making process is distributed, which is rendered as cognitive intelligence in the network. It enables equilibrium between total centralization, which promotes a global optimization, while restricting individual demands, and total distribution, where stability of the whole is at risk, as a result of the potential for excessive signaling and consumption by individual users. The QoS allocations that correspond to the available system resources, under equilibrium conditions, translate to a maximization of QoE across a variety of mobile users that are being served by the network. The QoS markers provide objective measures—delay, packet loss, prioritization, etc.—that enable favorable QoE conditions while being insufficient to provide guarantees from a broader-perspective QoE, which is subjective.

3.9.8 Self-Organization, Distribution, and Interdependence

The hybrid approach to the decision-making process, with respect to a distribution of RRM functions that include a delegation of the decision-making process to mobile devices, promotes a reduction in network complexities, in terms of signaling and processing demands. This implies that a mobile device is infused with cognitive capabilities, with respect to its local environment, which includes sensing and inferring information concerning the status of the radio subsystem and the core network subsystem. These capabilities are aspects of an SON framework.* It is a framework that harnesses both the interdependence and the distribution of entities within the system, to enhance the experience of mobile connectivity, while also optimizing the operational overhead. The SON framework offers a significant pathway for enhancements in interoperability and complexity reduction. It provides an automation of network management and operations to lower recurring costs. The three main components of SON are self-configuration, self-optimization, and self-healing. The architectural framework for SON may be centralized, distributed, or a hybrid of centralized and distributed models. An example of self-configuration is a dynamic plug-and-play base station, which automatically configures the initial parameters, such as physical cell identity, transmission frequency, transmission power level, etc. Automation of configuration simplifies and quickens the operational processes, such as planning and deployment. The interface between the core network† and the base station‡ (S1 interface), and the interface between base stations (X2 interface) are automatically configured through the use of the automatic neighbor relations (ANRs) procedure.§ As part of the self-configuration process, the IP addresses for connectivity over these interfaces are allocated automatically. The ANR procedure automatically discovers and configures the list of neighboring base stations for any additions of base stations or changes in installed base stations. The base station identifier is configured uniquely by the SON framework. These unique identifiers are used in the creation of a list of base stations that qualify as neighbors of any given base station. The identifiers of the neighboring base stations are acquired via mea-

---

* TS 32.500, "Telecommunication management, self-organizing networks (SON); concepts and requirements," 3GPP, v11.1.0, 2011.
† EPC—Evolved packet core, within the LTE EPS—Evolved Packet System.
‡ TS 36.300, "E-UTRAN and E-UTRAN—Overall description, stage-2," 3GGP, v10.3.0, 2011.
§ TS 32.511, "Telecommunication management; automatic neighbour relation (ANR) management; concepts and requirements," 3GPP, v11.2.0, 2012.

surement reports, provided by a mobile device being served, or via the interbase station interface. The initial configuration as well as the optimization of the configuration of a base station is a part of the automatic configuration procedure. A dynamic update and configuration of a base station neighbor list enables a superior handover performance and availability of mobile connectivity. The dynamic self-configuration process is governed by policies that are provisioned in the EMS* within the OSS.[†]

Automatic regulation of behaviors across any system may be characterized in terms of self-optimization and self-healing. The nature of self-optimization is dependent on the desirable behaviors of the system. In a mobile connectivity fabric, self-optimization is broadly characterized in terms of coverage, capacity, congestion, handover, and radio-frequency interference. For example, base stations that experience congestion conditions are enabled to offload traffic to other base stations with resource availability, through load-balancing procedures[‡] that have an awareness of traffic load distribution conditions across multiple base stations. This awareness is renewed through periodic exchange of information—over the X2 interface[§]—that reveals the traffic load distribution and available capacity profiles over a network of base stations across a coverage area. The thresholds for triggering load balancing procedures are based on the limit of the number of mobile users that can be served using the radio resources available at the base station while sustaining the QoS demands of these users. To meet the QoS demands associated with a service, the scheduler in the base station allocates the appropriate number of PRBs.[¶] The load balancing threshold limit for a given base station that triggers an offload to another base stations, with available capacity, is reached when the configured PRB limit is reached for serving a set of mobile users. The configured thresholds for triggering an offload could be set to a range of limits, from normal to increasing traffic load levels, to enable a dynamic adaptation of traffic load conditions in the vicinity of a given base station. This would allow look-ahead load balancing methods for a graceful offload of traffic to preserve the user experience while optimizing the radio network resource utilization.

In conjunction with traffic load balancing, mobility robustness is another aspect of self-optimization. The quality of a continuum of connectivity across the islands of coverage governs the experiential profile of mobile services. The performance of handovers of connectivity across these islands, both in the idle mode (information traffic absent) and in the active mode (information traffic present), shapes the robustness of mobile connectivity. Parameter settings for handovers affect the experiential profile of mobile services, where nonoptimal settings may result in a range of adverse impacts to the experiential profile, from increased latencies to radio-link failures. These adverse impacts also translate to a decreased utilization efficiency of the mobile connectivity system resources. Latencies in the handover process serve as markers of multimedia service quality. Delay sensitive services, such as VoLTE,** are relatively more susceptible to handover-induced latencies, especially at the edges of the coverage areas, where the uplink transmission power is likely to exhibit deficiencies. Feedback signaling across the source and target base stations adjusts the handover parameter settings to avoid both late handovers and early handovers. Load balancing handovers are steered to suitable base stations with available capacity, after radio signal strength conditions are deemed to be satisfactory. A reduction of the coverage area in terms of a

---

* Element management system.

[†] Operations support system.

[‡] TS 36.902, "(E-UTRAN); Self-configuring and self-optimizing network (SON) use cases and solutions," 3GPP, v9.3.1, 2011.

[§] X2: interbase station (eNodeB) interface.

[¶] 1 PRB = 12 subcarriers times 7 OFDM symbols = 84 resource elements.

** VoLTE, voice over LTE, a technology specific rendition of VoIP (voice over IP).

reduction in signal strength is used to steer the mobile device away from a loaded base station. Under low or no load conditions, the suspension or energy-saving mode operation of a serving base station triggers a handover of mobile traffic to the remaining active base stations that have sufficient capacity to serve the mobile users in a given coverage area. An increase in mobile traffic demand triggers a wake-up signal to the suspended base station, followed by a handover of mobile users to this base station.

The distribution of the self-organizing algorithms across the base stations provides a cognitive and dynamic scattering of mobile connectivity intelligence to optimize both the experiential profile of mobile services, from each user's perspective, and the system resource utilization. Distribution allows a significant reduction in signaling load and latencies in the exchange of mobile load balancing and handover optimization information. On the other hand, interworking and interoperability across different configurations and/or implementations are facilitated through the coordination function of a centralized network management system (NMS), where the signaling information for self-optimization is transferred from the distributed base stations containing the element manager system (EMS) to the NMS. Minimizing the participation of centralized coordinative functions, beyond an initial configuration of a multivendor, multidomain environment, avoids single failure point scenarios and latencies, and promotes outage isolation under dynamic conditions. Distributed mobile connectivity architectures, with an awareness of connectivity conditions, provides the foundation for a heterogeneous connectivity environment of small cells and macro cells in a technology-independent manner.

Reduction and optimization of operational complexities are invaluable objectives in a mobile connectivity fabric, both from resource utilization and user-experience perspectives. Self-healing capabilities are pivotal, within the SON framework, in the automation of fault recovery or recovery from performance-degraded conditions to optimize the robustness and availability of mobile connectivity. The detection of defined and configured metrics that characterize the nature of a degraded or fault condition related to mobile connectivity is performed through a monitoring of the associated alarms. The affected network elements generate alarms based on a triggering of the configured performance thresholds for each of the defined metrics for a given mobile network coverage area, at the resolution of a base station. The triggering of these alarm thresholds is independent of whether the automatic detection is automatically cleared or manually cleared* through appropriate recovery actions. The conclusion of a self-healing procedure results in notifications being sent to the integration reference point (IRP)[†] manager that contain a log of the recovery actions taken and the results of those actions. The levels of performance degradation range from partial to complete outage at the resolution of the base station coverage area. The recovery actions that are triggered include corrective measures such as reconfiguration of base coverage areas (larger or smaller to adapt to capacity demands or load conditions), software updates, fallback software activation, isolation of fault conditions to minimize propagation of unstable or connectivity outage conditions, activation of standby resources, connectivity context transfer, handover, traffic offload, etc. Preventive and look-ahead self-healing actions include an intelligent learning and adaptation to dynamic connectivity conditions, such as load and service quality demands. These self-healing actions are a part of the monitoring, detection, diagnosis, and prescription processes to optimize mobile connectivity, in terms of user experience, capacity, coverage, and resource utilization in a dynamic

---

* TS 32.541, "Self-organizing networks (SON); Self-healing concepts and requirements," 3GPP, v11.0.0, 2012.
[†] TS 32.522, "SON policy network resource model (NRM) integration reference point (IRP); information service (IS)," 3GPP, v11.7.0, 2013.

and stochastic wireless landscape. The essence of a self-healing framework consists of look-ahead condition detection logic, where the logic operates using a set of KPIs* with configured thresholds or profiles—static or dynamically adaptable—that trigger appropriate actions to heal proactively or to recover gracefully from a fault condition. For example, a KPI could be the base station throughput, where a configured profile is a long-term throughput average for the base station, based on six-sigma† principles applied during an inference and learning phase. These baseline profiles could be established for each base station to optimize the mobile connectivity experience. Deviations of a current KPI profile from the baseline KPI for a given base station are utilized as markers toward a configured value of deviation steps, measured as the number of standard-deviation increments. When the deviation of the current KPI reaches this configured value, the corresponding healing actions are executed to compensate for the throughput degradation. The self-healing actions are multifaceted and include cooperation from neighboring base stations in terms of coverage expansion through coverage shaping to bolster mobile connectivity for mobile devices served by the base station with degraded throughput. The adjustment of configurable parameters, such as handover thresholds, antenna tilts, addition of a secondary cell for throughput enhancement using the CA feature, traffic payload context-aware (real-time or non-real-time content) handover, and small-cell offload thresholds, reflect the nature of self-healing action types. Baseline profiles for coverage shaping, capacity balancing, and interference conditions are established automatically through measurement data collected via the mobile device, using the minimization of drive test (MDT)‡ feature.

The survivability and the robustness of the mobile connectivity fabric demand an embedding of self-healing capabilities, as a result of the stochastic nature of wireless links, interference variations, mobility patterns, and environmental conditions, such as weather and topology (outdoor and indoor structural profiles) and the intrinsically limited availability of high-quality radio-frequency spectrum.§ These characteristics are stark contrasts with wireline systems, which utilize extraneous dedicated resources to accomplish self-healing capabilities. In this sense, the embedded self-healing capabilities have a high level of autonomy and are distributed, much like the self-healing functions within the human body. The distributed self-healing functions act in concert through automatic measurements, coordination, and collaboration to preserve and optimize the system behaviors through an alignment with the baseline KPI profiles for a stellar user experience. This interdependent self-healing framework is reflective of the human "mind," which according to Dan Siegel¶ is defined as an "embodied and relational process that regulates the flow of energy and information."

3.9.9 Performance Enhancements through Traffic Offload

Wireless access systems that provide mobile connectivity require self-organizing capabilities to effectively manage connectivity at the inherently dynamic and probabilistic edges of the

---

* KPI—key performance indicators.
† Six-sigma: This is a process that is based on the idea that if the behavior of a function is within six standard deviation increments, between the average behavior profile and the nearest limit of specified behavior profile, then the behavior of the function will not fail to meet the specified behavior profile.
‡ TS 37.320, "Minimization of drive tests (MDT); Overall description; stage 2," 3GPP, v12.0.0, 2014.
§ Kant, L. and Chen, W., "Service survivability in wireless networks via multi-layer self-healing," *IEEE Wireless Communications and Networking Conference*, vol. 4, pp. 2446–2452, 2005.
¶ Siegel, D.J., *The Mindful Brain—Reflection and Attunement in the Cultivation of Well-Being*. New York: W.W. Norton & Company, 2007.

network—edge of chaos. This is reflective of the behaviors of world around us, such as beehives, weather patterns, cloud formations, ocean currents, social networks, societies, interest groups, and markets. They exhibit complex renditions of self-organization for order and sustainability. Self-organization is elemental for managing and simplifying the adverse impacts of systemic complexity, which is an artifact of widespread interdependence and cooperation, both of which are vital in the evolution of value.

Complexity along the dynamic and moving edges of connectivity is characterized by the randomness and the variability in the quality of the radio-access links that are available in the vicinity of mobile device. The notion of heterogeneous networks broadly extends beyond a single RAT to a coexistence of multiple RATs, over both licensed and unlicensed spectrum. The continuing prolific expansion of information pathways over mobile connectivity demands opportunistic capabilities to harness assorted wireless transport mechanisms that enable an attractive menu of choices that augment the user experience. Topologies of coverage and the type of access technology govern a range of mobility speeds, providing opportunities to optimize resource utilization, by intelligently distributing or offloading connectivity load. The E-UTRA technology, which is optimized to handle a wide range of mobility speeds, is well suited for mobile connectivity, independent of coverage footprint sizes and can be leveraged in both licensed and unlicensed spectrum. This technology can coexist with Wi-Fi technology, over unlicensed spectrum, to further extend the potential for connectivity load distribution, via the EPC framework, especially for best-effort type of service quality, with session continuity. The choice of different RATs promotes a selective and resource efficient utilization of access, for the transport of rich communication services, with multimedia content and a mixed range of service quality demands.

Langton* and Kauffman† characterized complexity as being the boundary between order and chaos, or in other words at the edge of chaos. Langton explored the conditions under which cellular automata (discrete, abstract, computational systems) could function as computational primitives for the transmission, storage, and the modification of information. The impact of order and chaos on these computational systems was modeled using a "lambda" value, which indicates the probability that in a given neighborhood or local configuration, of these computational systems, their internal state will lead to either a stable state (order) or an unstable state (disorder or chaos). When *lambda = 0*, all the neighborhood states move the cell automata to a quiescent state, and the system is completely ordered or stable. When *lambda = 1*, none of the neighborhood states move the cell automata to a quiescent state and the system becomes chaotic or unstable. As the value of "lambda" increases from 0 to 1, the state of the cellular automata transitions from order to chaos. Langton established a critical value for "lambda," between 0 and 1, where the average amount of mutual information—a measure of system complexity—within a given cellular automata is maximized. It was found that if this critical value is exceeded, the average amount of mutual information is depleted and that the cellular automata undergo a phase transition from order to chaos. It then follows that computation associated within the system at the critical value of *lambda* has the maximum amount of mutual information to effectively orchestrate state transitions toward sustaining order and self-organization in a dynamically changing environment.

---

* Langton, C., "Computation at the edge of chaos: Phase transitions and emergent computation," in *Emergent Computation*, Forest, S. (Ed.). Cambridge, MA: The MIT Press, pp. 12–37, 1991. Langton explains his investigation of one-dimensional cellular automata and examines the "lambda" parameter.

† Kauffman, S., *At Home in the Universe: The Search for Laws of Self-organization and Complexity*. New York: Oxford University Press, 1995. The views explore how complexity, including life itself can evolve from simple tenets. The notion of The "Edge Of Chaos" is pivotal in his perspective.

Such a system is required to navigate astutely between too much order, implying limited value as a result of limited flexibility or capability, and too much chaos, implying no value as a result of instability or dysfunction in a dynamically changing mobility environment. This is akin to navigating between Scylla and Charybdis.*

The density of mobile device users, the diversity of multimedia content, the business models, the diversity of coverage topologies, and the various RATs constitute a complex and dynamic computational system. Self-organizing system behaviors are naturally suited for mobile connectivity in a heterogeneous landscape of RATs and diverse coverage footprints. Collaboration between the serving networks and the mobile device in the neighborhood of mobile connectivity can be modeled as cellular automata, where the amount of collaborative information is maximized for a stable and optimized self-organizing system. Offloading and load balancing of information flows to and from the mobile device to different coverage footprints and RATs, via dynamic and learned behaviors, are among the capabilities of self-organization in a heterogeneous mobile connectivity system.

The enabling interoperable entity for interfacing the E-UTRA system with different RATs is the access network discovery and selection function (ANDSF).†,‡ The ANDSF resides within the EPC of the E-UTRA system. The ANDSF serves as a framework for orchestrating the discovery and selection of a radio-access network in a diverse landscape of heterogeneous technologies and coverage footprints to intelligently distribute mobile connectivity demands. The strategies for distributing traffic across heterogeneous access allow a flexible approach to enhance the combined mobile connectivity experience together with an optimization of resource utilization in the vicinity of the mobile device user. The use of geospatial coordinate information can be configured and dynamically updated to guide the mobile device about the available access technologies and the associated available capacity for optimized connectivity opportunities based on service quality and pricing levels. The ANDSF serves as a mediator in the selection of a suitable RAT for mobile connectivity using configured and dynamically changing coverage and capacity condition. The functions of the ANDSF can be distributed and virtualized to optimize the mobile connectivity experience across a dynamic landscape of heterogeneous connectivity, where both radio and network resources are effectively utilized. Intersystem mobility policies, including preferred access, dynamic access, connectivity pricing, etc.—within the ANDSF framework of virtualized functions can be dynamically configured, using self-organizing principles that account for dynamic changes in information flows and changes in mobile device user densities across time and topologies. Multimedia services with different combinations of mixed media (real time and non–real time) and hence with different characteristics and QoS profiles can be routed via IP flows over different RATs, with different bandwidth demands, via ANDSF framework policies.

3.10 Strategic Considerations

Communication and content—anytime, anywhere—imbued with a customizable experience, where the human is the centerpiece and the essence of recipes that promote the evolution and

---

* Odyssey—Book 12: In this epic, Odysseus and his men explore a passage between Scylla and Charybdis. Scylla is a six-headed monster that lives on a sharp mountain peak and Charybdis is a giant whirlpool.
† 3GPP TS 23.402, "Architecture enhancements for non-3GPP accesses," v12.2.0, September 2013.
‡ 3GPP TS 24.312, "Access network discovery and selection function (ANDSF) management object (MO)," v12.2.0, September 2013.

sustainability of mobile connectivity. Wireless connectivity serves as the cornerstone of mobile information transport, which is the closest to natural forms of human interaction—untethered and mobile. Technology-mediated forms of diverse wireless connectivity technologies that are designed to interoperate through the use of well-defined global standards are pivotal enablers in ad hoc and flexible mobile connectivity. The attributes of flexibility and scalability are significant ingredients of mobile connectivity together with a viable performance-to-cost ratio potential. The characteristics of connectivity technologies, assisted by technology innovation that promotes the performance-to-cost ratio potential, include high information transfer rates, flexible provisioning capabilities, minimized latencies, contextual awareness, and QoS to suit individualized and customizable usage scenarios—nomadic and mobile. Nomadic scenarios allow global roaming, while familiar metaphors and the QoE are seamlessly orchestrated.

The reliability of connectivity handovers across coverage areas, while optimizing coverage and information throughput, is vital for mobile broadband connectivity while providing robust levels of access availability, independent of the terrestrial environment—rural or urban. The utilization efficiency of radio-frequency spectrum requires simultaneous optimizations in radio and network resource management to provide a high QoS experience. The affordability and availability of ubiquitous mobile connectivity technologies are critical factors in the shaping of viable connectivity provider business models. These models in turn influence the sustainability and quality of mobile connectivity services. The design and evolution of compelling business models require an astute and strategic blending, leveraging, and management of the primary mobile connectivity assets—radio-frequency spectrum (licensed/unlicensed), subscriber base/demographic, base station/backhaul, and core network entities. This underlying fabric of considerations requires to be complemented with novel capabilities that embellish the connectivity experience.

The ubiquity and the spread of mobile broadband connectivity are a function of the total cost of ownership (TCO) and the trends in cost reductions associated with operations, connectivity networks, and the mobile device portfolio. Reusability and extensibility of base station sites and core network elements enhance the potential for a lowering investment risks and the TCO. A balance between enabling widespread coverage, while enhancing capacity utilization, is elemental in the minimization of deployment costs. The average and the peak information throughputs affect both the type and quality of services that are delivered over mobile connectivity. Flexibility in the growth of capacity to support dynamic market demands and the necessary initial and subsequent incremental investments are required to be aligned strategically for the evolution and the sustainability of the mobile connectivity provider business investments. A customizable combination of multistandard, multiaccess technologies imbued with interoperability provides new vistas of forward-looking, market-aligned, and market-making business models. These models selectively adapt different access technologies that cooperate and collaborate across different geographic topologies (terrestrial and satellite) to promote seamless mobile connectivity. Choices in mobile connectivity are not limited by geographic topologies but are attractively open to selection based on the available QoS, and pricing profiles to suit any given mobile service experience.

Interoperable interfaces across the heterogeneous access environments, glued together by virtualized core architectures, serve as an enabling fabric and framework to suit investment strategies and business models. Leveraging the commonality of the baseband processing modules that plug-and-play with radio antenna front ends and RRHs provides a platform for optimizing operational and capital costs, across the network and mobile devices. From the perspective of mobile connectivity, the radio frequency of the information transport channel affects the coverage area of connectivity in an inversely proportional manner—the higher the radio frequency, the smaller the connectivity area of the serving base station. On the other hand, higher radio frequencies of information transport channel permit higher rates of information transfer. The wireless segment of mobile connectivity is characterized by the nature of the radio transmission and reception mode (FDD or TDD). In contrast to the FDD mode, which uses different uplink and downlink frequencies for information transport, TDD utilizes time differentiation over the same frequency for information transport. The latter mode avoids the need for paired frequencies for information transport—one for the downlink direction and the other for the uplink direction. For a given frequency bandwidth, TDD accommodates asymmetry in the volume of information transport between the uplink and the downlink directions. Adaptive modulation coding (AMC) techniques allow flexibility in the information throughput, by enabling higher information transfer rates through higher orders of modulation, when the radio-frequency environment is accommodative in terms of improved signal quality such as relatively better signal-to-noise radio conditions. The use of OFDM leverages this concept through the use of an increasing number of orthogonally distributed (mutually noninterfering) frequency subcarriers (lanes of information transport). This concept effectively harnesses the performance complexity trade-off, by realizing information throughput efficiencies as the frequency bandwidth increases, since the signaling overhead is relatively constant, allowing a larger proportion of the frequency bandwidth to be available for mobile service transport.

Along the wireless segment of transport, SDRs (inherently reconfigurable) provide the recipes for carving out the desired trade-offs between performance and cost, dictated by mobile service experience. For example, in the case of machine-to-machine communications, the absence of the experiential factor allows an astute application of design simplifications (half-duplex transceiver) that allow a reduction of investment costs. Performance metrics across the wireless and wired information transport segments serve as measures of compatibility between the characteristics of connectivity and service components of information transfer, which in turn shape the connectivity and service provider strategies. These collaborative ingredients represent a confluence of technology and art in the delivery of experiential mobile multimedia services.

The wide range of potential business models range from providing managed mobile connectivity, which serves as an information bit pipe along the radio-access segment, the core network segment, or both. Other strategies for business models may include mobile connectivity with a variety of innovative mobile services. In all cases, the strategies for forward-looking business models include mobility as a service as the central theme in the evolution of the mobile network. From a user perspective, with the ubiquitous availability of a rich plethora of Internet services, mobile connectivity with its highly personalized nature is a distinguishing facet that complements Internet service access. Experience includes the attractiveness of the brand associated with the services, as much as the innovative nature and the quality of mobile connectivity provided to suit any given service. The perception of the connectivity provider brand is shaped predominantly by user trust, which is directly proportional to the authenticity of the provider and is the service being provided agile and adaptable to individualized needs, while being accommodative to user interests. Strategic and stewardship-oriented investments build sustainable user relationships that promote innovative business models and business health as markets evolve. With the rapid progress of technological innovation and the decline in the cost and energy demands of computing resources, the potential for frequent disruption of rigid and non-user-centric business models is on the rise. Non-user-centric strategies in business models are not likely to survive the disruptive nature of the ebb and flow of the market tides and the evolution of mobile communication technologies. Strategies that are user centric realize the benefits of innovation beyond the underlying technologies that are vulnerable to commoditization and opportunity contraction.

Such considerations naturally opt out of a zero-sum game and embrace the power of stewardship within the business model for a sustainable revenue potential, as a result of an implicit propagation of well-being.

Mobile connectivity awareness across an assortment of RAT is pivotal and complementary to service context and content experience. Backward compatibility of evolutionary directions provide the stepping stones to optimize the operational investments, while the direction of evolution remains open to radical change. In other words, strategic directions must embrace seemingly disruptive ideas that are necessary to invent and reshape sustainable business models. Technology evolution through standardizations and innovation are essential ingredients that create the "je ne sais quoi" for creating a stellar user experience that fuels business initiatives and sustainability.

3.11 Virtualization Perspectives

The two distinct attributes that have directionally influenced choices in system design and implementation are "interoperability" and "proprietary." Each of these attributes has their intrinsic benefits and limitations. The former attribute enables a heterogeneous ecosystem of participants that harness interoperable interfaces—hardware and software—to cooperate, collaborate, and promote an ever-expanding, forward-looking, interdependent, and sustainable symphony of mobile connectivity and services. It demands widespread collaboration and imagination to inspire and accommodate a myriad of ideas toward a consensus view of logical models of architecture, interfaces, and procedures. The objective is a minimalistic suite of interoperable building blocks while not encumbering innovation in product or service design and implementation. It demands a holistic view of forward-looking design imagination, existing market conditions, regulations, cultural nuances, and potential for market creation.

On the other hand, the proprietary approach embraces the perspective where a unique vision driven by a few is viewed as a potentially attractive and viable technology for either a broad array or a selective array of users. The advantage with this approach is that the realization of such a vision is relatively more rapid since the complexity of requiring a broad consensus is avoided. The weakness with this approach is that it relies on the uniqueness and robustness of the vision to transcend the typical barriers associated with shifting market demands, investment cost of research supported by few or a single entity, execution, and competition. This then also implies that given a unique and robust vision, the approach has a high potential to impress and influence market trends and evolution. The approach must also provide the impetus for advocacy through experiential inferences and conversations for an exploration of specific aspects of a given technology that lead to open and expanded perspectives. The proprietary approach, while directly at odds with the interoperable recipe for propelling evolution, serves as beacon for potential areas of exploration for the crafting of consensus-driven interoperable enabling building blocks that provide an expansion of evolutionary implementations and continuing innovation.

Innovation through imagination, ideas, and implementation shapes the leading edge of the evolutionary impulse to promote the creation of technology-mediated lifestyle enhancements, where communication is a central theme in the human journey. Proprietary approaches in the expansion of technology to augment and influence market demands serve as a cooperative ingredient in the envisioning of interoperable capabilities. The advancement of knowledge through these cooperative mechanisms of innovation allows progress beyond the stepping stones of customized hardware and software, to flexible and agile frameworks, independent of usage or deployment scenarios. The agility and adaptability of mobile connectivity frameworks allow a dynamic movement of functions and capabilities to suit space–time separated mobile connectivity demands. Virtualization embodies an agile and adaptable mobile connectivity network. The notion of virtualization has been adopted widely by Amazon—among the pioneers of the virtualization vision—in its rendering of services over a web of universal connectivity. Jeff Bezos,* speaking at the 2009 Annual Awards Dinners in New York City, characterized the elusive nature of forward-looking perspectives: "Invention requires a long-term willingness to be misunderstood." Virtualization in the arena of mobile connectivity system has profound systemic implications, such as technology, strategy, business, markets, and user experience. Interoperable specification directions, with universal advocacy, cooperation, collaboration, and consensus, are elemental ingredients in the realization of virtualized mobile connectivity frameworks.

The accumulation of customized hardware and software computing platforms that interwork through interoperable interfaces require continuing investments, while the commoditization of mobile connectivity continues unabated. The proprietary and customized design realizations, driven by innovation and competitiveness across infrastructure producers, are akin to the proprietary approaches to technology evolution. To invigorate the cycle innovation, realization, and evolution, a shift is required toward an enablement of experiential elegance and ease of mobile services, over generalized hardware and software computing frameworks. Such frameworks would be generalized as mobile connectivity resources that are invoked on-demand by connectivity functions rendered in software. This type of generalization of mobile connectivity resources requires a virtualization of these resources from the perspective of mobile connectivity functions, such as connection establishment and mobility management. The functions implicit in the proprietary customizations of hardware and software entities are explored and extracted while not compromising experiential performance objectives, namely, connection robustness, availability, and mobility. The functions are partitioned as interoperable software functions that execute over virtualized mobile connectivity resources. This provide a high level of elasticity and scalability while harnessing the benefits of an interoperable and interdependent, consensus-driven technology evolution. The focus of mobile evolution then naturally shifts to mobility-enabled service context that is pivotal in rendering memorable and convenience-oriented multimedia service within the virtually unbounded ecosystem of the IoT.

The challenge associated with proprietary and/or customized realizations of computing resources is their inherent complexity associated with the rendering of widespread mobile connectivity. This complexity associated with specialized realizations of computing resources leads to unsustainable investments. The required computing resources should instead simply serve as a generic fabric for a universal distribution of information. Apart from catering to the integration demands of the specialized realizations of infrastructure equipment, the skills associated with specific installation and maintenance are both narrow and limited in terms of extensibility, reuse, and evolution. In other words, the skills are of limited value since they are tightly coupled to the specifics of an implementation. The value is ephemeral and undermines the potential of investments, since the value is bound to the span of requirement–design–integration–deployment cycle, while extensibility is limited. Further, in specialized realizations, the lifecycle of hardware is short while the demand for service continues to proliferate through continuing innovation and increasing demands for functionality. The underlying reasons fall into two categories, namely, alternative forms of untethered connectivity technologies such as Wi-Fi, sufficient for a variety of broadband

---

* Amazon founder and visionary—Aspen Institute's 26th Annual Awards Dinner http:// www.aspeninstitute.org.

connectivity usage scenarios, and the perceived value of content/information that triggers the demand for new functions. The latter category is crucial from the perspective of forward-looking strategy formulation and business models. The exponential demand for service innovation and sophisticated functionality demand growth requires a shift away from specialized realizations of hardware/software.

In an internetwork of humans and things, mobile connectivity with its natural appeal bears the potential for boundless growth. From a service perspective, this natural appeal holds the potential for evolution and revenue generation for service and connectivity providers. A virtualized, generic, and interoperable fabric of computing resources for enabling widespread and efficient mobile connectivity becomes a natural evolutionary direction. This is a crucial step for optimizing costs for the providers of connectivity, where the infrastructure entities and the related resources are virtualized to provide mobile connectivity. The benefits of virtualization can be synthesized into two categories, namely, agility to enable and deliver innovative mobile services, and the optimization of resource utilization. The former category allows the network infrastructure to be strategically ready for the deployment of innovative and yet-to-be-envisioned mobile services while avoiding additional infrastructure investments. The latter category follows the principle of optimized resource sharing while promoting an effective scaling of information throughput. For example, a mobile connectivity serving network point of presence (POP) allows a scaling of aggregated information throughput, through effective and intelligent resource sharing, while not requiring a tight coupling between the physical topology of the resource and its utilization point, where a mobile device attaches to the network infrastructure.

Thinking differently, the crafting of business models that are aligned strategically with the system-wide implications of a virtualization of the mobile connectivity network is critical. The shifts that are required to adapt to the forward-looking vision of virtualization are deep, broad, and interdisciplinary. The impacts of the "long tail" in terms of the service uniqueness—relatively low cost, high levels of customization, high volume of differentiation, infused value creation—are essential consideration for long-term business sustainability. These considerations include the evolution and adoption of interoperable mobile connectivity specifications that harness the potential of computing resource virtualization. The benefits of a virtualization of computing resources are influenced by a market environment that is incessantly reshaped by the cycle of innovation and commoditization. In such an environment, strategies for profit- and nonprofit-oriented business models require an alignment with the incremental shifts that serve as harbingers, in an evolving market environment, where a potentially profound change may not be immediate but near. The incremental shifts in the virtually limitless vista of possibilities in the mobile service landscape include the appearance of service quality classifications such as premium, advertisement-supported free services like Pandora, YouTube, etc., crowd-sourced information services, location-independent virtualized work environment, and social networks such as Twitter, LinkedIn, etc. These innovative shifts are indicative of the enormously profound yet subtle changes that continue to reshape and reinvent the context and content of value creation. It is here that thinking differently from the status quo and popular beliefs is an imperative in the formulation of forward-looking strategies. Mobile connectivity customization and flexibility serve as significant attributes along new vistas of virtualization. The operation of generic computing resources becomes table stakes through technology innovation and automation. Along these directions, measures of user experience, through user conversations and feedback, is a cornerstone in the realization of an agile framework of design, development, distribution, and deployment of virtualized mobile connectivity. Virtualization is a strategic enabler in the propulsion of these 4Ds—design, development, distribution, and deployment.

3.12 Autonomic Perspectives: Nature's Theme

Themes of decentralization and coordination abound in the natural world. Swarm intelligence[*] harnessing appears widely through decentralized individual behaviors that influence the collective behavioral pattern. The decentralized individual behaviors are self-organized, autonomous, and loosely coupled with one another, via control signaling. Examples of autonomic themes in nature are reflected in various renditions, such as a school of fish, a colony of ants, or a flock of birds.

Such systems are also agnostic to notions of hierarchy while tuned into cooperation and coordination to realize intended collective behavioral patterns. This model is embodied in the starfish, which has a decentralized nervous system, where each of the five limbs has an autonomous nervous system, loosely coupled with one another through control signaling to coordinate an integrated behavior such as movement. Localization—a consequence of decentralization—limits the scope of interactions among autonomous cooperating entities while influencing the behavior of the whole.

In the natural world, ant colonies[†] exemplify a decentralized and distributed system, where a cooperating individual ant serves as building block for a colony of ants. Although the contributing behavior of each individual ant is relatively simple, the behavior of the colony as a whole is complex, much like a dynamically structured and coordinated social organization—flexible and adaptable. The swarm intelligence within ant colonies is evident in examples such as foraging for food, labor distribution, cooperative transport, etc. Harvester ants, such as the Messor (moniker for the Roman god of crops and harvest), exhibit intelligent labor distribution, where the small and large ants intelligently cooperate to transport and store grain. The coordination, collaboration, and cooperation among the individual ants to affect intended collective behavior are enabled through the use of stigmergy—mediation through environmental triggers, in the form of biological substances such as pheromones.[‡] The convergence toward a selected behavior is affected by the concentration of pheromones. The concentration levels are altered over time based on the balance between renewal and depletion through evaporation, which implicitly provides a high level of dynamism, and self-organization. Artificial enablers that embrace the notion of stigmergy, derived from the natural world, serve as catalysts to self-organize mobile connectivity and to optimize capacity, coverage, and information flow patterns and integrity. Distribution and decentralization are foundational tenets within the model of swarm intelligence.

Autonomic[§] computing, ubiquitous computing, and semantic web[¶] techniques reflect the concepts revealed within swarm intelligence. These techniques allow significant adaptability and flexibility while effectively equipped to enable diverse design and deployment scenarios. Such techniques, while generally applicable to a plethora of systems, are eminently suited for application within the mobile communication ecosystem. Collaborative computing techniques within mobile communications augment both the functional and the experiential aspects of interactions among humans and machines. Swarm intelligence techniques across distributed mobile network entities

---

[*] Beni, G. and Wang, J., "Swarm intelligence in cellular robotic systems," in *Proceedings of NATO Advanced Workshop on Robots and Biological Systems*, Tuscany, Italy, June 26–30, 1989.

[†] Dorigo, M. and Stützle, T., *Ant Colony Optimization*. Cambridge, MA: MIT Press, 2004.

[‡] Karlson P. and Lüscher M., "Pheromones: A new term for a class of biologically active substances," *Nature*, vol. 183, no. 4653, pp. 55–56, 1959. doi:10.1038/183055a0. PMID 13622694.

[§] IBM. Autonomic computing: IBMs perspective on the state of information technology technical report, IBM Research, October 2001.

[¶] Fensel, D., Wahlster, W., Lieberman, H., and Hendler, J. (Eds.), *Spinning the Semantic Web: Bringing the World Wide Web to Its Full Potential*. Cambridge, MA: The MIT Press, 2002.

and mobile devices facilitate a self-organized overall intended behavior. The self-organization benefits promote dynamic configuration, optimization, and healing, both at the machine level (mobile infrastructure) and at the human level (social communities). Howard Rheingold's book *Smart Mobs* provides the following illustration of swarming:

> Expect startling social effects after mobile P2P achieves critical mass - when the 1,500 people who walk across Tokyo's Shibuya Crossing at every light change can become a temporary cloud of distributed computing power.

The behavior of ant colonies inspires the principles of self-organization within the architecture, design, and operation of a mobile connectivity network. The attributes embedded within the swarm intelligence model shed light on the desirable guiding principles for architecting, designing, and operating mobile connectivity systems. These attributes include the following:

- Decentralized
- Distributed
- Autonomous
- Loosely coupled
- Fault tolerant
- Self-configuring
- Self-optimizing
- Self-healing

The loose coupling and autonomy allow each of the cooperating entities to coordinate intended collective behaviors while minimizing the control signal complexity among the interacting entities. Flexibility, robustness, and adaptability to dynamic conditions are then a natural consequence of the guiding tenets.

With the expanding complexity of heterogeneous mobile connectivity networks, there is a corresponding rise in the operational complexities, such as network planning, maintenance, deployment, and optimization. The proliferation of information consumption and creation, over next-generation connectivity frameworks, requires effective strategies to optimize operational and performance costs while minimizing system complexities. These strategies imply the adoption of autonomous behaviors within next-generation connectivity frameworks. The principles of self-organization, self-configuration, and self-healing that embody an SON framework provide a tool chest of capabilities that enable autonomous behaviors. These behaviors and attributes provide operational and deployment flexibility, through heterogeneity, coordination, distribution, and decentralization of NGN frameworks. Autonomous behaviors, infused with SON principles, provide both automation and dynamic adaptation to mobile service load conditions that are aligned with simultaneously optimizing the user experience and the operational efficiencies. These principles enhance plug-and-play opportunities to improve the usability and availability of mobile connectivity and services, by avoiding manual intervention, improved resource utilization, accelerating the speed of network deployment, and rapid recovery potential from fault conditions. Improvements in spectral efficiencies ensue, from a radio-access perspective, through a dynamic allocation of radio resource allocation that tracks and adapts to coverage area specific capacity demand. Together with an adaption to capacity demand, the automation of radio-access network management and mitigation of radio reception interference serve as markers of QoS improvement, which is a building block for QoE. Dynamic reconfigurability of network entities—wired and wireless network elements—within NGN frameworks demands the embodiment of autonomic behaviors, characterized by SON guiding principles.

The distributed and decentralized nature of NGN frameworks serves as a shift away from the notion of fixed point-to-point interactions to a more generalized and dynamic many-to-many interactions. This shift motivates a cloud model for distributed and decentralized connectivity, computing, sensing, and storage resources. It motivates a virtualization and abstraction of physical resources within an autonomy enabled NGN framework. The complementary capabilities of connectivity, computing, sensing, and storage provide pervasiveness—contextualization and location—in an interconnected Web of Things that include humans and machines. Intelligent allocation of resources in the cloud and at the edges of mobile connectivity, within a virtualized framework of functions, would be among the ingredients that maximize the user experience and optimize resource utilization. These directions have the potential to maximize investment returns across developed and developing economies on a global scale. A mash-up of wired and wireless connectivity then becomes a flexible fabric for seamless information creation and consumption. It opens new horizons where connectivity of content, things, and services shapes innovative bridges between supply and demand over unbounded vistas of human endeavor and existence.

Mobile connectivity serves as a natural component of widespread information flows, characterized by a dizzying array of services that encompass personal, social, and professional well-being. The intelligence of the features and capabilities that enable these rich and value-added services over mobile connectivity continues to proliferate at the wireless network edges. Autonomic capabilities, such as those based on the SON guiding principles, provide the tools to manage both operational efficiencies and performance in the shifting nature of complexity toward the edges of the network, where mobility and functionality intersect. Self-organization, self-configuration, and self-healing capabilities that reflect an autonomic style of design and deployment complement one another to enable a dynamic environment for flexible and nonintrusive introduction of new functionality, to diagnose the connectivity system conditions, or to heal a malfunction. This approach allows a graceful and scalable extensibility in the management of NGNs, while the sophistication of mobile connectivity continues to evolve, expand, and diversify, to deliver user-centric services.

Chapter 4

Service

> We are all inventors, each sailing out on a voyage of discovery, guided each by a private chart of which there is no duplicate. The world is all gates, all opportunities.
>
> **—Ralph Waldo Emerson**

Service universality and service personalization must coexist to meet the demands of human behavior, choices, and interests (unique to every user) over a foundation of service interoperability and service expectations (nonunique to every user). Historical product and service trends have typically adopted a monolithic perspective of experience; if it is good for one user, it should be good for all, an inflexible and limiting view. Rigid objectives, behaviors, and benchmarks, for service experience, in this context, are insufficient.

The interfaces between the human usability, namely, intuitive, ergonomic, and attractive attributes, and the technology-powered service mobility capabilities are vital ingredients that must effectively conceal any non-usage-related nuances of a technology implementation.

Reflections in universality are a pivotal attribute of service, where the semantics of evolution appeal to the constancy of change. Flexible options for customization, in this universality, to cater to the preferences of each user are equally significant for enabling experiences that are as diverse as the richness of the human fabric.

Experiences in the human journey are manifested in individual and collective forms. It is this diverse and common themes of experience that define the value of service, which is especially significant in the nature of mobile service. Preferences for the types of mobile services are recognizable, differentiated, and experienced in both personal and social themes. Studies performed by the British Sociological Association* revealed that the flexibility and choices afforded by mobile services outweighed the potential for stress resulting from communication overload, while being connected to the Internet. The choices afforded by the ability to connect and communicate, independent of space or time, promote adaptable user preferences. These choices are increasingly relevant to simplify, plan, discover, and experience information and its application in a globally

---

* Bittman, M., Brown, J.E., and Wajcman, J., "The mobile phone, perpetual contact and time pressure," *Work, Employment & Society*, vol. 23, no. 4, pp. 673–691, December 2009.

connected world. Communication services, in its rich array of modalities, namely, notifications, presence, availability, conversational, scheduling, etc., are paramount capabilities that offer the potential for balanced lifestyles.

Service encompasses both the quantitative and qualitative aspects of a variety of human endeavors: education, health, home, travel, commerce, well-being, social, relationships, art, self-improvement, etc. Mobile connectivity, as a dominant fabric of the next-generation Internet, is a global stage for a proliferation of both the quantitative and qualitative elements of service. In these shifting winds, mobile service is a natural pathway for service ubiquity rooted in a sea of imagination. It is one that is intrinsically aligned with human behaviors, beyond cultures and geography.

Mobile broadband connectivity subsumes the elemental components of *always-on*, *speed*, and *mobility*. Each of these components is distinct and complementary building-blocks that serve as essential ingredients, in varying degrees—driven by user-driven choices—in a virtually unbounded ocean of service recipes. This implies a seamless blending of mobile service into assorted lifestyle choices, enriched in a canvas of familiar and emerging metaphors of experience. The traditional notions of capacity and coverage of connectivity morph into heterogeneous modalities, from near-area range to wide-area range, as voice vanishes into one of the media types among voice, video and data, within the larger context of multimedia service. This latter context is where *always-on* and *mobility* form the elemental pillars of experience in mobile service. Experience, in mobile service, is the essence of a significant and emerging paradigm shift in a variety of changing frontiers affected by a technology-mediated communication evolution.

A sense of being present and being in touch has been a pivotal motivator in the evolution of telecommunications. Over 100 years ago, Arthur Mee* mulled, "If as it is said to be not unlikely in the near future, the principle of sight is applied to the telephone as well as that of sound, earth will be a true paradise, and distance will lose its enchantment by being abolished altogether." To paraphrase Mee's foresight further, in his magazine article, he envisioned a vista of infinite charm which few prophets of today have dreamed of, and who would dare to say that in future, the electric miracle would not bring all the corners of the earth to our own fireside. This vision of personalization has proliferated beyond the wired communication era to the wireless domain, with enormous potential for information richness, personalization, and mobility. The augmentation of the physical realities around us draws the user within through the window of the mobile interface into a personalized experiential fabric. The portability and convenience offered by this moving window into a world of rich media experience, embellished by personal choices and interests, are among the compelling attributes of mobile service offerings. For example, handheld mobile devices can be used by doctors to monitor specific aspects of the human body, through ultrasound procedures, which are visually augmented by overlaying contextual graphical information, enabled through related mathematical models. Remote collaboration opportunities in similar endeavors are mediated by intuitive mobile device interfaces through an appropriate rendering of rich multimedia services.

The collapse of separated spaces, together with instant or archived communications, is the essence of technology-mediated communications. Beyond this essence lies the experiential aspect of human communications, which provides the attractiveness of mobile services and is a part of the richness that is beyond technology mediation: one that is inherent in the fabric of face-to-face human communications.

---

* Mee, A., The pleasure telephone, *The Strand Magazine*, pp. 339–369, 1898.

4.1 Shifting Horizons

A rhapsody of perceptions pervades in a nomadic world textured in a plethora of expectations and experiences. So what are the expectations, desires, in these mobile orchestrated experiences? The access to communications, in its various incantations, in the shape of multimedia services and a multitude of social interactions, evokes the central theme in a human story. Communications is rendered in a pocket, promoting a virtually real experience, instantaneously for information and relationships, across the barriers of time and space. It is exciting—with endless possibilities. It is untethered—to connect here, there, and everywhere—to create, consume, and choose information.

A paradigm of mobility fosters a natural sense of independence in an interdependent world, where personal choices to explore connections are enabled for enhanced lifestyles and experiences, in an intrinsically collaborative and cooperative information age: indeed, an alluring realm beyond the assemblage of hardware and software toward a symphony of capabilities cast in form and function, with a distinctly human touch, where the hardware and software are only discernible in the venues of experience, and form, function, and behaviors intuitively adapt to each other—a magnum opus.

The scaffolding—enablers, hardware, software—blends invisibly in the wrapping of form and function as the architect, and artisan crafts the intelligence behind the compositions to unfold the promise of a compelling service.

The measure of attractiveness of a service must be linked to the uniqueness of identities—woven in the distinct descriptors of behaviors and choices that identify each human being, a notion with compelling implications in service mobility.

It is the breadth of usage and adoption, across diverse segments of interests and ideas, that are far more relevant, beyond a silo-oriented categorization hinging on rigid demographic boundaries and stereotypes. A service is a framework of capabilities and features, from configuration to customer care, that renders the functionality of an application or collection of applications to elicit a favorable experience for the user, while a specific user desire is being fulfilled. It is in the experiential realm that the user desires and technology mediation blend to foster new behaviors that shape the edge of innovation. It is a technosocial evolution, where mobile communications is an emergent paradigm with its inherent handiness and amenability to virtually unbounded levels of personalization. Familiar metaphors influence behaviors and embellish experience, while the ease of universal interaction continues to evolve as technology evolution—Long-Term Evolution (LTE)-augmented mobile broadband connectivity together with assorted local area access, natural human interface enhancements, and rich multimedia—indistinguishably blends into new stories in the moving frontiers of lifestyles. These then become foundational in the veneer of cultural diversity, around the globe in the evolution of human consciousness, knowledge, and understanding, through the enormous proliferation and exchange of information, ushered by the evolution of mobile communications.

Mobile communications through its various incantations of multimedia content and multiinterface presentation have a natural propensity to be central to the emergent theme of enormous amounts of data creation and consumption. Most of these seas of information embedded in data are unstructured, in the sense that the information comprises of various combinations of voice, video, text, and files that are dynamic and volatile in nature, and are not intended for storage or further processing. These then contain impressions of communication patterns and behaviors that could serve as catalysts in understanding behaviors, as well as in the derivation of insights for continuing technology innovation and advancements, while not compromising individual privacy.

Artificial intelligence techniques, such as pattern recognition and inference algorithms, are pivotal in a harnessing, understanding, and research of human behaviors over the changing and multi-faceted horizons of untethered communications. Intelligent voice-recognition engines, such as the Siri,* in the iPhone, are harbingers of machine learning, cognition, and natural interface advances that are vital in a technology-mediated experiential ecosystem.

According to studies at IBM, about 2.5 quintillion bytes of data are generated each day, of which about 90% is accounted for in the last couple of years. Sources of data proliferate around us from sensors, weather, online commerce, e-learning, telemedicine, metering, online banking, social media, pictures, videos, location services, etc.; the list is virtually endless and continues to be propelled by imagination and innovation. This is Big Data.† Among its main defining attributes are: *Volume*—large, beyond terabytes or petabytes of information; *Velocity*—time sensitive information, such as streaming, interactive, or conversations; *Variety*—structured or unstructured information, such as audio, text, video, files, and mixed-media types. Among the various prongs of initiatives that hinge on Big Data is the United Nations'‡ endeavor to explore sentiment analysis of social networking information transactions, using natural language recognition algorithms to promote human well-being around the globe. It mirrors a plethora of opportunities for data-oriented information discovery, inference, and decision making. The aspects of well-being encompass proactive measures and guidance in the improvement of health, education, economy, and lifestyles.

The smartphone- and tablet-engineered expansion of the social web, in concert with the likes of Google, Twitter, and Facebook applications, is an incubator of innovation, where technology mediation highlights the interdependence between the individual user and the world around. This underscores the significance of context in its shaping of experience and imagination in the dynamic frontiers of mobile communications. Context brings relevance and appeal in the fabric of mobile service. For example, an overlay of contextual information, on Google Maps, associates the abstract location coordinates with user preferences, such as area attractions, namely, restaurants, theaters, businesses, fitness centers, friends, coffee shops, etc., prompted by user interests, choices, and desires. The possibilities are endless and the notions' niche contextual embellishments are profound in the Long Tail of mobile communications, in the tapestry of Big Data. It reflects an emergent paradigm, whose time has come, with mobility and wireless as its dominant actors.

4.2 Data Ubiquity

The profundity of mobile information communications in the human realm is a function of the convenience of connectivity and the attractiveness of service offering. The availability of information over separated spaces, through data transactions, is a function of the underlying technologies. Mobile data binds and sorts different categories of information across networked devices and sensors around the globe, allowing a proliferation of services, which continue to create and augment personal and commercial endeavors. Data streams that encapsulate the distribution of information serve as containers of multimedia content to characterize virtually unbounded service potential:

---

* Siri is a spinoff of the CAL—Cognitive Assistant that Learns and recognizes—project, sponsored by DARPA—Defense Advanced Research Projects Agency.
† http://www-01.ibm.com/software/data/bigdata/.
‡ http://www.unglobalpulse.org/.

through assorted human and machine interfaces, such as smartphones, consumer devices, sensors, tablets, and laptops. The ubiquity of data transport is a formless mesh in the proliferation of information that is accelerated by the widespread adoption of social media and networking services, a reflection of the intrinsic human desire to connect with one another to express, share, understand, entertain, consume, and create in a journey of knowledge, well-being, and consciousness advancement. These directions promote enormous oceans of data—the alphabet and words of information—which pose both challenges and opportunities in the crafting of mobile services that are attuned to the experiential domain, while adapting to the constraints and capabilities of the connectivity fabric.

Data serves as the universal fabric of information exchange, which unveils opportunities for mankind to continue on the path of a global understanding and endeavor, beyond the boundaries of separate realities—geography, cultures, behaviors, language, and traditions—that have shaped historically shaped human evolution. The emergence of the mobile web* and usage patterns echo this shift in terms of a coexistence of a singular *me*, within a plural *we*. This is central to the theme of service in social media, which engenders notions of individual experience in the web of universal entanglement through technology-mediated-awareness. Individual service experience is modulated by various attributes that include convenience, efficiency, personalization, learning, leisure, entertainment, and well-being in a virtually unbounded vista of possibilities.

Interconnected computing resources, distributed across global networks that accommodate wireless mobile access, provide the foundations for a highly scalable and flexible environment for cloud-enabled mobile service. The ubiquity of shared resource networks, with IP as the unifying glue, continues to promote innovative ideas and modalities of mobile services for information exchange and sharing. This is a pivotal catalyst in the availability of mobile service for people to participate, create, and consume in an ever-expanding sphere of untethered communications, embellished through a dizzying array of devices and sensors, interconnected through global networks. The utilization of data, creation and consumption, spans across various traditionally segregated sectors, such as industrial, utilitarian, automotive, retail, etc., as well as across emerging sectors of virtually unlimited personal, business, and social models of usage behaviors. The unfolding of Moore's law,† and its generalized applicability in a variety of arenas, such as information storage, continuing commoditization of technologies, cloud computing, etc., promotes a lowering of technology barriers and the related costs. This fosters a cycle of innovation and a corresponding potential for market expansion in an evolutionary convergence of assorted technologies.

The emergence of smartphones and tablets continues to motivate the exchange of mobile data in the realm of human communications on an ever-increasing scale of connectivity and service. These mobile interfaces to multimedia service serve as subjective portal of experience, which are intrinsically beyond the artifacts of devices and networks, while the latter serve as advancing scaffolding in the fabric of technology evolution. The significance of experience in its unbounded forms of diversity and distinctiveness in technology-mediated communications is immutable, and is incubated in the constancy of ceaseless change. QoE (Quality of Experience) is a subjective measure of the quality of multimedia service, viewed through the lens of the mobile user. On the other

---

* Kelly, K., *Web 2.0 Expo and Conference*, March 29, 2011. http://www.web2expo.com/webexsf2011/public/schedule/proceedings.
† Moore's law—postulated by Intel's cofounder Gordon Moore—projected a doubling of transistor component densities within an integrated circuit, approximately every two years. This translates to an equivalent doubling of computing resources, for the same cost, approximately every two years.

hand, QoS (Quality of Service) is a technology-centric control scheme, which is tuned as needed to deliver a desired level of quality for a multimedia rendered to the mobile user.

Corruption of a wireless data link, resulting from a loss of signal integrity, which is probable in the implicit uncertainties associated with the nature of a radio waves, or congestion resulting from system resource constraints have an adverse impact on the expected flow of data between a mobile multimedia service function and the delivery of the service function to the mobile user. Information loss, delay, or delay variability, characterized by corresponding data packet loss, delay, or delay variability, are symptoms of corruption or congestion of information over a data link between a source and a destination of information. Resource constraints, within the mobile system—network or device—are manifested in terms of contention for resources to allow a desired propagation, which is characterized by bounds associated with loss, delay, or delay variability of data packets. The desired propagation characteristics of information hinge on the nature of the mobile multimedia service being rendered, such as interactive, real time, and non–real time. Mobile services, belonging to each of these categories, are affected differently, as a result of corruption, congestion, or both. In the propagation of information, intermediary routers of information and the endpoints of information creation and consumption are affected by the availability of resources and the contention for the underlying connectivity, storage, and processing resources by the mobile services participating in the transfer of information. The techniques for information propagation include temporary buffering of information and/or a graceful degradation of a given service, within the prescribed experiential constraints. For example, the buffering of information has the potential to increase or add latency variations, if they adversely affect the anticipated service experience. Since the temporary storage of information is a function of memory resources, congestion or corruption could result in an overflow of an information buffer, which leads to packet loss, and in turn potential information loss. The throughput of information and information transfer latencies are inversely affected by packet loss, and are managed by information transport mechanisms (e.g., TCP [Transmission Control Protocol]). The attributes of information throughput and latency are common measures that influence experience across a variety of services. With improvements, in these common measures at the wireless mobile edges, the evolution of smartphones will continue to accelerate the utilization of the Internet, toward a web of mobile Internet services.

4.2.1 Information Archetypes

Information exchange in its various forms, expressed as multimedia, forms the backbone of mobile service in its assorted costumes of creation and consumption. Data, which serves as the building-block of information, represents the constituents of mobile service and is exchanged through the conduit of mobile connectivity networks. The archetypal categories of service information may be viewed as being either accessed within a mobile connectivity network, or accessed beyond the mobile network, over the Internet. In the latter category, the mobile user access to service information is not mediated by the mobile connectivity network, in contrast with the former category. This distinction has ramifications in terms of usage patterns and preferences from a human perspective.

From a mobile connectivity network view, the information conduit over the wireless mobile connection is common for both categories of mobile service. When the information mediation, within the mobile connectivity network, occurs for service provisioning and delivery, the nature of the mobile service is customized and unique to a serving mobile connectivity network. On the other hand, when the information mediation within the mobile connectivity network is limited to signaling for control purposes, the mobile service provisioning and delivery occurs remotely over the Internet, and the mobile network simply serves as an edge segment of the related information conduit. The attractiveness of mobile network mediated services is a function of their appeal, customizability, and the enthusiasm of the mobile user to leverage the uniqueness of a given service offering. The ubiquity of a service offering, independent of a mobile connectivity network, has its attractiveness embedded in retention of familiar usage metaphors across the edges of wireless mobility. In the latter case, the attractiveness spans across a variety of mobile users—savvy and simple. The advent and ongoing advances in smartphones, with their human-centric interfaces of vision, audio interaction, touch, gestures, etc., promote the consumption of mobile services hosted anywhere in the Internet, or within the mobile connectivity network. Convenience, enjoyment, and natural interfaces are among the significant attributes of attractiveness of mobile service. The mobile device, with natural interfaces, is pivotal in the evolution of smartphones. Natural interfaces provide a seamless portal to mobile service to unveil rich possibilities in the landscape of human experience.

The mobile service usage modalities influence the adoption rates differently among classes of people, namely, innovative and noninnovative, in a relatively broad sense.* The subjective influence of others, with respect to new mobile service features or capabilities, is not a consideration among users with an innovative lens. On the other hand, subjective views are likely to tangibly influence those that have a noninnovative inclination. The nature of evolving technology-mediated mobile communications hinges on innovation and invention. It follows that the mobile service users with an innovative orientation serve as a catalyst in the propagation of emerging mobile services in the populations, where they live, work, and play.

The conduit of mobile connectivity is an essential element in the presentation and use of information being created or consumed. The mobile device in its various designs, interface elegance, form factors, and appeal is a frontline in the landscape of user perception in the language of convenience and customizability. These aspects tangibly affect service delivery and usage behaviors motivated through a palette of intention. As the mobile device capabilities advance in terms of appeal, convenience, and customizability, the motivators for an adoption of new and innovative services will follow and expansive trend. The application space associated with the iPhone and Android categories of smartphones and tablets echoes a reflection of this trend. The appeal of smartphones and tablets naturally serves as a bridge between the types of services offered by a mobile connectivity provider, or those offered over the Internet with mobile connectivity. This vanishing gap between the tech-savvy and non-tech-savvy categories of mobile users is a market expansive trend that promotes mobile communications. It is an invitation for mobile connectivity providers to explore and embrace innovation with a holistic approach: one that shapes disruptive business models, through a confluence of technology and art, through partnerships rooted in cooperation and collaboration.

The influence of utilitarian values on the one hand and the enjoyment factor on the other are the significant prongs in the attractiveness of emergent services. These are twin motivators in the case of the inherently curious or innovative types of users of mobile service–oriented information. Services may be viewed as catering to emotional human needs of social interactions, entertainment, health, leisure, etc., and functional requirements such as research, business, home, banking, commerce, and education, among others, or a combination of sorts. Renditions of these diverse categories include e-mail, web browsing, social networking, presence, instant messaging, content, etc. Social influences and stories of value and appeal are more likely to influence the adoption of

---

* Rogers, E.M., *Diffusion of Innovations*, 5th edn. New York: Free Press, 2003.

services offered under the auspices of mobile connectivity providers. In the latter case, customer care and a tailored portal of information provide the tools for service adoption, avoiding the higher entry barriers, where self-directed research is warranted. Compelling customer care and intuitive portals together with a holistic view of service, encompassing the mobile device, mobile connectivity, uniqueness, valuation, and QoE, serve as markers of adoption uniformity and market expansion, across the average users and early adopters. In this mix, the noninnovative category of user is likely to be attracted to utilitarian types of information services, relative to the exploratory minds of early adopters and enthusiasts—a category that is vital for moving the wave of advances in mobile evolution.

4.2.2 Subjective Realm

The nature of information communications, in the personalized realm of mobility, is inherently driven by human factors that are universal, beneath the veneer of social and cultural costumes around the world. The choices in this arena are motivated by intention-driven behavior, which shape the degree to which preferences are explored and selected. TPB (theory of planned behavior)* is a widely established and researched model for examining the motivators of behavior. This model reveals the underlying dynamics of preferences and pursuits that pertain to the virtually unbounded and emergent possibilities in the rich fabric of mobile service. The model is particularly relevant in this context, since it is driven by personal choice and intention. Choice-driven control is the trigger for the behavioral expression of an intention, revealed through research and the application of the TPB model. Intention synthesizes the motivational factors that influence behavior. An application of the TPB model, in terms of multimedia service adoption, is shown in Figure 4.1.

The choice of control or decision making, influenced by motivational factors, influences a corresponding behavioral expression. At the same time, nonmotivational factors, such as affordability, skills, social cooperation, time, etc., which are elements of appropriate opportunities and resources, are also among the factors that to some extent affect behavioral expression of an intention. Stated another way, behavioral expression appears to be a combined function of intention and ability, as revealed by TPB. Further insights from the TPB model reveal three kinds of considerations, pertaining to human behavior: behavioral beliefs, normative beliefs, and control beliefs. *Behavioral beliefs* translate into either favorable or unfavorable attitudes that affect behavior. *Normative beliefs* shape the subjective nature of behavior. *Control beliefs* affect the perception of behavioral control. These three elements, according to the TPB model, craft behavioral intention. The stronger the influence of perceived behavioral control, the greater is the likelihood that an individual will either strengthen an intention and/or perform an intended behavior, when the opportunity arises.

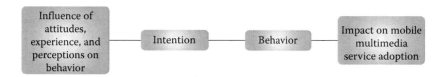

Figure 4.1 Elements of influence on mobile service adoption.

---

* Ajzen, I., "The theory of planned behaviour," *Organizational Behaviour and Human Decision Processes*, vol. 50, pp. 179–211, 1991.

Service mobility is particularly relevant, with respect to the various considerations that are revealed by the TPB model. Mobile access to the rich variety of a multimedia service landscape is abundant in innovative horizons that span utilitarian, entertainment, education, and wellness categories of service offerings. The nature and texture of these types of service offerings inherently motivate and inspire their adoption, through a variety of behavioral expressions driven by intention. The ability to communicate and collaborate with ease fuels the utilitarian and well-being-related aspects of the fabric of human existence.

Mobility may be broadly categorized* as: *traveling* (wide-area mobility), *wandering* (local-area mobility), or *visiting* (wide-area or local-area mobility), culminating in a period of stay at a given location. Behavioral expressions, related to mobile service adoption, are applicable within each of these categories. The mobile device as a personal companion serves as an easily accessible portal of being touch in a social context. It is a vehicle that invites various modalities of communication for maintaining relationships or for creating new ones. Service mobility is a function that embellishes "weak ties"† through sustenance of relationships, whether personal or professional, through customizable and individual behavioral expressions. The ensuing experiences, as compared with the expected relational and functional outcomes, affect the shaping of emerging individualized behavioral expressions that in turn impact the nature of mobile service adoption.

Intention-driven behavior is primarily influenced by the contextual nature within which a mobile service is used, which is a conditional value. The degree of commitment and cost aspects are auxiliary factors that affect usage. The crafting of a mobile service that carves out a compelling experience augments a commitment for adoption through the pivotal ingredients of conditional benefits in concert with a high emotional appeal.‡

4.3 Contextual Awareness

Face-to-face communications are rich in media content that includes ambient information and expression information, which combine to create a sphere of perception, which interplays with this holistic array of information. The sphere of perception provides a contextual awareness, which embellishes the quality of communications. This is the essential fabric that resides in the experiential domain of communications: one that technology mediation strives to attain through user-oriented mobile services that adapt to personalized styles, across the chasms of time and space.

The contextual aspects, in addition to technology-mediated embellishment, subsumes a symphony of notes associated with the user, service content, service provider, and technology mediation.

The influence of technology mediation, with the experiential domain of a user, is an element within the contextual landscape, together with the nature of service content, and the quality of personalized care offered by the service provider, as shown in Figure 4.2. The contextual aspect is enhanced as the awareness of context is augmented, together with the associated perceived value, which is characterized by the ability to invoke a mobile service conveniently—anytime, anywhere.

---

* Kristoffersen, S. and Ljungberg, F., "Mobility: From stationary to mobile work," in *Planet Internet*, K. Braa, C. Sorensen, and B. Dahlbom (Eds.). Lund, Sweden: Studentlitteratur, pp. 137–156, 2000.
† Granovetter, M., "The strength of weak ties: A network theory visited," in *Social Structure and Network Analysis*, P. Marsden and N. Lin (Eds.). Beverley Hills, CA: Sage, pp. 105–130, 1982.
‡ Pura, M., "Linking perceived value and loyalty in location-based mobile services," *Service Innovation Management*, vol. 15, pp. 509–538, 2005.

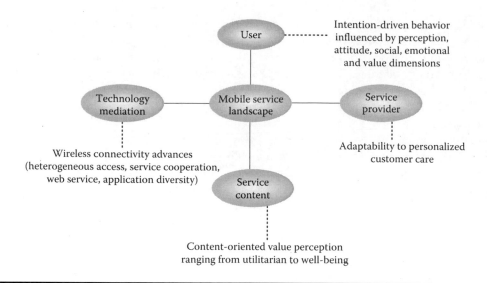

Figure 4.2 Influencers of mobile service adoption.

The contextual landscape, illustrated in Figure 4.2, includes automatic service content augmentation, in terms of context-awareness through the overlay of related information, such as pictures or videos. Such a holistic augmentation lends itself to the melioration value in terms of both perceived utilitarian and well-being aspects. Context-awareness is multifaceted. The physical aspect of context may be defined in terms of time and spatial orientation, such as geolocation or a descriptor, such as a locality, a town, a city, or a country. The physical context serves as a trigger for a variety of contextual enhancements that pave the path toward virtual reality and augmented reality-oriented mobile service opportunities. Social linkages and attributes define a social context, through the use of services such as Twitter, LinkedIn, Facebook, etc. Lifestyle preferences, personal or business related, define a functional context. Each of these categories of context is an exemplification of the synergistic potential of context-oriented embellishment of mobile service. A holistic view of context consists of compelling recipes of information aggregation and presentation that are relevant and customizable, to situational interactions among humans and mobile services.[*] The flexibility and adaptability of mobile services to serve personalized choices and behaviors further improves the perceived value that a contextual landscape brings to mobile service.[†]

4.3.1 Interdependence

In a woven complex of human communications, technology mediation requires an inclusive lens. It is one where artistic and human factors—together with utilitarian aspects—become essential ingredients within an expanded technology paradigm. It is a creative ensemble of these ingredients that requires infusion into the evolutionary considerations within the realm of mobile communications. This view evokes considerations of interdependence across assorted elements in the

---

[*] Pura, M., "Linking perceived value and loyalty in location-based mobile services," *Managing Service Quality*, vol .15, pp. 509–538, 2005.
[†] Bouwman, H., Haaker, T., and de Vos, H., "Mobile service bundles: The example of navigation services," *Electronic Markets*, vol. 17, pp. 20–28, 2007.

arena of mobile communications. Convergence across the nature of multimedia service experience over the fixed Internet and the mobile Internet, where mobility is an implicit attribute of an internetworked service usage, is one which demands an evolving vision for service orchestration. A coming together of the experiential domain within the mobile domain requires a realization of a cooperative and collaboration service framework that bridges service creation, delivery, and consumption over assorted networks, mobile or fixed. The all-IP connectivity fabric is universally binding and allows an integrative mobile service pathway, while accommodating heterogeneous and autonomous access technologies. While this emerging paradigm imbues a new vision for a mobile Internet, advances in computing power, smaller form factors, storage densities, cognitive user interfaces, such as touch, gestures, voice, etc., and commoditization are ushering a convergence at the device level through appealing mobile interfaces—laptops, tablets, and smartphones. These twin advancing paradigms fuel the motivators for a leveraging of a common framework for service orchestration, whether it is performed over a fixed or mobile network. The common framework approach is powerful, from a service provider perspective, since the opportunities for mobile service offerings then become independent of the source, fixed or mobile, of a service offering. This independence through a leveraging of interdependent convergence fosters innovation in service offering, such as contextualization, flexibility, and adaptability to user-level customization of a mobile service. From a user-level perspective, contextualization and a trust relationship with the service provider are vital, since mobility, unlike fixed access, requires specialized care with respect to the dynamic nature and quality of connectivity to any mobile service. The trust relationship involves a user identity, for wireless connectivity together with the associated unique descriptors of the related user profile and mobile service preferences. This unique relationship can be leveraged to enhance service contextualization, whether the service orchestration occurs over a fixed or a mobile network.

The mobile device is an individualized portal to the changing seas of information and interaction, over the Internet. It serves as a source of contextualization, which is a function of user behaviors that are qualified by input information acquired through location, touch, gestures, voice recognition, etc. The realm of mobility as a service involves opportunities for a contextualization of information,* situational and user-initiated, that are inherently dynamic and offers the potential for enormous enhancements to mobile service experience. Innovation that leverages this interdependent web of contextual information, in the form of collaborative architectural models and functional capabilities, is the underlying canvas for service augmentation. Sources of contextual information include GPS (global positioning system) information, cameras, proximity sensors, health monitors, home environment, in-vehicle interaction, etc. Lower levels of context from sensors could be used to infer higher level contexts of user behavior and preference for enhancing individualized adaptability, flexibility, and customization.

A framework to extract and deliver contextual information is intrinsically interdependent and dynamic in nature. The mobile device collects information from its surroundings and usage behaviors, utilizing user preferences. Within a context management framework, this information is sent to the network for grooming and embellishing multimedia information streams that are destined for the mobile device. Applications that are resident in the mobile device can then leverage the relevant contextual information to augment the service experience. The processing of dynamic contextual information in the network alleviates the scale of processing requirements and power consumption with the mobile device. At the same time, an awareness of the

---

* Dey, A., Abowd, G., and Salber, D., "A conceptual framework and a toolkit for supporting the rapid prototyping of context-aware applications," *Human–Computer Interaction*, vol. 16, no. 2, pp. 97–166, 2001.

contextual information at the network level allows a convenient and scalable processing of contextual information across different mobile devices, such as those that may be demanded by peer-to-peer applications: social networking or gaming, etc. Cooperation between a mobile client and a network server is the foundational entity in a context management framework. The context information is automatically and dynamically refreshed to reflect the mobile service flavor, as the information derived from the ambient conditions changes. The conceptual aspects of the framework are independent of the types of operating systems (OSs)—iOS, Android, etc.—resident in a smartphones, tablets, or laptops. An illustration of a context management framework is depicted in Figure 4.3.

The mobile device serves as a source for contextual inputs, from its orientation and location with respect to its ambient conditions, which is provided for further processing over the upper layers of the protocol stack, which includes application-level processing. In the abstract interdependent contextual framework depicted in Figure 4.3, the mobile network gateway, which acquires the contextual information, through the context information wrapper, redirects the contextual information toward the context server.

The context data server processes the contextual information into formats that are relevant for inference and use by the application layer. As an example for location-oriented applications, the context server collects the GPS coordinate information together with other metadata, using a context data model, together with temporal context information, and transfers the corresponding information for use by a mobile location-oriented service. The context server correlates the applications and the different streams of contextual information based on the corresponding mobile applications that are registered with the context server. Different context providers, resident in a mobile device, publish contextual information streams and subscribe to context servers, to provide contextual information streams to context servers.

The richness and diversity of context augmentation are elemental in the pursuit of enabling advances in experiential appeal of mobile service. Along these directions, the enabling capabilities offered by the IMS (IP Multimedia Service) and the RCS (Rich Communication Suite) frameworks are foundational. These frameworks, in concert, provide a platform for convergence—fixed, mobile, and assorted applications—and augmentation of mobile multimedia service offerings. These augmented technology-mediated capabilities have enabled innovation in the context of

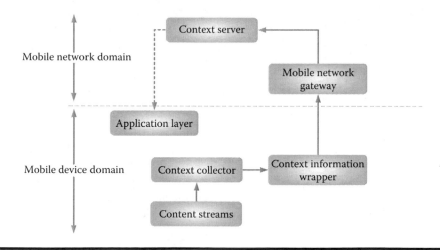

Figure 4.3   Interdependent frameworks for context management.

mobile Internet-propelled application-centric service horizons. It is one where service offerings are enriched through various combinations of flavors for human social interaction, as well as for virtually unbounded types of information availability. Web 2.0 applications are reflective of the ability to carve various service offerings with assorted types of applications that are influenced by personalized motivating factors: utilitarian to well-being. This reveals the attractiveness of meaningful cross-linkages that mirror the behavior of the natural world. Interdependence then extends beyond the technology fabric of convergence enabling frameworks to business, social, and personal aspects of cooperation and collaboration for unique and contextualized mobile service offerings.

4.3.2 Ambient Intelligence

The cooperative and collaborative exchange of context associated with people, places, and events creates a web of awareness among the actors, enabling an experiential service space for people. The interactions nonintrusively meld into the ambient, and the service appears as one that is experiential. The distribution of multimedia applications, sensors, and information repositories, in both mobile and fixed scenarios, is a pivotal paradigm in the progression toward an intelligent ambient ecosystem. In such an ecosystem, the shift is away from the traditional menu-driven, command and control interfaces toward intuitive and perceptual interactions. The integration of behaviors and information in such an ecosystem is enabled through ubiquitous or sentient computing models.

Mark Weiser* postulated: "The most profound technologies are those that disappear. They weave themselves into the fabric of everyday life until they are indistinguishable from it." The proliferation of the state-of-the-art of computing, in its various revisions, such as mainframes, desktops, and laptops, over the past decades, has been much about augmenting the same model: enhancements of an entity distinctly separated from the user, and linked through a series of artificial command and control accessories, such as a keypad. This is analogous to the challenges faced by scribes and authors in early times, when good penmanship was just as important as the creative pursuit of writing. A thought-provoking observation reveals that although there is much more information exchange, during a walk in the woods, relative to that available via a computing intermediary, we are unaware of any information overload*. The former is relaxing, while the latter is intrinsically unnatural. This reveals a profound motivator toward inherently intuitive intermediary interfaces between a user and an ambient enhanced mobile service. The vanishing of the enabling aspects of technology is fundamental to a user appeal. Such an attribute serves without any noticeable intrusion or impediment—an attribute of human psychology, where a user is unaware of technology mediation, while enjoying its benefits. Michael Polyani† coined the term "tacit dimension," where "we can know more than we can tell." This is a precursor to knowing and is referred to as "tacit knowledge,"‡ which is a collection of conceptual and sensory information that can be exploited to make sense out of something or to create something new. Users are then empowered to pursue and realize their intentions, without being explicitly distracted by the specifics of a service.

The mobile device, with its personalized dimension, is situated naturally in the realm of service offerings that integrate the richness of ambient information with the user. Ubiquitous computing

---

* Weiser, M., "The computer for the 21st century," *Scientific American*, vol. 265, pp. 66–75, 1991.
† Polanyi, M., *The Tacit Dimension*. New York: Anchor Books, 1967.
‡ Hodgkin, R., "Michael Polanyi: Prophet of life, the universe and everything," *Times Higher Educational Supplement*, p. 15, September 27, 1991.

through its inherent pervasiveness orchestrates interactions among the actors, including the user in an ambient space. The broad themes of interactions among the actors mediated through applications span natural interfaces, contextual awareness, and automatic information exchange. Natural interfaces reflect the modalities of human expression between the user and the ambient service environment. Contextual awareness and automated information exchange embellish ubiquitous applications to enrich the service adaptation and service experience. The inclusion of cognition promotes intelligence in the ambient space, in terms of the emotional dimension of human experience, which further promotes the service appeal by adapting the service behavior to a user's emotional state.[*] Situational—temporal and spatial—context, which includes an inference of emotional states, is derived through cognitive agents using artificial intelligence techniques, such as fuzzy logic and inference engines, embedded within a pervasive computing environment of services. For example, a cognitive agent could recognize the difference between morning, evening, or location in an ambient space and render different modalities of the same service, such as different styles and/or volumes, related to a music collection, in concert with a user's behavioral preferences.

Cognitive agents could utilize a timed FCM (Fuzzy Cognitive Map),[†] to infer the impacts among related concepts, such as the temporal and spatial attributes of an ambient space—a dynamic living environment of a mobile service user—on the experiential aspects of a given mobile service. A timed FCM is a fuzzy signed graph with feedback, where the sign ± at the edges of the graph indicates a causal increase or a causal decrease of the significance of a represented concept, among a collection of concepts and relationships among concepts. The timed FCM utilizes an approach that is akin to manner in which neural network dynamically infers causal relationships. It enables cognitive agents, associated with a service with dynamic behavior information, service concepts and user concepts, within the ambient ecosystem, to adapt its decisions to the spatial, temporal, and emotional situations. The significance of a concept or a relationship across concepts, at a given time, is revealed by the timed FCM, which is then utilized by a cognitive agent to make decisions that optimize the mobile service experience. An illustration of a model that harnesses the ambient intelligence, to augment the experiential aspects of a service, is depicted in Figure 4.4.

Multifaceted sensing capabilities, proximity, light, positioning, voice, gestures, accelerometers, gyroscope, touch, camera, QR (Quick Response) codes, assorted radio access, etc., within smartphones and tablets provide pathways for enhanced cognitive and contextual awareness capabilities. Collectively these versatile sensing faculties provide crowdsourcing opportunities for collaboratively enhancing personalized lifestyles through a rendition of mobile services enriched with these enhancements. For example, iCartel[‡] is an application that enables motorists to receive position-oriented real-time traffic conditions, using a collection of information detected from other motorists in the vicinity. Another scenario, among virtually unbounded innovative possibilities, is Transitgenie,[§] which cooperates with the sensed ambient

---

[*] Zhou, J. and Kallio, P., "Ambient emotion intelligence: From business awareness to emotion-awareness," in *Proceeding of 17th International Conference on Systems Research, Informatics and Cybernetics*, Baden-Baden, Germany, 2005.

[†] Acampora, G., Loia, V., and Vitiello, A., "A cognitive multi-agent system for emotion-aware ambient intelligence," in *2011 IEEE Symposium on Intelligent Agent (IA)*, Department of Computer Science, University of Salerno.

[‡] http://www.iCartel.net.

[§] Thiagarajan, A., Gerlich, T., Biagioni, J., and Eriksson, J., "Cooperative transit tracking using GPS-enabled smartphones, in *SenSys 2010*, Zurich, Switzerland, November 3–5, 2010.

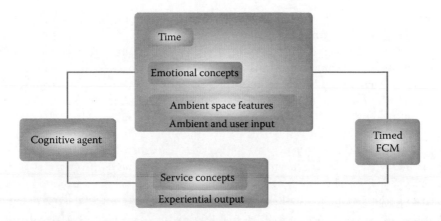

Figure 4.4   Experiential enhancement through ambient intelligence.

information consisting of the status of trains, buses, etc. equipped with advertising their real-time location-related context, to provide a value-added mobile service that enhances lifestyle and well-being.

Interdependent information streams enhance the richness of context awareness, associated with a mobile service, in a personalized fashion. The elements of context, which act in concert for a specific service, provide the building blocks for an attractive QoE , within a dynamic framework of mobility, and they are depicted in Figure 4.5. The model for an enhanced context awareness composes a holistic view of the context, from the interdependent streams of information available from a user's internal state of choice-making, behavior, and emotions, together with the ambient conditions such as a conference gathering, emergency situation, attributes of light, sound, touch, etc., and crowdsourced information such as traffic patterns, web of social information, etc. Each of these factors has an effect on contextual awareness, which is dynamic in nature and uniquely affect the experience of the mobile service user.

The greater the improvement of perception and a resolution of uncertainty, through a holistic contextual awareness, the greater is the potential for a favorable user experience.

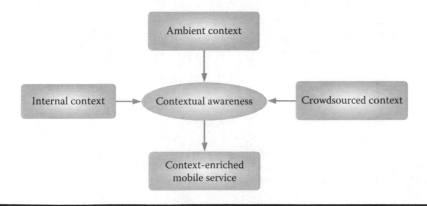

Figure 4.5   Factors that affect the QoE in mobile service.

Through social networking interactions[*], such as micro-blogging via Twitter or haptic[†] sensing capabilities via instant messaging, a mapping of comfortable or stressful environments can be created. These types of technology-mediated communications utilize personalized and crowd sourced information for rendering context embellished content. In a global economy, where cultural and geographic boundaries are no longer relevant in a social, leisure, or business context, stress factors are a side effect of these emerging lifestyles, which then demands therapeutic regimens, such as exercise, and healing techniques through sight, sound, and touch. It is in the realm of touch that multimedia mobile services can be augmented through the inclusion of haptic components. Technology mediation, through haptic mobile services, can then become a tool for well-being, in emergent lifestyles, where the barriers of time, space, and the comfort zones of familiarity are often breached. Both physiological and immunological benefits are a function of touch triggered well-being.[‡] Service augmentation, where touch effects are an integral component of remote social interaction, enriches the audio-visual component, such as that experienced, when someone includes a physical "nudge" to complement a greeting: "Hello, there." Such touch-oriented, encoded information streams could be classified as "affective communication," which is ambient, crowdsourced, and ubiquitous to create a perceptually attractive experience. Online interactions, such as through the various modalities of social networks, represent the enormous potential for affective communications. Expressions, feelings, and aspirations pervade the web of crowds, around the globe for variety of purposes, yet converging to a common existential theme of well-being, with significant personalization that is afforded uniquely through mobility-powered connectedness. Patient therapy, through affective communications, has been implemented at a North Carolina hospital[§] to promote healing and well-being.

The highly multidimensional and multimodal nature of haptic perception leverages a variety of sensory information, which is then aggregated in the human brain to enable perception.[¶] The context of intention, emotion, and imagination, together with the sensory information, provides the ingredients of a personalized and differentiated experience. Haptic perception opens avenues for immersive, yet remote interactions for compelling, collaborative, and memorable experiences. It enables augmentation in telepresence service scenarios, consisting of a remote immersion and interaction, in both human-to-human and human-to-machine telepresence situations. Elements of tactile information are composed of attributes such as pressure, touch, temperature, comfort, and pain for neuromuscular sensory feedback for an articulation of perception. This aspect of an affective mobile communication service could either augment sight and sound or may be a substitute, where visibility or hearing may be impaired as a result of individual or ambient conditions. Haptic service capabilities are central to the theme of TPTA (Tele-Presence and Tele-Action)[**] systems,

---

[*] Blom, J., Viswanathan, D., Spasojevic, M., Go, J., Acharya, K., and Ahonius, R., "Fear and the city: Role of mobile services in harnessing safety and security in urban use contexts," in *Proceedings of the 28th International Conference on Human Factors in Computing Systems*. New York: ACM, pp. 1841–1850, 2010.

[†] Chung, K., Chiu, C., Xiao, X., and (Peggy) Chi, P.-Y., "Stress outsourced: A haptic social network via crowdsourcing," in *Proceedings of the 27th International Conference Extended Abstracts on Human Factors in Computing Systems*, Boston, MA, pp. 2439–2448, 2009.

[‡] Zeitlin, D., Keller, S., Shiflett, S., Schleifer, S., and Bartlett, J., "Immunological effects of massage therapy during academic stress," *Psychosomatic Medicine*, vol. 62, pp. 83–84, 2000.

[§] Noguchi, Y., "Cyber-catharsis: Bloggers use web sites as therapy," *Washington Post* article, October 12, 2005. http://www.washingtonpost.com/wp-dyn/content/article/2005/10/11/AR2005101101781.html.

[¶] Ernst, M.O. and Bülthoff, H.H., "Merging the senses into a robust percept," *Trends Cognitive Sciences*, vol. 8, pp. 162–169, 2004.

[**] Steinbach, E., Hirche, S., Kammerl, J., Vittorias, I., and Chaudhari, R., "Haptic data compression and communication," *IEEE Signal Processing Magazine*, vol. 28, no. 1, pp. 87–96, January 2011.

where an immersion into ecosystem of people and things is enabled through ambient intelligence, context-awareness, and crowdsourcing. These capabilities can be used as dynamic, therapeutic, and well-being databases for service innovation in the experiential dimension.

The design of such applications entails an adaptation of sensory information capture, with different OS platforms, such as iOS, Android, etc. This implies a leveraging of the relevant API (application programming interfaces) that hide the underlying access network complexities and nuances—by serving as an abstraction layer that encapsulates application-level protocols and data formats—from the users of APIs, and an optimization of wireless bandwidth, QoS, and power consumption. Crowdsourcing cooperatively mines, filters, and serves pertinent information to a mobile user, to suit individualized preferences. Haptic feedback for social, therapeutic, and well-being is realizable through innovative service-driven haptic actuators, such as wearable transducers that add the dimension of touch, within a multimedia stream of information. Haptic actuators enable collaboration for remote immersion, through anthromorphic limbs and gripping artifacts that promote a TPTA service: a synergistic collaboration, among the various attributes of emotional and ambient categories, of context and crowd sourcing.

Through a resolution of uncertainty, produced by the separation in time-space and dynamically changing situations, such as traffic conditions in the vicinity, a motorist benefits from crowd-sourced information for alternative directions to avoid congestion. This type of service augments and optimizes a motorist's driving experience. At the same time, the information used in such as service employs innovative techniques to extract the embedded wisdom of the crowds, where the quality and reliability of the information is improved through a diversity of information sources.

The selective mining of a web of information embedded in the crowds,[*] coupled with contextual and ambient awareness, transcends the traditional boundaries of information types, namely, home, office, well-being, healing, manufacturing, learning, education, etc., and holds universal promise of experiential advancement of mobile service. The harnessing of information that permeates the crowds serves as a dynamic, real-time repository of energy and intelligence that offers new paradigms for mobile service, where the user, the service, and the ambient operate as a symphony to usher new perspectives, behaviors, and innovation toward new horizons in mobile service evolution.

4.4 Personalization

Among the most widely consumed forms of content appears music, which transcends cultural and geographic boundaries, while being amenable to individual choices. Consumption of music, in mobile usage scenarios, stored locally or in the cloud, is especially compelling with flexible user selectable preferences in terms of music genres, playlists, and intelligent discovery of music derived from a likeability history associated with the user.

Similarly, book content, delivered via mobile e-readers, namely, iPad, Kindle, etc., provides the user with the convenience of a mobile library of books, driven by individual choices and customizations. These choices for individualization are metaphors, in an expanding vista of possibilities in the realm of technology-mediated communications, whether the storage of content is local or in the cloud, where the access is enabled from any authenticated mobile device.

---

[*] Afridi, A.H. and Gul, S., "Method assisted requirements elicitation for context aware computing for the field force," in *Proceedings of the International Multiconference of Engineers and Computer Scientists*, Hong Kong, vol. I, March 19–21, 2008.

These denote examples of conveniences in assorted communications that incubate shifts in lifestyles and behaviors that in turn serve as catalysts to propagate the evolution of mobile services that hinge on the dominant theme of experience. Among the most compelling aspects of the fruits of the information age, with mobility-powered communications as its crown jewel, is the notion of individualized or narrowly customized artifacts of experience-oriented services. In other words, it implies an evolution toward a Long Tail paradigm. This archetype of service creation and delivery exemplifies a shift characterized as one where "Our culture and economy is increasingly shifting away from a focus on a relatively small number of 'hits' (mainstream products and markets) at the head of the demand curve and toward a huge number of niches in the tail. As the costs of production and distribution fall, especially online, there is now less need to lump products and consumers into one-size-fits-all containers. In an era without the constraints of physical shelf space and other bottlenecks of distribution, narrowly targeted goods and services can be as economically attractive as mainstream fare."*

In late 1993, the world was introduced to the World Wide Web, with the arrival of the Mosaic web browser. It marked the advent of widespread access, creation, and consumption information and a practical realization of a digital era through technology mediation, from a wired to a wireless mobile world that continues to see unprecedented shifts through the third millennium. The distributed nature of information through the inherent interconnectivity of the web allows the capabilities of social networks and cloud computing to be leveraged by mobile communications, where this information continues to be pushed inexorably to the edges of creation and consumption—the individual. It is at the edges of interconnectivity, where uncertainty proliferates, that information possesses the maximum potential for experiential value generation through the resolution of uncertainty. Consequently, technology-mediated innovation at the edges of interconnectivity is profound in the realm of mobile services. This implies a compatible shift in the paradigm for technology mediation and in the related user behaviors. A depiction of this shift—STEP (shift to edge paradigm)—is illustrated in Figure 4.6, which depicts the corresponding changes in the associated attributes, going from one that has been traditionally hierarchical to a new model of distribution, autonomy, and self-organization. This shift is symbolic of the evolution of the digital

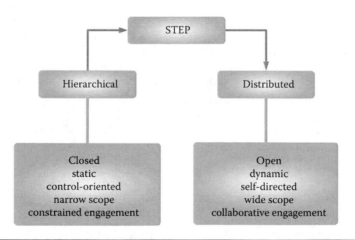

Figure 4.6 STEP (shift to edge paradigm)—reflections in a technology-mediated evolution.

---

* Anderson, C., *The Long Tail: Why the Future of Business Is Selling Less of More.* New York: Hyperion, 2006.

era of an information web, where advancements in mobile communications continue to evolve to the edges and then into the experiential domain of human communications.

The experiential domain is uniquely subjective, while being simultaneously interdependent. Functional compliance and performance quality are inadequate in of themselves to guarantee a favorable or attractive mobile service experience. Since the service experience is intertwined with the user and the related context, it cannot be examined separately from the context or the user.* The personalized nature of mobile service adds to the significance of the embodied nature of experience, within the user and the context, which is shaped by the function, performance, and usage scenarios. Subjectivity in the experiential domain is *emergent*—one which is a process that is shaped by a variety of ingredients, although not easily understood or predictable through a knowledge of characteristics of the ingredients.† The emergent nature of subjective experience results in its expansion and advancement over time, while remaining elusive to external or internal, user level, descriptors.‡

Subjective perception that shapes experience is depicted in Figure 4.7, as consisting of horizontal and vertical dimensions,§ to enable both a sequential and a simultaneous awareness of features, respectively—such as those within a mobile service—within the conscious state of a mobile service user. Sequential awareness spans perceived features over short periods of time enabled by memory, while simultaneous awareness is a momentary snapshot of perceived features. These dimensions in their aggregate create and shape a mobile service experience in a uniquely personalized manner. The modalities of sensory input, in concert with relevant multimedia information content that influence experience, include orientation enabled through accelerometers, gesture, gaze, speech, and touch. As a part of the Architecture Machine Group at MIT, Richard Bolt⁵ demonstrated the enormous appeal of the natural modalities

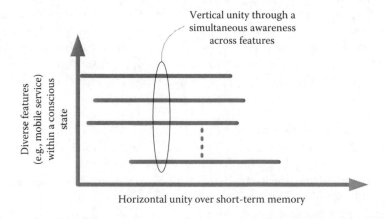

Figure 4.7 **Interplay of the orthogonal dimensions of features in the experiential domain.**

* Searle, J.R., *The Rediscovery of the Mind*. Cambridge, MA: MIT Press, 1992.

† Cerf, V.G., "Emergent properties, human rights, and the internet," *Internet Computing, IEEE*, vol. 16, no. 2, pp. 87–88, 2012.

‡ Swanson, E. and Sato, K., "Structuring for subjective experience: The contextual scenario framework," in *2012 45th Hawaii International Conference on System Sciences*, Maui, HI, pp. 589–598, 2012.

§ Searle, J.R., *The Rediscovery of the Mind*. Cambridge, MA: MIT Press, 1992.

⁵ Bolt, R.A., Architecture Machine Group, and Massachusetts Institute of Technology, "Put-that-there: Voice and gesture at the graphics interface," in *SIGGRAPH '80: Proceedings of the 7th Annual Conference on Computer Graphics and Interactive Techniques*. New York: ACM, 1980.

of human communication, which is inherently entangled and multimodal. The central theme of this experiment illustrated the interdependent nature of multimodal content and sensory information: "how voice and gesture can be made to interorchestrate actions in one modality amplifying, modifying and disambiguating actions in the other." The modalities may be user intentioned, namely, deliberate inputs of touch, speech, gesture, etc., or they may be cognitive, namely, computer recognition of facial expressions, image, movements, orientation, ambient, etc. This type of interorchestration is evident in smartphones that are embellished with multimodal inputs and outputs as the advances in cognitive technologies allow practical implementations in the design of attractive user interfaces. These directions in evolution and innovation are aligned with an adaptation of technologies to suit the innate aspects of natural human behavior, sensory perception, and input.

The mobile device, smartphones, tablets, etc., is representative of the various aspects of an interactive and personalized system, such as its features and capabilities, which are influenced by the usage scenarios that shape a contextual profile. The expectation of the performance of the mobile device is colored by the context of a usage scenario: personal, business, social, temporal, spatial, location-oriented, preference-oriented, interest-oriented, etc. It then follows that the user experience is crafted as a function of the contextual flexibility served by the suite of features and capabilities offered by smartphones, tablets, etc. Complex combinations of contextual profiles that span a variety of usage scenarios, within an ecosystem of mobile service operation, are pivotal in energizing the experiential domain, where the subject and object blend indistinguishably. The embodiment of the mobile service user, within an assortment of contexts that vary with situation, shapes the mobile service experience. This contextual awareness of mobile service weaves the situational ecosystem with the subjective usage scenario, while the mobile-oriented technology mediation remains nonintrusive in the experiential landscape. It is an essential ingredient in the design and implementation of the mobile device interfaces and applications. Contextual awareness reinforces the narrative that the mobile service user is both the actor and the audience, where the enabling technologies serve as intuitive mediators. While passive mediation enables through an attentiveness of human behavior, active mediation enables through the intentionality of human behavior: explicit user inputs such as gestures, touch, voice, image, etc.

The appeal of a mobile service may be estimated as a function of its alignment with user intention, using the TAM (Technology Acceptance Model).* According to TAM, the user's attitude concerning a mobile service is a significant factor in the adoption of its use. The user's attitude is a function of two belief components associated with a mobile service, which are PU (Perceived Usefulness) and PEOU (Perceived Ease Of Use). The latter belief influences the former, which then implies the significance of innovative designs that yield intuitive and artfully attractive mobile device user interfaces. These belief components must then be an intrinsic consideration of mobile service design approaches from a holistic, system level perspective. The mobile service model encompasses functions in mobility and applications, in distributed user-centric landscape, which is strongly aligned with the Web 2.0 paradigm of collaboration, interdependence, and interaction through social relationships, blogs, wikis, mashups, content sharing, and applications. The impacts of PU and PEOU are influenced by emotional motivators of attitudes toward mobile services, which in turn affect its adoption through intention. It has been observed that positive emotional attitudes toward a mobile service, in terms of PU and PEOU, tend to favorably

---

* Davis, F.D., "User acceptance of information technology: System characteristics, user perceptions and behavioral impacts," *International Journal of Man–Machine Studies*, vol. 38, pp. 475–487, 1993.

influence its adoption.* Design, implementation, and customer care approaches should then be tuned according to the nature of a mobile service—utilitarian or well-being-oriented—to optimize the PU and PEOU aspects of the mobile service to enhance its appeal for a mobile user.

TAM considerations, enhanced with holistic and enjoyment attributes[†] of a mobile service, are a significant determinant of the appeal and adoption of the mobile adoption, which in turn provide insights into the underlying technology-mediated designs and innovation. The holistic[‡] attributes of a mobile service view the context of usage as a whole, where the objective and the subjective dance together as an indistinguishable aspect of the whole, in an experiential field. Cultural nuances[§] of user behavior that are not evident from observation may be inferred from verbal and nonverbal expressions that are valuable in determining other behaviors that influence service adoption. On other hand, there is a commonality in the cognitive methods, within the human brain, that transcends variations across cultural boundaries, but is affected by individual belief systems.[¶] A collection of these attributes, in a holistic sense, influences the nature of abstract, logical, and analytic thought that pervades the building-blocks of the underlying technology-mediated enabling functions, within the innards of a mobile service, which are comfortably non-intrusive, with respect to personalized usage scenarios. The identification and consideration of elements that promote a holistic experience, within technological aspects of mobile service design, attract the engagement of users, within the domain of a mobile service execution. This synthesizes an enjoyable experience,** which is a compelling motivator, which is distinct and yet a part of the underlying technology mediation and the related performance that it offers.

Application stores have a virtually unbounded potential for the adoption of applications and services influenced through personalized and intention-driven belief patterns. Since applications are the building-blocks of services, they provide a specific function or a capability that appeal to a user. These building-blocks conform and thrive in the Long Tail landscape of possibilities—the rendering of a few building-blocks, in the creation of different service offerings, conforming to a virtually unlimited adaptation to user personalization. In turn, these attributes promote openness, mobile device, and mobile service differentiation, where innovation and incremental value generation abound.

4.5 Social Ties

A technology-mediated social fabric, with mobility as a natural ingredient, opens pathways to an immersive communication ecosystem, where a remote physical presence embodiment, separated in time-space, is virtually replicated for a local sensory perception and interaction. Virtual co-location of communications has been central to the theme of human interactions that underscores

---

* Lee, Y.K., Lee, C.K., Lee, S.-K., and Babin, B.J., "Festival scapes and patrons' emotions, satisfaction, and loyalty," *Journal of Business Research*, vol. 61, no. 1, pp. 54–64, 2008.

† Agarwal, R. and Karahanna, E., "Time flies when you're having fun: Cognitive absorption and beliefs about information technology usage," *MIS Quarterly*, vol. 24, no. 4, p. 665, December 2000.

‡ Nisbett, R.E., Peng, K., Choi, I., and Norenzayan, A., "Culture and system of thought: Holistic versus analytic cognition," *Psychological Review*, vol. 108, no. 2, pp. 291–310, 2001.

§ Hofstede, G., *Culture's Consequences*, 2 edn. Thousand Oaks, CA: Sage Publications, 2001.

¶ Noguchi, Y., "Cyber-catharsis: Bloggers use web sites as therapy," *Washington Post* article, October 12, 2005. http://www.washingtonpost.com/wp-dyn/content/article/2005/10/11/AR2005101101781.html.

** Davis, F.D., Bagozzi, R.P., and Warshaw, P.R., "Extrinsic and intrinsic motivation to use computers in the workplace," *Journal of Applied Social Psychology*, vol. 22, no. 14, pp. 1111–1132, 1992.

the intrinsically social in nature of human behavior and existence. Subjectivity dominates in the communication realm, where both responses and perceptions are intertwined. Technology-mediated communication services attempt to transform the state of mind or emotion such that space and time gaps are transcended. Such services strive to mimic the ingredients and attributes of face-to-ace social interactions, through assorted applications: social networking, voice, video, text, pictures, etc. On the one hand, technology mediation, such as mobile services, allows instant communication modalities, around the planet to satisfy business and social demands.* On the other hand, technology mediation to effectively capture the subjective nature of human experience is an ongoing endeavor in research and evolution. While information transfer rates and capacity over mobility-augmented communication continue to improve, the experiential quality dimension has subjective attributes, such as ambient immersion that is nonintrusively present in face-to-face communication that requires consideration in mobile services. Nonverbal content, such as position, gesture, eye contact, attentiveness, touch, posture, and appreciation cues, can be both displayed and detected in face-to-face communications. Such tacit expressions and acknowledgements are what make the communications experientially impressionable in a social context.† The tacit dimension afforded by face-to-face communications is the prime motivator for co-located activities: conferences, trade-shows, family gatherings, entertainment, sports, etc. It then follows that service augmentation through contextual awareness provides a framework for immersive communications. The elements of such a framework are to simulate the co-location of spatially and temporally distributed environments, for an immersive QoE. Since nonverbal content is predominantly subjective, the objective nature of technology-mediated effects is amenable to innovation to create an illusion of immersion. Social signal mediation through innovative mobile services opens the horizons for integrating personalized behaviors, in a social context, while bridging time-space barriers, in the experiential domain.

Innovation in the experiential realm in the form of ambient immersion has its roots in the art-form of animation and stories that create a fabric of sensory experience. Animating the social web through intuitive and instinctual information, mapped as sight and sound, is among the pursuits of the entertainment industry pioneer DreamWorks Animation SKG, Inc.‡ For example, Apple's talking mobile companion, Siri,§ is illustrative of the innate appeal of intuitive and instinctive interfaces to human sensory perception and response. The QoE of a technology-mediated immersive communication session may be characterized in terms of objective performance metrics—latency, jitter, throughput, audio-video quality and coordination, etc.—together with subjective metrics such as psychometric and psychophysical measures. The experiential domain merges the artificial boundaries that tend to split objective and subjective measures, where technology innovation subsumes the subjective in a symphony of oneness. The integration of multimodal descriptors of a communication—beyond the basic audio-visual components of multimedia—to include and correlate complementary sensory information, such as touch, gestures, ambiance, etc., enriches the quality of an immersive telepresence. Contextual awareness is a common theme, for promoting immersive communications ranging from interaction to collaboration, as depicted in a model shown in Figure 4.8, for illustrating the archetypes of communication in an immersive framework.

---

* Friedman, T., *The World Is Flat: A Brief History of the Twenty-First Century*, 3rd edn. New York: Farrar, Straus and Giroux, 2007.
† Vinciarelli, A., Pantic, M., and Bourland, H., "Social signal processing: Survey of an emerging domain," *Journal of Image and Vision Computing*, vol. 27, no. 12, pp. 1743–1759, 2009.
‡ http://www.dreamworksanimation.com/.
§ http://www.apple.com/iphone/features/siri.html.

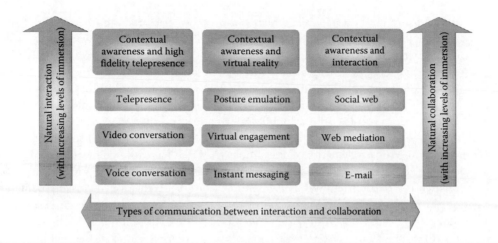

Figure 4.8   Contextual awareness for an immersive experience.

Advances in LTE-enabled wide-area, broadband access, generic all IP connectivity, computing capabilities within the small form-factor of smartphones, and tablets are pivotal ingredients in the evolution of immersive communications. All IP connectivity implicitly contains the flexibility and adaptability for innovation at the application layer for the transport multimodal sensory signals that shape a desired QoE. The psychometric component* of QoE includes a measure of the experiential dimension, enabled through a contextual awareness embedded in a mobile service. It attempts to calibrate how one feels during the experience of a mobile service. It embodies a collection of sensory signals that create an experience of *being there* through a holistic rendering of the collection of sensory signals that are transported and processed across spatially and temporally distributed locations. Robustness and reliability coupled with wider bandwidths are integral components in a vivid orchestration of experience associated with an immersive mobile service.

The psychometric component of QoE may be illustrated through a study of the use of an ESP (Embedded Social Proxy),[†] within a hub-and-satellite collaborative team, consisting of a mix of face-to-face and remote participants. Spatial and temporal co-location in the hub naturally enables a high-level of contextual awareness through the modalities of physical presence and proximity nonverbal and social cues that augment the audio-visual content streams. On the other hand, the satellite or remote participants are constrained through an absence of physical presence, proximity, and social cues. The ESP terminal located within a co-located hub represents a dedicated embodiment of a remote participant enabled through an audio-visual contextual awareness. Such an embodiment reflects a localized presence of a remote participant, in terms of collaborative activities and availability. The ESP-mediated contextual awareness provides different types of contextual information, associated with the remote participant, such as "awareness of topics of collaborative interest," "awareness of perspectives," "awareness of priorities," "awareness of likeability," "awareness of social aspects, such as behaviors and styles." These attributes of contextual awareness

---

* Apostolopoulos, J.G., Chou, P.A., Culbertson, B., Kalker, T., Trott, M.D., and Wee, S., "The road to immersive communication," *Proceedings of the IEEE*, vol. 100, no. 4, pp. 974–990, 2012.

† Venolia, G., Tang, J., Cervantes, R., Bly, S., Robertson, G., Lee, B., Inkpen, K., and Drucker, S., "Embodied social proxy: Connecting hub-and-satellite teams," in *Proceedings of Computer Supported Cooperative Work*, Savannah, Georgia, 2010.

are measurable, using the Likert* scale, which is a psychometric gradation of the experience of *being together*. At this level of technology-mediated audio-visual contextual awareness, attributes such as trust, understanding, and creativity thrive.

According to Weber's law,[†] there is a logarithmic relationship between perception and stimulus, is a measure of the psychophysical aspect of QoE. A change in perception is proportional to the amplitude of the stimulus, which is inherently subjective. Digitization and compression have provided efficiencies and scale, from a mobile information transport perspective. On the other hand, the multimodal (audio-visual, tactile, gaze, gesture, temperature, etc.) sensory experience has a continuous and subjective nature. For example, the ITU[‡] has explored measures of subjectivity that characterize a mobile service experience in terms of parameters, such as MOS (mean opinion score), which is a mean listening-quality opinion score. With the ever-increasing demand for information throughput, over the relatively limited wireless access spectrum resources, compression of audio, picture, and video information content is an essential evolutionary consideration, for an efficient utilization of these resources. The objective of compression of perceptual information content is to lower the information transfer bit rate, while minimizing the perceived distortion of information. This implies that the associated compression algorithms for a low bit rate coding of the digital representation of information content must be resilient to information distortion impacts over the probabilistic nature of a wireless link. If the noise distortion, introduced by the information compression algorithms, is properly distributed over the actual information, then the noise distortion is imperceptible to human sensory perception. This is referred to as JND (Just Noticeable Distortion),[§] where the noise distortion introduced by information compression is imperceptible from a spatial and temporal locality perspective of the actual information content. Aside from prefiltering and postfiltering for information signal conditioning, the removal redundant and irrelevant information are integral aspects of information compression, while preserving an attractive perceptual experience as depicted in Figure 4.9.

Perception provides an experience that hinges on a temporal snapshot of sensory information captured from the ambient. In audio-visual terms, high-resolution perception occurs, within a field of view or gaze, and audio with high amplitude or fidelity obscures other audio content. Inferences from these inherent sensory perception patterns are vital in the design of mobile services that provide a naturally favorable experience that is the essence of immersive communications. Subjectivity shapes perception to differentiate the QoE on an individual basis. In this differentiation, there are rough common measures that affect the QoE across individual users. For instance, the human sensitivity to audio information with frequencies beyond 15–20 kHz[¶] diminishes. It has been observed that for lip-synchronization, without perception impairments,[**] the audio lag behind video is limited to 100–125 ms, while the audio lead ahead of video is limited to 25–45 ms. In the area of the fovea of the retina containing the cones for visual acuity, the sensitivity degrades beyond 60 cycles per degree.[††]

---

* Spector, P.E., *Summated Rating Scale Construction: An Introduction.* Thousand Oaks, CA: Sage Publications, 1992.

† Ross, H.E. and Murray, D.J. (ed. and transl.), *E.H. Weber on the Tactile Senses*, 2nd edn. Hove, U.K.: Erlbaum (UK) Taylor & Francis, 1996.

‡ Methods for Objective and Subjective Assessment of Quality, ITU-T Recommendation P.800, 1996.

§ Jayant, N.S., Johnston, J.D., and Safranek, R.J., "Signal compression based on models of human perception," *Proceedings of IEEE*, vol. 81, no. 10, pp. 1385–1422, October 1993.

¶ Moore, B.C.J, *Psychology of Hearing*, 5th edn. New York: Academic, 2003.

** Relative Timing of Sound and Vision for Broadcasting, ITU-RBT.1359-1, 1998.

†† Pirenne, M.H., "Visual acuity," in *The Eye*, vol. 2, Davson, H. (Ed.). New York: Academic, 1962.

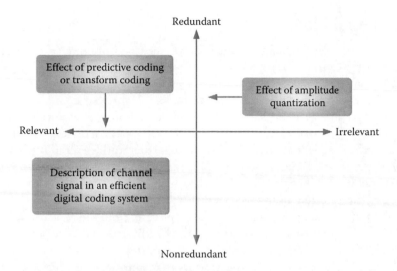

Figure 4.9 **Perceptual experience enhancement (Schouten diagram).**

An accurate expression of eye gaze is a significant aspect of spatially and temporally distributed mobile communications that approaches one of the attributes inherent in the richness of face-to-face communications. Studies in videoconferencing[*] have revealed that the 90th percentile of the sensitivity of perception of eye contact is roughly within an error of 7° downward, but only 1° upward, left or right.

In these unfolding horizons, mobile services infused with Web 2.0 capabilities cater to a variety of human factors around the theme of social ties. The conceptual building-block of Web 2.0 consists of (1) open participation that fosters coordination and collaboration of information, (2) instant availability of shared, collaborative information, (3) leveraging of open standards, and (4) leveraging of open source. These capabilities underscore a fading of the boundaries between mobile information content creation and consumption, toward interdependent styles of innovative frontiers that hinge on human experience. The advent of *intelligence* within a mobile service, such as natural language processing, learning, and reasoning capabilities, marks an evolution of Web 2.0 to Web 3.0. Virtualized, multifaceted mobile service connectivity—to the world of things around us—and a three-dimensional representation of content on smartphones, tablets, etc. are a few examples of Web 3.0-mediated, immersive communications.

The mobility paradigm mimics the naturally dynamic nature of the observer, within an ecosystem, where both perception and response change with the changing scenes. This understanding is foundational in the design of mobile services, where an immersive awareness is an integral component of technology-mediated innovation. It is one where the user is imbued with a sense being uniquely recognized, while being a part of the ecosystem at the same time. Adaptation then becomes mutual—entanglement between the mobile service user and the local environment— much like the dynamic relationship between a stage performer and the audience. The proliferation of mobile services and information share continues to shape the evolution of social ties, interactions, and behaviors.

---

[*] Chen, M., "Leveraging the asymmetric sensitivity of eye contact for videoconferencing," Computer Graphics Laboratory, and Interactive Laboratory, Stanford University, 2001.

4.6 Experiential Plane

The rich diversity of social interactions demands the inclusion of contextual content, which is vital in the discovery, expansion, and nurturing of social ties. The inclusion of subjective considerations then becomes an integral component of evolutionary mobile service horizons, through innovation and enhancements in technology mediation. Nonverbal contents, such as posture, proximity, touch, gesture, gaze direction, eye contact, intention, and attention, etc., are the signatures of co-located or face-to-face interactions. This type of articulation of content, which is subjective, lies in the experiential plane. In this arena, virtual worlds exemplify technology mediation possibilities to embed the experiential plane in mobile communications. Virtual worlds represent a persistent virtual environment within which people experience others as being with them, and where they are able to interact with them.* The experiential dimension is expanded, beyond the relatively static artifacts of voice, text, pictures, and video, through animation of peoples and their environments that are remote in the physical plane. A prominent harbinger of virtual worlds was *Second Life*.† It established a generalized notion of a *second life* in a virtual context. It provides technology-enabled ingredients that simulate an immersive experience in a cost-effective manner, for personal and commercial applications. For example, NOAA (National Oceanic and Atmospheric Administration) utilizes *Second Life* to socialize and evolve its Earth System Laboratory research through access to more people through visualization, interaction, and spontaneous collaboration.

The creation and sharing of knowledge, through the artifacts of virtual worlds, promotes innovation. A virtual world platform serves as an accessory in the access and exchange of information for an enhancement of knowledge‡ and understanding. Knowledge management entails user-centric and culture-centric attributes that shape the progression of knowledge and innovation.§ This progression is a function of cognition and knowing through participatory interactions,¶ such as sharing, contributing, collaborating, aligning perspectives, and learning through virtual worlds that complement face-to-face communications. In this realm, multimedia mobile service enables rich content mediation. Three-dimensional virtual ecosystems, enabled by Web 2.0 applications, are among the emerging frontiers of mobile service that promote collective discourse and research in the shaping of new ideas, approaches, and imagination. An awareness of social presence is a significant aspect of fostering and promoting remote cooperation and collaboration across personal and professional groups of people around the planet. Mobile access and service, with a variety of sensory interfaces, will serve as a convenient, personalized, and distributed platform for a natural harmonization of the virtual and physical worlds. The usage and presence** of avatars will serve as augmented virtual modalities of representation of physical presence. The augmented virtual

* Schroeder, R., "Defining virtual worlds and virtual environments," *Journal of Virtual Worlds Research*, vol. 1, ISSN: 1941–8477, July 2008.

† http://secondlife.com.

‡ Leonard-Barton, D., *Wellsprings of Knowledge: Building and Sustaining the Sources of Information*. Boston, MA: Harvard Business School Press, 1995.

§ Hazlett, S.-A., McAdam, R., and Gallagher, S., "Theory building in knowledge management: In search of Paradigms," *Journal of Management Inquiry*, vol. 14, pp. 31–42, 2005.

¶ Orlikowski, W.J., "Knowing in practice: Enacting a collective capability in distributed organizing," *Organization Science*, vol. 13, pp. 249–273, 2002.

** Lombard, M. and Ditton, T., "At the heart of it all: The concept of presence," *Journal of Computer-Mediated Communication*, vol. 3, no. 2, 1997.

modalities, TSI* (Transformed Social Interaction), through avatars in interpersonal communications are at the heart of innovative vistas in collaboration and cooperation in an interdependent evolution. The expressiveness of avatars in the form of conversations, text chat, gestures, demeanor, etc. symbolizes physical representations that endear and endure relationships in virtual worlds. They reflect a presence that provides a state of immersion within the context of personalized mobile services that elevate the sense of being there, which in turn enhances the experiential landscape. The enhanced levels of experience naturally lend themselves to advanced and sophisticated levels of user customization evolution, while enabling dynamic, diverse, and memorable communication services. It is the experiential aspect that will serve as a measure of the effectiveness of immersion provided by contextually significant information that wraps a mobile service.

As the technology mediation supplies contextually rich multimedia content, the social demands for a compelling mobile service—anytime, anyplace—are anticipated to evolve to provide a dynamic feedback for mobile service evolution. Improvements in the QoE for each user are directly correlated with the ability of the contextual information and representation to induce a compelling experience. Improvements in the experiential domain are enabled through widespread collaborative capabilities fueled by Web 2.0 and cloud-assisted technology mediation and human factor considerations in mobile service.

The use of web-oriented technologies, Web 2.0 and beyond, promote the technologies that are vital for the delivery of customized and niche mobile services for many, as depicted in Figure 4.10 in the flat portion of the Long Tail curve. For instance, virtual retailers, such as Amazon.com, utilize web-oriented technologies to leverage and facilitate service value for a potentially unbounded range of specialized products and features in the Long Tail to a large number of consumers. The attractiveness for engagement in the Long Tail, for both the producer and the consumer, hinges on revenue potential sustainability for the producer and customized experience for the consumer. Business sustainability accrues from web technology–mediated customization, which spurs incessant innovation for a few specialized and less popular items, which caters to a large number of consumers leading to a robust revenue potential. The use of social media is another example, among many, where the Long Tail model is significant.

The demand for products and services on a global scale continues to shape and mold the evolution of web-oriented technologies. These technologies, when applied in the realm of mobile

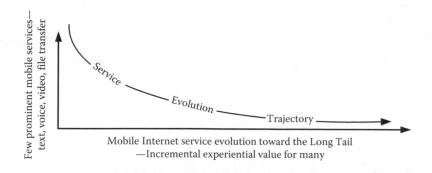

Figure 4.10 Shift toward incremental experiential value for many.

* Bailenson, J.N., Beall, A.C., Loomis, J., Blascovich, J., and Turk, M., "Transformed social interaction: Decoupling representation from behavior and form in collaborative virtual environments," *PRESENCE: Teleoperators and Virtual Environments*, vol. 13, no. 4, pp. 428–441, 2004.

services, promote a widespread distribution of products and services, together with the required logistics for delivery. The categories that span the landscape of products and services include books, movies, music, news, commerce, publishing, and design. The awareness and the creation of specialized mobile services that cater niche demands leverages ideas around natural metaphors for user-centric participation and information management across the entire web of usage, which includes the tail of participation and engagement, not just in the head or in the dominant themes of mobile service.

Mobile connectivity, an integral component of the Internet, provides innovative opportunities in the arena of the Long Tail, mediated by web-oriented technologies. Open APIs serve as a catalyst in the transformation of creative ideas into an unlimited rendering of mobile services. The mobility paradigm is naturally distributed and therefore is aligned with the notion of service creation and consumption in the Long Tail. The appeal of niche mobile services leads to a rich variety of features in terms of both supply and demand. The specialized nature of niche mobile services, experientially attractive, propels their usage frequencies,* which not only promotes the revenue potential, but also serves as a harbinger for innovation.

The adoption and integration of Web 2.0 ideas allow the expansion of the flexibility and capabilities of technology-mediated communications toward the natural modalities of human communication that are intrinsically experiential. These directions are manifested through the capabilities of social networking, LinkedIn, Twitter, Facebook, etc., widgets, RSS (Really Simple Syndication), Wikis, Widgets (an application or interface that enables a user-centric functional capability), and Tags (metadata that enables user-centric information access through labels for content categorization for intuitive access, such as a Twitter "Hash Tag"). Intuitive techniques for accessing, identifying, analyzing, creating, and processing the massive amounts and diversity of information in the dynamic and evolving web of information are the primary motivators for the ingredients of Web 2.0. While the modalities of physical transportation remind us of separation in time and space, the modalities of multimedia content–rich mobile service bind humankind through a variety of endeavors, namely, well-being, personal, social, cultural, and professional, that occupy life. Since face-to-face communications provide the ultimate benchmark of experience in human communications and interaction, the evolution of the type and nature of technology mediation in mobile service in the experiential direction is profound.

The advent of these directions was marked by a proliferation of information publication and access to a variety of content types through search engines, over the Internet via web browsing user interfaces. This was a relatively static user interface in that it was primarily a read-only paradigm, and was referred to as Web 1.0. As the usage, interests, and learning evolved, the user interfaces evolved toward those of an intuitive nature derived from experience. The social aspects emerged as a dominant theme, where user behaviors continue to transform toward one that is intrinsically participative and engaging. It marked a shift toward an inclusion of subjective considerations in technology mediation. In this collaborative and participative landscape, users are enabled to create, share, and publish ideas and perspectives through the artifacts of social networking, blogs, etc. It creates a spontaneous cycle of information discovery, usage, creation, and transfer that lends itself to continuous innovation and enhancements in lifestyles, technology evolution, and well-being. An assortment of these value-added capabilities serves as an incubator for new mobile service strategies and business models. These improvements and capabilities, characterized by a read-write paradigm, are collectively labeled as Web 2.0.

---

* Verkasalo, H., "Open mobile platforms modeling the long-tail of application usage," in *2009 Fourth International Conference on Internet and Web Applications and Services,* Venice/Mestre, Italy, May 24–28, 2009.

These capabilities of Web 2.0 enabled a creative and participatory ecosystem of personal and professional lifestyles. It is an evolution of the read-only web version Web 1.0, to the read-write capabilities, realized in the form of wikis, blogs, social networking, etc. of Web 2.0. On the other hand, this ecosystem generates enormous amounts of information that are not automatically correlated. An automatic semantic correlation and linking of information types and sources would further enhance the value of information for both producers and consumers of information. The evolution of the capabilities in Web 2.0 is realized in Web 3.0, which includes the semantic correlation and linking of information types and sources. It represents an evolution toward an intelligent organization and availability of information. These advances in capabilities shape improvements—information collection, correlation, interpretation, ease of use—in the experiential domain, while the volume and scale of disparate and diverse nature of information continues to proliferate. These capabilities allow the expansion of information dissemination, while enabling a preference-oriented, intuitive mining, content relevance, and understanding of raw information, ignited by interests, creativity, and imagination. It serves as an incubator for experiential, innovative mobile service creation and delivery.

Collective collaboration and information management are the cornerstones of the Web 2.0 foundation, which promote the proliferation of intelligence and assorted mobile services, including niche services in the Long Tail. The foundational concepts of Web 2.0 can be realized through the use of technologies, such as AJAX (Asynchronous JavaScript and XML). The prominent theme of mobile services within the Web 2.0 context is to facilitate a distribution and advancement of the knowledge within a group or a community, through a participative and mutual engagement: one-to-many and many-to-one transfer and embellishment of thoughts, ideas, and services. The formalization of interactions among users, while allowing user-centric forms of interaction, is a prominent aspect of the Web 2.0 paradigm. Information in this paradigm is considered in a generalized manner, which is particularly attractive in multimedia mobile services.

For instance, the generalized information patterns include various types of service augmentation content through appropriate arrangements of voice, video, picture, data, or text. Further embellishment of content is realizable through the use of metadata. Examples of metadata include ratings, reviews, annotation, history, etc. associated with a service. The dynamic information sharing and distribution, afforded by the Web 2.0 paradigm, allows a decoupling of applications, metadata, and interfaces, in the creation and delivery of mobile services. An exemplification of this type of decoupling is reflected in the rendering of Amazon.com cloud services, where the metadata, such as ratings and reviews, is provided by users, the data is provided by the sellers, and the developers utilize APIs to build the required interfaces.

The experiential value of the loosely coupled and coordinated fabric of the Web 2.0 paradigm is shaped over time and the frequency of interactions among the service providers and users distributed over time. A significant characteristic of this paradigm is that the techniques are not formulated a priori, but are facilitated to evolve, governed by the nature and appeal of the supported services: as reflected by the extent of user engagement, in local and global communities. The evolution of Web 2.0 toward Web 3.0* is characterized by a shift of a network of content and application silos toward an interdependent, interoperable, and a collective oneness, where connectivity—anytime, anyplace—is indistinguishably blended, whether it is mobile or fixed access to services. In this emerging dimension, some of the dominant enabling facets include cloud computing, open APIs, open technologies standardized by global standards organization, open source

---

* Hendler, J., "Web 3.0 emerging," *Computer*, vol. 42, no. 1, pp. 111–113, 2009.

software platforms, distributed databases, machine learning, intelligent applications, and natural language processing.

Web 3.0 is often characterized as the Semantic Web.* The vision of the Semantic Web, as described by W3C (World Wide Web Consortium), extends the underlying concepts of interlinking documents to the interlinking and the correlation of data. A common thread of interest binds assorted data across disparate and assorted repositories of information. Natural language processing and semantic relationship procedures indexes data based on a common thread of interest to find, interpret, and identify relationships across the disparate and assorted databases of information to configure user-centric data extraction and search choices. The vision consists of two dominant notions: common formats for integration, and the combination of various types of semantically related types of data extracted from different silos of information repositories.

This direction of semantically related data enables a mobile user to conveniently navigate across a potentially unlimited array of information databases. The semantic relationships are characterized by virtual, not physical network connections, linkages to a given object. For example, an organizer may link a calendar to office meetings, linked to a picture of people, linked to a video of one among many events, linked to a production studio, linked to a mobile payment entry, linked to an e-mail message, which then may be linked to a seminar presentation, and so on. The semantically linked list could be vast and contain access to rich content, providing enormous experiential value, within a mobile service. Figure 4.11, illustrates the functional elements of the Web 3.0 framework.

The foundational technologies in the framework of Web 3.0 include RDF (Resource Description Framework), OWL (Web Ontology Language), and SPARQL (Simple Protocol and RDF Query Language). The RDF provides information description for flexible interpretation and expression. The OWL describes the semantic relationships across diverse types of information. The SPARQL is utilized to query RDF data repositories, which are natively represented graphically, or contained within traditional databases. The Web 3.0 framework allows data sharing and reuse across a multitude of domains—across community, enterprise, application provider, and service provider boundaries, etc. It is based on RDF (Resource Description Framework), which is W3C standard for the interchange of data, using a directed, labeled graph to model a linking structure, where a variety of data is allowed to be blended, shared, or exposed across the boundaries of information silos.

The OWL builds on the RDF, by using URIs (uniform resource identifiers) for naming and the common framework for semantic linkages provided by the RDF. More vocabulary is included,

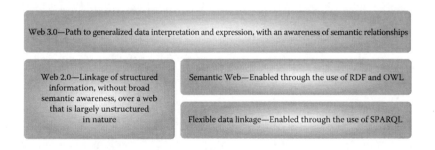

Figure 4.11   Elements of the Web 3.0 framework.

---

* Berners-Lee, T., Hendler, J., and Lassila, O., "The semantic web," *Scientific American*, May 2001.

within OWL, relative to RDF, for describing properties, data classes, relationships between the data classes, etc. Additional features for ontologies, enabled by OWL, include extensibility, openness, and ability to be distributed across different domains, interoperability, and scalability over the web. The prominent underlying concept of OWL is information content processing, which facilitates widespread machine interpretability as well, instead of limiting the content presentation only for human consumption. The use of URIs, within the RDF for the representation of data, allows the blending of information from a variety of sources that may be globally distributed, for the development of rich, multimedia mashups. This has the potential to significantly augment the experiential appeal for innovative mobile services—especially in the evolution of services in the Long Tail.

Figure 4.12 illustrates an evolution of the functional attributes in the shift from Web 2.0 to Web 3.0. This evolution is inherently user-centric, and interdependent with usage patterns and behaviors that leverage the semantic relationships of data, enabling and molding the shape of business, entertainment, innovation, information exchange, communities, and cultures.

The experiential aspect of mobile service is a barometer of appeal, where the perceived value of service is characterized by the dimensions of functionality, adaptability, attractiveness of user interfaces, usage convenience, performance, and cost. These directions suggest a convergence of mobile evolution toward the holistic and immersive nature of human interactions with the world around, as well as among one another.

It is a movement in technology mediation, where the objective modalities of protocols and procedures blend with the subjective of nature of the perceptual world around us. In the evolution of the blended patterns of the objective and subjective, mobile service usage behaviors evolve as well toward natural modalities. Technology mediation and its evolution, aligned toward enabling the natural modalities human interactions, has a virtually unbounded potential in the landscape of innovative mobile services, since it would then be in the arena of mitigating

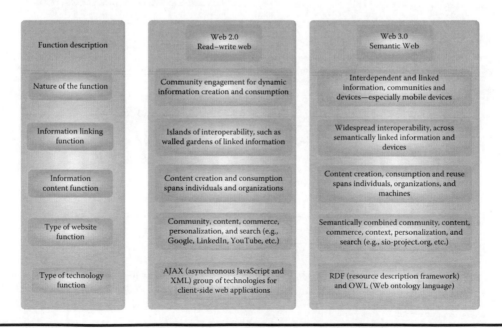

Figure 4.12 Functional attribute evolution from Web 2.0 to Web 3.0.

the time-space gap impairments, inherent in natural human interaction. At the same time, immersive, multimedia information presentation and transaction would embellish the natural modalities of human interactions. In these compelling directions, where both expression and reflection are richly represented, the dominant theme in mobile service evolution hinges on the nature and relevance of the content, relative to the presentation user interface. In this paradigm, the interplay of expression, reflection, and social dynamics are blended, to enrich the mobile service experience.

The QoE is shaped both by the expectations of the type of service and the performance of the service within each type of service. This in turn unlocks a variety of revenue opportunities* in the mobile service value chain orchestrated by the flexibility and adaptability of web service as illustrated in Figure 4.13. The vertical axis depicts the type of revenue potential ranging from *direct* to *indirect*, where the direct revenue sources include service subscriptions, and service subsidization, and examples of indirect revenue sources include collaborative partnerships, mobile payments, attracting service traffic to a website, brand augmentation, premium service, free service, and advertising. The horizontal axis depicts the type of web service categories ranging from private partnerships on left toward openness and individual mobile service users. The open category represents the vast potential of services creation and consumption through the use of standardized and well-defined APIs, opening the collaborative opportunities to an ever-expanding base of service

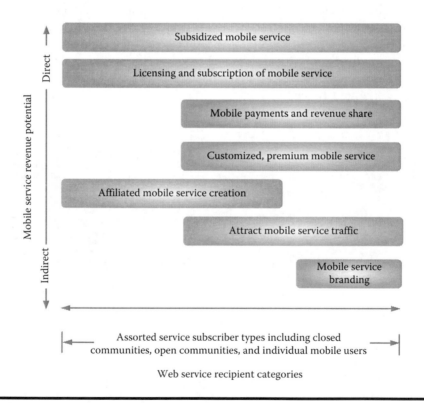

Figure 4.13   Fabric of revenue potential through QoE catalysts in mobile service.

---

* Brinker, S, "Business models for linked data and Web 3.0." http://www.chiefmartec.com/2010/03/business-models-for-linked-data-and-web-30.html.

development communities. The individual mobile service user category represents a diverse and expansive category of end-users, with unlimited potential for service customization demands and usage scenarios. The mobile payments service is illustrative of the customization opportunities at the level of the mobile user that is a vital ingredient in the evolution toward QoE enhancements. This leads to innovative models to harness revenue streams from a variety of payment scales, including micropayments for assorted content—music, video, e-books, on-demand viewing such as movies, sporting events, etc.

Online banking and commerce—anytime, anyplace—enables flexible choices and convenience in scheduling and organization, to suit different lifestyles in a globally interconnected information ecosystem. In addition to unveiling new revenue streams, these QoE-centric mobile services serve as opportunities for an optimization and reduction of operational costs, through an avoidance of paper or physical transactions, such as checks, receipts, etc. From a mobile user perspective, the online availability of personal profiles and choices augments the QoE through a simplification of the service request and billing process, increasing the frequency of usage of a content and context-aware website, which are enabled by the Web 3.0 paradigm. The potential for an increased frequency of usage of Web 3.0-augmented service portals provides attractive opportunities for advertisers to augment their brand visibility and awareness among mobile service users. This in turns paves the way for new revenue streams, and a dynamic and selective embellishment of the service or information presented at the web portal. In these directions, new avenues for serving content based on contextual information, derived from mobile user behaviors, are enabled, which in turn enhances the potential for new revenue streams that accrue from a corresponding augmentation of QoE.

In the shift toward enabling complex interactions that are reflexive in nature, the content interdependence and its evolution process becomes a dominant actor relative to the distribution system of networks over which the content and service is transported. The capabilities of the information distribution system serve as a catalyst for users to interact through the deliberate actions of selection, creation, or consumption of content. An evolution of this initial step, on an improving experiential scale, is where the creation or consumption of content becomes a process of reflection and interaction among the users. A core enabling capability to foster experiential improvement—convenience, intuitiveness, well-being, and enjoyment—of mobile services is the use of semantic relationships across disparate pieces of distributed information. In other words, this enabling capability enhances the expressiveness of distributed information, rendered via mobile services. This essence of the Web 3.0 paradigm to harness the power of semantic relationships and content interdependence is an enabling platform for revenue opportunities that are both direct and indirect.

From a mobile service vantage point, ubiquitous connectivity together with user-centric and context aware applications unveils evolutionary directions that are aligned with the notion of the Internet of Things (IoT), where machine learning and distributed autonomous computing agents further enhance the potential for user-centric interactions and services to effectively harness information and its relevance individually and collectively. Semantic relationships of content and context are pivotal ingredients* in blending the virtual and physical worlds through enhancements in knowledge representation, machine learning, natural language search, information mining, and artificial intelligence.

---

* Madhu, G., Govardhan, A., and Rajinikanth, T., "Intelligent semantic web search engines: A brief survey," *International Journal of Web & Semantic Technology*, vol. 2, no. 1, pp. 34–42, 2011.

4.7 Pervasiveness

A throttling of information flow, in the creation, transmission, and consumption of information, has been examined widely, with differing perspectives. The advent of packet data communications was marked by ARPANET in the late 1960s, which utilized packet switching technology, powered by computing nodes, to transport information.

In the cases, where the quality of information is unaffected by the transport framework, as a result of a deterministic nature of the transport, or is sufficiently over-provisioned, where one path does not differ from another, the type of information is clearly independent of the transport framework. In such scenarios, the transport resources may be throttled, using a minimalistic approach, to the extent that the throttling is not explicitly biased by the nature of the information content.

On the other hand, where the quality of the information is affected by the nature of the transport framework—wireless transport, where the transport is probabilistic—the boundaries between transport and information are blurred, since cooperation is essential. The QoS provided for the information, such as multimedia content, demands an awareness of the nature of the associated service being transported for ensuring an appealing user experience. Both a scarcity of wireless transport resources, spectrum-enabled wireless highways, and the probabilistic nature of wireless highways demand a collaborative and cooperative processing of information and transport. In such scenarios, any throttling of information flows to manage the availability and allocation of the transport resources is a function of subscription levels and service level QoS.

Whether or not the information and the transport are naturally decoupled—such as in the case of overprovisioned wireline networks—the information itself must be allowed to flow gracefully, within the constraints of resource availability and subscription levels to promote unfettered usage, innovation, and investment potential. In the world of multimedia services, social media continues to emerge as a transformational arena in a virtual world, where individuals, groups, and communities are empowered to create and consume mutually beneficial ideas and services. These continue to shift incrementally, with ideas and experiences fostering a fertile landscape for ongoing enhancements providing value in the human journey. The new horizons appear to be shaping an incessant change, with movements from competition, to collaboration, to cooperation.

Aligned with this movement, the shift of cultures and economies toward increasing volumes of niche services at the tail of the demand curve is evident with the continuing evolution of mobile connectivity. This shift toward services at the tail of the demand curve is a significant departure from the models of service creation and consumption that cater to a few popular or mainstream markets and usage behaviors. A few illustrative examples of service providers that continue to harness the ever-expanding, distributed, and niche nature of mobile services in the tail of the demand curve include Amazon.com, Google, iTunes, Netflix, third-party application development communities, content developers, etc. These evolutionary mobile service directions are reflective of the value and attractiveness of the uniqueness of user-centric modalities that mirror the nature and diversity of human behaviors, while availability and interoperability are preserved. The nature of this shift in service value toward the tail of the demand curve is exemplified by an observation* of the sale of books at Amazon.com: "We sold more books today that didn't sell at all yesterday than we sold today of all the books that did sell yesterday."

---

* Anderson, C., http://longtail.typepad.com/the_long_tail/2005/01/definitions_fin.html, quote from Josh Peterson, January 10, 2005.

The pervasive experiential potential of mobile services is incubated in innovation-driven technology evolution that promotes a continual reduction of the costs of production and distribution of these services. As a result of these increasing efficiencies in mobile service creation and delivery, new opportunities that further augment the QoE in terms of niche services are enabled, beyond a rudimentary one-size-fits-all approach service models. Ubiquitous mobile connectivity, with the degree of interconnectivity—proportional to the square of connections, as postulated by Metcalfe's law—is a vital fabric for a pervasive distribution of QoE, anytime and anyplace. Increasing levels of pervasiveness and QoE create tipping points,* where "the momentum for change becomes unstoppable" in the evolution of mobile services, which in turn affects the proliferation of ideas, knowledge advancements, and social well-being.

The significance of social and perceptible ambient intelligence is advanced through the proliferation of information types and transactions, over mobile and wireless connectivity of web-enabled devices. Both context awareness and social networking features and capabilities are at the core of this technology-mediated information distribution. In this realm, the term "context" represents a descriptive information collection that characterizes the environment that envelopes the user either virtually (preference and profiles) or physically (device interfaces and the ambient). The management of contextual information is delineated in terms of information capture, formatting, representation, mediation, and inference. The distributed connectivity of sensors and smartphones provides a fabric for shaping and processing dynamically changing contextual information. Context-aware systems harness the widespread relevance and the distributed nature of web-oriented technologies, such as REST† (REpresentational State Transfer) APIs for interoperability and consistent service behaviors. The architectural ingredients that facilitate a pervasive contextual awareness, in a distributed fabric of connectivity, include a collection and aggregation of contextual information, and its delivery to requesting entities in the Web of Things—machines and humans. The ingredients also include the capabilities to correlate a service and an associated user, together with an inference of social relationships by parsing the related social graphs that represent a web of social relationships.

To facilitate the consistent behaviors of social network applications, embellished with context-aware information, an interoperable layer consisting of a collection of frameworks that handle a variety of service attributes such as content, policy, profile, identity, etc. serves as an intermediary between mobile connectivity network layer and the service layer. These frameworks, based on open standards initiatives,‡ enable a social web of relationships, which are not limited to memberships to single site, but transcend site boundaries across a plurality of websites that represent the entire web. The decentralization of open standards paves a path toward the development of specifications that harness the power of distributed individual and collective intelligence. The establishment of open interfaces and specifications across heterogeneous radio-access technologies and assorted radio-coverage footprints—ranging from small to large—acts in concert with open APIs to expand the participation of third-party application development communities. These cooperative components are elemental, for an overlay of the social web, with its pervasive relationship linkages and context. Service enablers within SOA (service-oriented architecture)

---

* Gladwell, M., *The Tipping Point: How Little Things Can Make a Big Difference*. New York: Back Bay Books, 2000.
† Fielding, R.T. "Architectural styles and the design of network-based software architectures," PhD dissertation, University of California, Irvine, CA, 2000.
‡ Federated Social Web Community for standardization studies in W3C, http://www.w3.org/community/fedsocweb/.

building-blocks provide an exposure of standardized network interfaces to third-party application development communities. The collective space of the social web at the application layer, and the objects of the physical world, such as sensors, appliances, transducers, mobile devices, etc. at the Internet layer, is referred to as the Web of Things.* The notion of IoT is subsumed within the umbrella of the Web of Things, and is representative of the web of life in a technology-mediated sense. In this paradigm, the objects of the physical world are virtually characterized over applications and services that are realized through the use of web-oriented languages, such as HTML, Perl, Python, PHP, JavaScript, etc., and capabilities such as search, browsing, linking, caching, bookmarking, etc. It then follows that information and service mashups can encompass the virtual characterization of physical objects enriching the contextual possibilities in a user-centric mobile service evolution. The availability of open APIs, for Web 2.0 type of services and beyond, based on the simplified styles of expression inherent in the REST-oriented API definitions, is pivotal in the creation of web mashups. Web 2.0 service types include the social web, rich Internet applications, and web-oriented architectures, where the applications expose functionality to other applications for the augmentation of functionality through a selective combination of functionality from different application sources. Web mashups, RSS feeds, etc. are illustrative of web-oriented architectures that augment service functionality through a selective combination-assorted applications. In this Web of Things, sensors, health monitors linked via wireless connectivity, may offer their services in terms of sensed markers of well-being, such as heart rate, blood pressure, stress, etc., to lifestyle and fitness guidance agents, which in turn provide advice for improvements in the markers of well-being.

Much like the manner in which an OS of a computer acts as an intermediary between a process and the various distributed hardware and software resources, a mobile communication session serves as an intermediary between the endpoints engaged in the session, across one or more distributed networks. In the latter case, the communication session in the generalized sense provides the various mobile connectivity resources, for the delivery and experience of a mobile multimedia service. This effectively amounts to a virtualization of all the distributed and intermediate information transport networks—wide-area and small cell forms of connectivity—via an associated communication session between a source and a destination. This architectural construct allows for pervasiveness in mobile service delivery, over mobile devices and networks diversely characterized in both form and function. Sensor-oriented information collection, which is inherently distributed, serves as an information network that spans a variety of practical applications—home appliances, control systems, proactive health and well-being, environmental quality, etc. The distribution and evolution of the Internet from being a networked information fabric of established computing devices—servers, desktops, laptops, notebooks, tablets, and smartphones—to a dizzying array of devices of all sizes, shapes, and capabilities ushers unbounded possibilities in the form of service proliferation, mobile or fixed wireless scenarios. This in turn implies the enormous potential of these services to augment the health and well-being of our planetary civilization.

Personalized health monitoring, both from prevention perspective as well as for the treatment of an existing health imbalance, provides new and innovative avenues for mobile service–mediated promotion of health and well-being. The nonintrusive nature of wearable wireless sensors that collect information on specific aspects of the dynamic functions within the human

---

* Guinard, D. and Trlfa, V., "Towards the web of things: Web mashups for embedded devices," in *Second Workshop on Mashups, Enterprise Mashups and Lightweight Composition on the Web* (*MEM 2009*), Madrid, Spain, April 2009.

body promotes unprecedented levels of convenience. For example, the BAN (Body Area Networks) standard* provides proximity wireless transport for personal health monitoring via wearable sensors. This is experientially attractive,[†] since it allows users to continue with their daily lives, without the impediments—scheduling of appointments, visits to diagnostic centers, visits to hospitals, or treatment centers—of the traditional doctor–patient relationship model. Figure 4.14, illustrates a telemonitoring and a teleguidance paradigm of mobile-mediated services, using an assortment of proximity, local-area and wide-area mobile connectivity, via open interface, interoperability. Organizations, such as the Continua[‡] health alliance, promote a globally emerging vision of personalized health and well-being that serves as a fabric of user-centric service requirements, in the landscape of experientially attractive mobile service evolution. These directions provide compelling motivators for both business model and interoperability innovation—directions that are aligned with the notions of decentralization, collaboration, and cooperation.

While the role of mobile services is firmly planted in the realm of human experience, as a result of its intrinsically personalized nature, it naturally thrives as a distributed capability. The client-server model of mobile service rendering is an elemental building-block. A pervasive service environment is naturally aligned with enhanced experiential capabilities, as a result of the integration of mobile connectivity and service in an unobtrusive fashion, where connectivity is by design always available. The functional components of a pervasive mobile service are dynamically orchestrated, utilizing a variety of user-centric attributes, such as user profile and user context, and a variety of ambient affecting service conditions, such as adaptability, suitability, and availability.

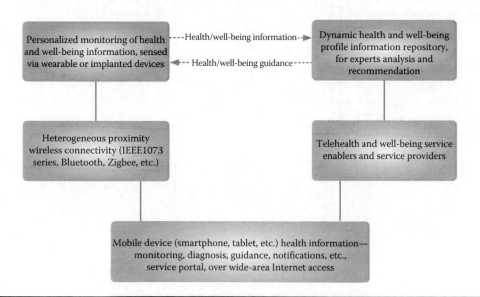

Figure 4.14 Proactive well-being directions in a mobile Web of Things.

---

* IEEE 802.15 WPAN™ Task Group 6 (TG6) Body Area Networks. http://www.ieee802.org/15/pub/TG6.html.

[†] Mearian, L., "Body area networks should free hospital bandwidth, untether patients," June 6, 2012, Healthcare IT, Computerworld.

[‡] http://www.continuaalliance.org/about-the-alliance.html.

The presence and availability of appropriate interfaces is among the vital ingredients for enabling a pervasive information environment. For example, speech and gesture-oriented interfaces provide a natural and intuitive modality of interaction with a pervasive information environment. The ease of user interaction is a marker of the attractiveness of mobile device–mediated services. Voice-enabled interactions allow users to perform other activities in unison, muck like walking and talking. Multimodal interfaces and applications provide a fabric of experientially rich mobile services, using a framework of assorted APIs and web services, in an immersive world of information. The discovery of available services and their composition, using the building-blocks of APIs and web services, is pivotal for exquisitely crafting unique and user-centric mobile service. These ingredients together with an exposure of the appropriate interfaces and resources, within a mobile connectivity framework, are essential for a realization of innovative mobile services. This approach is in contrast with the OS-centric approach of third-party service providers, such as Google, Apple, and others that leverage the social network web of information to promote user-centric services. On the other hand, this approach lacks the depth of information, and usage behaviors that can be derived from wide-area mobile connectivity subscribers, where the connectivity is managed over licensed spectrum. This gap between the two approaches translates into limitations in the experiential quality: whether the service delivery is uncoordinated/over-the-top (disintermediation), or over unmanaged wireless connectivity. It is the probabilistic nature of wireless transport and its integrity, which are subject to variability, inherently dynamic based on mobility and topology conditions, that warrants the need for specialized mobile connectivity management. The latter in turn affects the experiential QoS, perceived by the smartphone or tablet user.

The shift toward the notion of establishing the mobile network infrastructure is vital, in the leveraging of the network resources and capabilities to enhance the pervasiveness of mobile service availability and experience. It is a direction which enables a continuous enhancement of mobile service experience through a virtual imperceptibility of network boundaries, in terms of mobile service delivery. The essence of this shift is characterized by the use of open architectural principles, to lower the barriers for service creation and distribution, and the ease with which users communicate and consume mobile services. The realization of a Web 2.0 ecosystem of services, with semantic enhancements, is the raison d'etre for an interoperable exposure of the mobile network resources, and capabilities, as depicted in Figure 4.15.

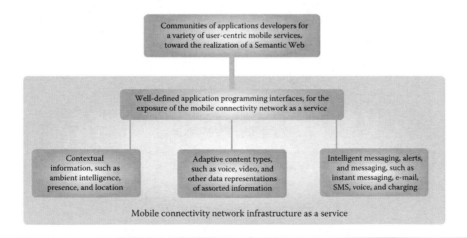

Figure 4.15   Mobile network exposure for enabling ubiquitous mobile services.

The utilization of applications in the creation of services, together with unique combination of services, unveils pathways to service attractiveness. The use of mashup* technologies with open web-oriented APIs intrinsically enables rich multimedia services that have the potential to be a composite of various constituent content sources, applications, and services. The combination of assorted content and services, with the use of mashup technologies, allows the mobile service delivery independent of mobile device clients, if the mashup is performed in the mobile network.

These notions embellish the distribution and collaboration of information, and its advancement in a social networking landscape. They promote the evolution of mobile services, through the related usage, in a variety of environments, in the people domain. For example, the social networking landscape is a mashup of personal, cultural, and professional interests, contexts, and interactions, that is, LinkedIn, Facebook, etc. According to Dave Snowden,[†] "social computing is not about selecting a tool based on predetermined criteria, it is about allowing multiple tools to co-evolve with each other, people and environments so that new patterns of stable interaction form, and destabilize as needed to reform in new and contextually appropriate ways." In such a dynamic ecosystem, experiences and interactions serve as iterative cycles of disruption and enhancement of mobile services through organic cooperation and collaboration. The participatory influence on the proliferation of information and knowledge are a function of service experience, where the semantic relationships offered via web-oriented frameworks, including Web 2.0 and beyond, are pivotal elements for popular services, as well as for services in the long tail.

4.8 Architectural Views

In the backdrop of the experiential domain, where both the user and the environment blend and interact, in virtually unbounded oceans of information, the enabling mobile service architectures and enabling capabilities must support these emergent paradigms of mobile service. The user-centric nature of mobile service demands an intuitive and choice-driven customization of services. This empowers users to acquire mobile services, anywhere, anytime, over any type of mobile connectivity. Technology mediation through the building-blocks of logical enabling functions promotes both interoperability and a management of the inherent business model, design, and implementation complexities. SOA is a model that allows the organization and utilization of capabilities that are distributed across different administrative domains and implemented in a variety of technologies. The creation of capabilities to satisfy certain technology or business requirements associated with an organization may be leveraged by another organization that may have the same requirements. In the SOA model, the design and applicability of elemental capabilities or functions promote ease of reuse and integration, while these capabilities and functions themselves remain autonomous and relevant across different usage scenarios: business, operational, or end-user-related services.

The relationship between a requirement and a supporting capability may be one-to-many (one requirement is supported by a set of capabilities) and vice versa. The framework of SOA then becomes a compelling approach to align requirements and capabilities, where different

---

* Maximilien, E.M., "Mobile mashups: Thoughts, directions, and challenges," in *IEEE International Conference on Semantic Computing*, Santa Clara, CA, pp. 597—600, August 4–7, 2008.
† Snowden, D., "Weltanschauung for social computing," March 12, 2007. http://cognitive-edge.com/blog/entry/4329/weltanschauung-for-social-computing.

capabilities may be combined and leveraged as needed. This framework denotes an architecture, where the emphasis is on the *services*, which occur on the boundaries between requirements and capabilities. As with any adaptable framework, its significance is established as a result of its logical expression, independent of any specific implementation. Each of these services has a service description that elucidates the functional implications of invoking the service. The service description specifies both the syntax and the semantics that allow the usage of the service by a human, machine, or application. The service accesses the mobile connectivity network resources and capabilities, through a service interface for supporting the needs of the service to function. The service description does not delineate or constrain implementations of the use of the service interface. The service executes its functionality through manual or automatic procedures that may utilize other services.

The architectural directions for mobile service enablement are driven by the guiding principles of the SOA framework. This allows interoperable building-blocks that are loosely coupled to orchestrate a rich suite of QoE-enriched forward-looking, innovative mobile services. Figure 4.16 illustrates a mobile service enablement context that serves as a basis for logical service architecture models, woven in the SOA fabric.

The tapestry of interdependent and experiential communications is woven in the enabling ingredients mobile service creation and consumption. These enabling ingredients are crafted with the objective of serving as interoperable and extensible building blocks of architectural and functional capabilities. The architectural and functional capabilities are envisioned and specified as logical enablers that permit innovative designs and flexible business models and choices, for virtually unbounded service rendering possibilities. The diverse service rendering possibilities allow for an adaptation to dynamic and changing market demands. The forward-looking nature of the enabling capabilities provides a design and rendering potential for both market enhancements and market evolution. The demonstration of the significance of QoE is underscored through the enormous variety of content types conveyed through the mobile services rendered over the

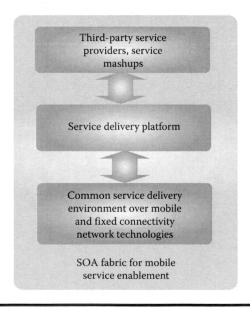

Figure 4.16 Mobile service enablement context.

Apple iPhone and the Google Android OSs. The role of mobile services, unlike other types of technology-mediated services for personal or professional segments of human existence, is naturally personal: highly individualized. The level of individualization demands both an attractive and a customizable experience of mobile service. The underlying segments of mobile connectivity involve a combination of both wired and wireless access, of which the wireless segment is characterized by a variety of radio-access technologies. The wireless access represents the mobile edge of connectivity, which enables the rendering of mobile services, over a QoS aware connectivity to serve as a foundation for QoE: an attribute that is experienced by the user, for any mobile service. The wireless signal robustness, bandwidth, network capacity, and network coverage are among the significant foundations for shaping the QoS demands of multimedia services over mobile connectivity. The incessant growth in the volume of mobile multimedia traffic is propelled by the insatiable appetite for mobile access to information and communications, fueled by the convenience and the ergonomic attractiveness of smartphones, tablets, and notebooks. This implies the need for advanced radio-frequency technology capabilities that were elucidated earlier, in terms of evolving next-generation mobile connectivity, namely, LTE-Advanced. The flexibility, adaptability, and cross-layer awareness of the radio-layer provide a tool-chest of connectivity modalities, through diverse topologies, such as heterogeneous networks, advanced radio-frequency modulation techniques, such as OFDM (Orthogonal Frequency Division Multiplex) schemes, antenna technologies, such as MIMO (Multiple Input Multiple Output) and three-dimensional radio-frequency beam-forming, and intercell interference coordination. These capabilities provide the underlying QoS aware transport for assorted and rich multimedia mobile service content and signaling, with dynamic resource allocation features to tune the network capacity and coverage to suit the mobile service demands. The LTE-Advanced radio-layer interfaces in concert with the EPC (Evolved Packet Core), within the EPS (Evolved Packet System), with the universality of IP as the binding glue that orchestrates the pathways of the mobile service layer multimedia traffic and signaling. This architectural framework utilizes IP as the thread of commonality across diversity represented through a multitude of RATs (Radio Access Technologies), the wired routing fabric of networks, and a variety of multimedia service environments.

The mobile services landscape continues to unfold with the movement propelled by user-centric behaviors motivated by experiential evolution in lifestyles and convenience, promoted via smartphones and tablets. While the volume of the information proliferation saga—*big data*—continues unabated, the commoditization of hardware and software continues rapidly. It then becomes apparent that the semantics and the usability of Big Data, rendered through the lens of mobile services, is a pivotal portal of evolution.

4.8.1 Service-Oriented Architecture

From a service consumer perspective, the service is opaque, implying that its implementation or inner details are hidden. The exceptions to this information hiding principle are as follows: (1) The service provides an exposure of information as a part of the service description. (2) The service provides an exposure of certain information, as required by the service consumer. The separation of roles between the service (capabilities) and its consumption (usage scenarios) is a central concept in the SOA framework. It is this clarity of expression of the distinct natures of these roles that highlights the widespread interoperability, relevance, applicability, scalability, and adaptability of the SOA framework. The SOA framework serves as an enabling foundation for a variety of service architectures and design patterns, such as Web 2.0. Some of the widely applied design patterns in the Web 2.0 paradigm of service evolution, such as Rich Internet

Applications (RIA), are compatible with the emerging directions toward cloud-oriented services. These design patterns are aligned with a cloud-computing style framework with dynamic allocation of resources for rendering service over a mobile connectivity network. In the cloud computing model, users are unaware of the underlying resources—infrastructure—in the cloud that are utilized by the services that are rendered for the user. The access to these services in the cloud is enabled through the use of API definitions, such as those used by social networking applications—Twitter, LinkedIn, etc.—to name a few in the Web 2.0 model of applications and services. A reference model of the SOA framework is shown in Figure 4.17. The abstract nature of the SOA framework provides insights into the prominent entities and the relationships between them, independent of the unbounded vistas of technologies and implementation nuances.

Collaboration and cooperation through an understanding and application of this unchanging template of entities and relationships promotes a mobile Internet scale of service creation and delivery of services, while minimizing the costs of realization and universal interoperability.

The power of abstraction lies in that it is based on minimum assumptions regarding mobile connectivity technologies and systems, which enables unbounded applicability and extensibility. The dominant themes of value within the principles of SOA are that they enable interoperability, scalability, manageability, delegation, integration, provisioning, distribution, flexibility, collaboration and cooperation across business systems, boundaries, and functional components to propel the evolution of mobile services. SOA principles invite a departure away from a model of point-to-point or pair-wise functional component interactions to a broader framework of cooperative functional components that promotes both agility and responsiveness in the process of service creation. It opens pathways that evolve from limited point-to-point functional component interaction, to a rich landscape of functional component interactions, built on portable, conceptual foundations to realize the demands of a diverse and changing marketplace of user-oriented mobile services.

The discovery of a service by a service consumer is dependent on the capability of the service provider to provide the service description and the policies, and on the capability of the service consumer to become aware of the service-related information. On the other hand, a service provider is able to discover a service consumer through a visibility of service consumers, which is

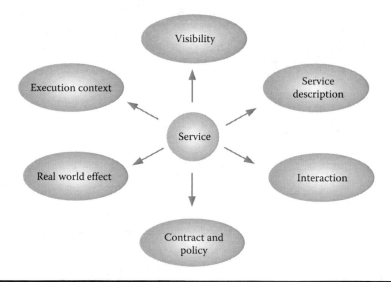

Figure 4.17   Reference model for SOA.

provided through the service consumer's description. The awareness of a service description and policy information are the vital building-blocks for a service consumer to be enabled to potentially utilize the service capabilities as needed. The provision or consumption of a service may be pushed by a provider, or pulled by a consumer, where the service availability is advertised or probed. The roles of a service requestor, service provider, and service broker in service delivery are modeled in Figure 4.18.

The distribution of software functionality and components is a prominent attribute of SOA. In this regard, it is vital to enable access to the functionality and components, beyond any single organization or domain, for the realization of value, while each of the participating parties operates autonomously. Open and globally standardized interfaces are intrinsic actors that allow various third-parties to participate in the access and in the provision of these distributed capabilities. Policies described within contracts among parties that are engaged across the interfaces serve to manage and enforce the necessary QoS requirements, together with technical and business-specific models.

The contractual descriptions include the considerations of interoperability among the distributed components, and the composition models for the functional components. The former is categorized within the web service specification layers, and the latter is associated with the type of model utilized for the functional component: for example, the SCA (Service Component Architecture) specified by OASIS.* These categories reflect the two prominent ingredients for a realization of SOA: consist of distributed software components and the policies that state the relevant behavioral characteristics of each of the distributed components.

Each distributed software component serves as a building-block, where the services they offer are available for a potential requestor of the corresponding services. These elemental ingredients for a realization of SOA, shown in Figure 4.19, are aligned with the model for service components, within the foundational concepts of WS-BPEL (Web Services-Business Process Execution Language) or BPEL, and the SCA. The service composition models in SOA, for combining services, are categorized as (1) process-oriented service composition—BPEL and (2) structure-oriented service—composition. The SCA framework represents a realization of the SOA principles, through its agile nature in the assembling and composition of services with loose coupling, while exploiting standardized service enablers and technologies. The attractiveness of the SCA framework, in creation and delivery of services, stems from its inherent neutrality with respect to vendors, applications, and programming languages. It serves as a framework for programming

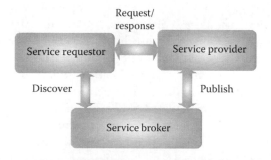

Figure 4.18 Model for service acquisition.

---

* OASIS, Service Component Architecture—Assembly Model Specification, Version 1.1, May 31, 2011.

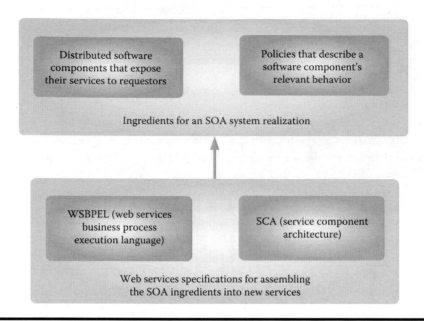

Figure 4.19 Ingredients and specifications for SOA realization.

languages and web services. The prominent motivator for these attributes of the SCA framework is rooted in versatility to allow and enable diverse business requirements, while avoiding encumbering constraints that limit the realization of a service to specific technologies, applications, languages, or vendors.

The framework is independent of system topologies, types or duration of interactions, client/server models, and peer-to-peer models. There is no implicit assumption with regard to the types of system resources that are exposed, system boundaries, or administrative domain boundaries. With this approach, the SCA framework is well-suited to address any complexities associated with the integration of the cooperative components for the realization of mobile services. It is a decentralized mediation framework for service realization, where the enabling capabilities may be leveraged by the service provider or by the service user. This leads to a simplification of mobile service creation and delivery, while reducing complexities associated with the service realization and avoiding the burden of any intermediary entity in the process. The complete specification of a software component consists of both internal and external specifications. The software component's internal specification characterizes its implementation and the execution environment. The implementation of the software component may be realized as a BPEL script, where the execution environment is a corresponding BPEL engine, which may be an open source product that embraces the open source philosophy, much like those of open standards. The latter is a catalyst toward broader participation and collaboration that fuels information dissemination and in turn innovation.

The interdependence and cooperation among the software components is archetypal, within the SOA framework. The external specification of a software component delineates the interactions between the software components, for a realization of SOA. The external specifications represent a contract description of the software component for interactions with other software components, and are also a measure of the software component's participative capability in the composition of services. The contract profile of a software component may include a variety of functional (software component interactions) and nonfunctional (SOA policy aspects) elements based on the context of interactions. The contract description serves a pivotal enabling role for third-parties, independent of administrative domains, to access the corresponding SOA software component. The business and technical requirements are revealed by the contract description, for the service accessed by a third-party, via the software component. The automatic integration of the SOA software components is facilitated through the use of contract descriptions that use machine-readable formats. Contracts may be encoded as XML documents, utilizing WSDL formats, which are leveraged for a variety of uses, such as querying repositories, code-generation, etc.

The dominant aspects within the SOA paradigm consist of the interoperability between the SOA software components, and the composition of SOA software components. The former is modeled by the web services* specification stack and the latter by composition models, such as the SCA. The encoded policies that are contained in a software component's contractual description include transactional behaviors, security, etc. oriented toward Semantic Web services. The potential of the SOA paradigm hinges on the richness and relevance of the software component's external contract. The latter is enhanced through a standardization of business-independent and common semantic definitions, for widespread applicability, adaptability, and flexibility.

A well-crafted utilization of the SCA framework lends itself to lower entry barriers, reduced complexity, and the reuse of building blocks, to allow a broader participation of third-parties in the expanding landscape of mobile service innovation and demand. Mobile OS platforms driven both by Google's Android and Apple's iOS continue to serve as a springboard in the proliferation and diversity of smartphone- and tablet-oriented mobile services, over a mobile Internet. Mobility-oriented time and space malleability promotes an expansion of mobile service possibilities, and unbounded horizons in the realm of mobile service experience, represented as QoE that is influenced by a variety of connectivity and service-related QoS attributes. The data volumes that required to be delivered in the form of rich multimedia application–driven services are virtually unbounded, with a corresponding unbounded appetite for information and imagination. While these directions are a reflection of the human propensity to explore, educate, and enrich, the underlying wireless mobile connectivity networks play a crucial role in enabling both scalability in connectivity resources and in the mobile service–related QoE. Enhancement in the radio-frequency connectivity techniques—modulation, topologies, and heterogeneous access—from a holistic perspective is vital in orchestrating a distributed and adaptive connectivity, while optimizing the utilization of wireless connectivity resources. The OFDM-MIMO radio-frequency techniques are foundational in the LTE EPS and its interworking with assorted radio-frequency technologies, over licensed and unlicensed radio spectrum. SON (self-organizing network) principles allow the distribution and coordination of autonomous wireless connectivity networks to realize for a dynamic allocation of resources, load balancing, fault isolation, and self-healing, while being aware of user mobility conditions to switch between wide-area connectivity and local-area connectivity; the latter modality being a wide-area connectivity off-load choice, for pedestrian mobility or fixed access to mobile services, within a local area, such as home, airport, office, shopping space, or other areas of attraction.

The QoE measure, which is inherently subjective, the human realm, is pivotal in the consumption of rich multimedia services, over a variety of smartphones, tablets, etc. with user interfaces. Enhanced radio-frequency techniques, OFDM and MIMO, inherent in LTE-oriented mobile connectivity technologies enable the handling of large volumes of multimedia information. It then

---

* Weerawarana, S., Curbera, F., Leymann, F., Storey, T., and Ferguson, D.F., *Web Services Platform Architecture: SOAP, WSDL, WS-Policy, WS-Addressing, WS-BPEL, WS-Reliable Messaging, and More.* Upper Saddle River, NJ: Prentice Hall, 2005.

follows that these enhancements enable improvement in the QoE for a mobile multimedia service. The LTE EPC is tuned to cooperate with the radio-layer, to promote the advancement of QoE, with the convergence of connectivity being enabled by the Internet protocol. The nature of the cooperation between the fixed IP core network, such as the LTE EPC, and the radio-layer consists broadly of mobility management and radio resource management. This cooperation is pivotal in enabling an attractive experience for streaming video services (YouTube, Hulu, etc.) over mobile connectivity. This type of multimedia service delivery is realized by distributed service networks, over the Internet, and is generally referred to as CDNs (content distribution networks). With the mobile connectivity segment, inserted as a part of the end-to-end transport, the radio-layer transport behavior plays a significant role in the QoE associated with mobile multimedia services. This is a consequence of the probabilistic nature of the radio-layer transport, which is a departure from the inherent assumptions, nonchanging availability of the allocated transport resources, of multimedia service delivery, over the fixed Internet.

4.8.2 Service Paradigm

The various facets of mobile communications that span a variety of business models and industry boundaries have a common theme of enabling next-generation mobile services. The notion of a service provider in this emergent theme has generalized and inclusive implications that cover the traditionally disparate business models labeled as network providers, service providers, content providers, and application providers. To leverage the common themes of requirements and functionality across disparate industries, in terms of services rendered and consumed, the SOA framework of abstract entities and relationships is a generalized template for service realization. This generalized template allows ways to develop, publish, and integrate application logic and resources as a service. The SDF (service delivery framework) provides an interoperable business and operational framework, using the SOA concepts and guiding principles. The SDF allows assorted service mashups, a mixing and matching of different pieces of applications and information, to create virtually unbounded service possibilities that ultimately augment a mobile service experience. It serves as a glue framework for a variety of SDPs (service delivery platforms), each with their own resources of information and functions or enablers. The next-generation SDP architecture facilitates mobile connectivity providers, traditionally referred to as operators, to collaborate with a variety of mobile service providers, for the creation and delivery of mobile services, independent of the underlying mobile connectivity architectures and the related radio-access technologies. This model paves an evolutionary path of widespread collaboration, not only among different connectivity or service providers, but also among providers and consumers—mobile service users. In this emergent paradigm, Web 2.0 and beyond, of virtually unlimited service possibilities, the mobile service user could serve as a content or service provider. The rigid roles of provider and consumer are allowed to be interchangeable, in a next-generation mobile service ecosystem.

In the NGN (next-generation network) vision, the IMS (IP Multimedia Subsystem) serves as a binding framework for a variety of traditionally disparate service creation, provisioning, and integration environment for different services. The identification and the specification of common and interoperable building blocks across different types of multimedia services—voice, video, and data—are foundational in the establishment of the IMS framework. The enabling capabilities of IMS serve as common thread of loosely coupled functions and capabilities, independent of the underling mobile connectivity or fixed connectivity technologies. It enables the delivery and management of next-generation multimedia mobile services. As the interdependence across the various domains, namely, business, industry, institutions, social, and personal, moves toward increasing synergies and benefits, the applicability of common themes in technology mediation, across these domains for mobile service creation, delivery, and consumption, will continue to evolve.

The harnessing of common themes for interoperability is a dominant component of standardization initiatives, for enabling heterogeneous building-blocks that serve as the essence of virtually boundless variety and possibilities, in the creation and delivery of mobile services. The distinguishing characteristics between the offerings of multimedia services, over the fixed Internet versus the mobile Internet, are mobility, high levels of personalization, resource limits, heterogeneity of wireless access technologies, discovery of attractive services, ease of interactions for service discovery, ease of service usability, and contextual enrichment. The enrichment of mobile multimedia services, through context-aware offerings, serves as a central characteristic in experiential value, since it enhances both service discovery and usability. Context embellishes usability through an offering of services or nuances that are relevant to a mobile user's dynamic usage conditions, while at the same time minimizing or avoiding user interaction or intervention. This implies that the mobile system—device and network cooperation—has an awareness across services. For example, Google Maps reveals directions, location, etc., while instant messaging services, such as Twitter, provide a social perspective, and a combination of these types of information augments the context for a mobile user.

The mashup of open APIs enables the availability of context augmentation. A framework of standardized open APIs, such as that based on REST, provides a broad array of mashup opportunities for context-enriched service experience. According to Dey,* context is defined as: "Context is any information that can be used to characterize the situation of an entity. An entity is a person, place or object that is considered relevant to the interaction between a user and an application, including the user and applications themselves."

The blending of context, with connectivity, anyplace, anytime, and over any personalized mobile device, exploits the enormous potential of cross protocol layer cooperation: access and service-enabling technologies. It is here, where the embedded intelligence is harnessed for a seamless extension of an attractive experience: with the evolution of connected things—personalized devices and machine-type devices—ranging from smartphones, tablets, home appliances, rich multimedia applications, messaging, social interactions, well-being monitors, well-being guidance, and a virtually unbounded web of distributed sensors and transducers, a customized creation and delivery of experiential services both appealing and viable. Along these directions, open platforms built on open architectural concepts and open API frameworks serve as a malleable fabric for a realization of the enormous potential that is intrinsic in a customizable and an experientially oriented mobile service ecosystem. The use of XML protocol–oriented documents to exchange information in an open and well-defined format—independent of the underlying arrangement of platform—over open interfaces is a foundational enabler that facilitates universal glue for the exchange of information associated with applications, networks, relevant data, and services. This is foundational to allow the proliferation of the richness of mobile connectivity and service, with cross protocol layer cooperation that is pivotal in the shaping of an experiential mobile world. Figure 4.20 exemplifies a cooperative environment that blends the traditionally dichotomous Interment and Telecom domains, through the mediation of an REST-oriented API framework,

* Dey, A.K. and Abowd, G.D., "Towards a better understanding of context and context-awareness," in *Proceedings of the Workshop on the What, Who, Where, When, and How of Context-Awareness Held in Conjunction with CHI'2000*, the Hague, the Netherlands, 2000.

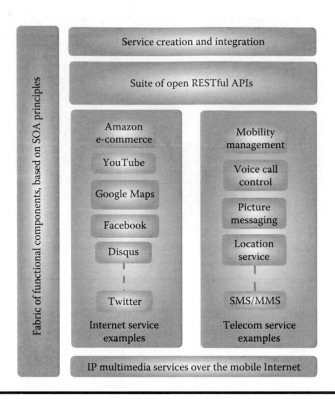

Figure 4.20 SOA framework of open APIs for mobile Internet services.

for a realization of contextually rich and media-rich mobile service potential, where service experience is the dominant theme.

The nature of mobile service is such that its inherent value and appeal are shaped by its various components and the manner in which it is orchestrated. These components that mold the service may be derived from a variety of interdomain sources, which may be internal or external to those of a mobile service provider. Service composition entails the integration of various disparate and potentially distributed applications, to satisfy the requirements of any given usage scenario. To accomplish the composition of a mobile service, the relevant service logic and data flows are specified, among the constituent functions that cooperate and collaborate to realize the objectives of the required composite service. The underlying business processes are instrumental in this realization, through a molding of the constituent functions toward a fulfillment of the objectives demanded by the composite service.

For instance, a mobile service, that provides consumer-targeted information concerning a product, its availability, the location of a retail outlet that stocks the product, together with proximity and directions to the retail outlet, requires a cooperative harnessing of relevant information from different sources. In this example, product description information, the direction mapping information, and the consumer's current location information are composed and delivered to the consumer, via a service that leverages a relevant mashup of applications.

The service orchestration utilizes SOA principles and concepts that leverage the implicit policy schemes, security, identity management, business models, etc. The underlying network and computing platforms over which service composition and delivery occur are likely to be heterogeneous and distributed, together with variety of administrative domains. It is with these practical considerations that the OSE (OMA Service Environment) model has been crafted, where the service enablers embrace an abstraction of the network functions, from the underlying network technologies, through an exposure of well-defined service-oriented interfaces.

Rachel Botsman* noted that for the first time in history, the age of networks and mobile devices has created the efficiency and social glue to enable the sharing and exchange of assets, from cars, to bikes to skills to spare space. This exemplifies the notion that it is the experiential aspects of a service that shape the appeal and value of a mobile service, beyond the technology mediation that blends gracefully in form, behind the intended function. The mobile Internet serves as a catalyst on a global scale to bridge time-space barriers, through the vehicle of mobile services, embellished with customized multimedia content to enable new horizons for innovative business endeavors and appealing lifestyles on an unprecedented scale.

With the widespread and global adoption of NGN architectures that embrace the broadband LTE RAT and the all IP EPS , new horizons for innovation in the arena of mobile services through creative business models, and rich multimedia content are virtually boundless. In these vistas of creative endeavor, interdependent collaboration and cooperation are enabled in dynamic leading edge of competition, where the experiential dimension shapes the attractiveness, appeal, and sustainability of a mobile service. These attributes that are modulated by user experience then naturally translate into an organic growth of users that adopt, learn, and retain the usage of a mobile service. The influence of the experiential modality of a mobile service is likely to extend and expand the nature of business models in the Long Tail of revenue propagation: services, and especially mobile services, with their elemental nature being in the experiential landscape of assorted lifestyles, personal or professional, they are implicitly aligned with the notion of "heuristics." According to the *Merriam Webster* dictionary, heuristics is defined as an aid to learning, discovery, or problem-solving by experimental and especially trial-and-error methods. In the context of mobile services, it pertains to the application and experience of a solution to a demand-oriented user-centric requirement, desire, or preference. This implies an exploratory dimension of problem-solving, where experiential feedback serves as guidance for mobile service enhancements, and innovation, including the creation of new functionality and experience. For example, heuristic system design, such as those built on the concepts of artificial intelligence, or expert systems, has the capacity to learn dynamically, and adapt to user-centric behaviors and requirements, through experience-driven feedback. In turn, this naturally promotes mobile service augmentation, where the service design is shaped to self-organize and self-adapt.

Loosely coupling and open interfaces, together with self-organization and adaptation capabilities, over heterogeneous access technologies, are among the pivotal attributes for a next-generation mobile SDP. The glue of the IMS bridges assorted flavors of connectivity, mobile and fixed, while the SDP facilitates a flexible environment for service creation, deployment, orchestration, execution, and management that hinge on SOA concepts and principles. The SDP framework functions as an overlay environment to complement the IMS fabric, to provide interoperable interfaces, which enable third-party application development communities to create, innovate, and distribute applications that are deliverable over heterogeneous broadband connectivity, such as LTE, Wi-Fi, cable, etc.

The coordination and cooperation of IMS capabilities with the SDP environment promotes the notion of *build once* and *use often* for mobile service creation. The *use often* paradigm is independent of the underlying connectivity technologies. The architectural and design concepts in the

---

* Botsman, R. and Roger, R., *What's Mine Is Yours: The Rise of Collaborative Consumption*. New York: Harper Collins, 2010.

next-generation SDP paradigm enable Web 2.0 capabilities, where users are enabled to function as service or content providers. The attractiveness of *build once* and *use often* paradigm, that is enabled through a diverse and interoperable suite of APIs, resonates with enhancing the user experience, service creation, service integration, and optimizing operational costs while harnessing service value through innovation. A framework of assorted, and interoperable, APIs opens pathways for service creation, across a multitude of development communities that span service and content providers, where service creation and deployment are independent of mobile or fixed connectivity, over all IP broadband networks. This opens new vistas for cooperation across service providers, content providers, and connectivity providers, where a shared vision and objective converge toward a continuous evolution of user experience.

At its core, the defining attributes of a next-generation SDP include the following aspects:

■ Interoperability with heterogeneous connectivity networks: mobile and fixed all IP networks
■ Utilization of SOA concepts and guiding principles
■ Adoption of Web 2.0 and Semantic Web capabilities, within mobile services
■ Widespread participation across various technology-enabling communities, such as service providers, connectivity providers, equipment providers, software providers, etc.
■ Cooperation and collaboration across the technology-enabling communities for revenue-sharing opportunities, optimization of costs, and service innovation

The dynamic nature of the experiential domain demands a time-variant adaptation of mobile technology mediation that reflects a changing landscape of life as it unfolds. Google mapping service is an example of the time-variant nature of user experience, where its serves not only as a static description of a map, but as dynamic descriptor of the user location, and as dynamic descriptor of the topology, such as New Orleans profile impact, after the Katrina hurricane disaster, as changes occur. Similarly, the events of 9/11 served as a harbinger of the significance of mobile multimedia services. This is representative of a widespread shift from the information distribution modalities of established media—TV, print, etc.—to mobile-mediated multimedia web-oriented services—anytime, anywhere. The intersection of the fixed Internet world with an all IP-enabled mobile world lies beyond the mechanistic nature of technology mediation, in the realm of human experience, since mobility is intrinsically personal.

4.8.3 Framework of Enablers

Interoperability at the service layer is realized through identification and specification of logical enabling components described in the architectural frameworks, reference points, interfaces, and functions. With the emergent directions of mobile connectivity, shaped by the assorted radio-layer techniques—modulation improvements, antenna techniques, interference management, access coordination, heterogeneous access technologies, and a variety of coverage topologies—and all-IP access networks, the scaffolding of logical service enablers complements the mobile connectivity with innovative service possibilities that weave the experiential mobile service landscape. The logical service enablers are envisioned and specified under the auspices of the OMA (Open Mobile Alliance). With the appropriate interfaces to the underlying mobile connectivity resources, through the IMS framework, the logical suite of enablers provide the elemental building-blocks for a seamless access to information and communications services. The service-enabling building blocks are relevant for the design and implementation of applications and services targeted for smartphones, tablets, notebooks, and machine-type communications. Context awareness, flexibility, and adaptability of service rendering are among the benefits facilitated by interoperable service enablers.

Interoperability, provided by the building-block service enablers, spans an assortment of devices, service providers, network providers, usage scenarios, and regulatory regimes across the globe. At the same time, the service enablers allow service innovation and differentiation for businesses to compete and collaborate in terms of specific mobile service offerings, designs, and implementation. A pivotal characteristic of OMA service enablers is their ability to interface with the IMS framework that hides the nuances of the mobile network resources and technologies. The IMS framework serves as an interoperable layer that facilitates the reuse of common capabilities for leveraging of the underlying mobile network infrastructure capabilities, across a myriad of distinct mobile services. This strategy obviates the need for repetitive vertical integration, an approach that adds complexity and cost, with the introduction of every new service.

The architecting of common capabilities—interfaces, protocols, and procedures—within the IMS framework promotes consistent mobile service behaviors across a variety of deployment and usage scenarios. The IMS framework provides a logical intermediary layer between service creation communities and assorted fixed and mobile network technologies: wide-area and local-area radio technologies, such as LTE, small cells, Wi-Fi, etc. The mobile service enabler building-blocks provide an interoperable service environment, where the service-enabling capabilities allow interactions among the building-blocks through well-defined interfaces. Openness through published specifications, based on global and evolutionary market and technology perspectives, serves as a catalyst for a broad participation among service and application developer communities. The OSE promotes an extensible and flexible architecture, for these communities to mine the potential and applicability of the mobile service enablers, specified by OMA.

The elements of OSE consist of the mobile service enablers, mobile service enabler components, and the interactions among the service enablers. A model for the exposure of the mobile infrastructure to services enablers is shown in Figure 4.21. This facilitates consistent mobile service behaviors, which are vital for promoting and preserving an attractive QoE. The positioning of the OSE between third-party application development communities and the IMS layer in mobile

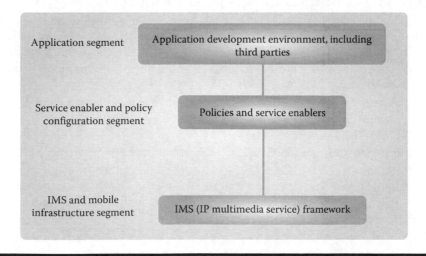

Figure 4.21   Model for mobile infrastructure for exposure to service enablers.

service and connectivity provider's infrastructure provides insights into a convergence of the traditionally isolated silos of the telecommunication services and Internet services.

Personalization and usage convenience are the twin attributes upon which experience in terms of its quality is perceived (QoE). Mobile devices (smartphones, phablets, tablets, etc.) with human interfaces provide the familiar Internet services, such as browsing, social networking, messaging, and e-mail services, together with voice communications. This is a departure from the legacy usage models, where a computer of sorts, desktop, laptop, notebook, etc., was used for Internet services and a phone was utilized for voice-oriented communications. The integration of these capabilities in mobile devices serves as a catalyst for an evolution of social behaviors, as well as for mobile service–oriented technology evolution and innovation.

4.8.4 Web Service Convergence

In the emerging frontiers of multimedia and multimodal ubiquity of services over mobile connectivity, interoperability of technologies, in terms of protocols and procedures, is the bridge that integrates the web services with mobile communication services. This convergence requires an integration of mobile real-time communication services, across the mobile network infrastructure and the World Wide Web, while preserving an attractive QoE. The IMS framework, within the mobile network infrastructure, provides a control plane architectural paradigm for harnessing and rendering the ubiquity of IP multimedia services in a mobility-oriented landscape, independent of the mobile connectivity–specific protocols. The generalized nature of the IMS framework provides new avenues for mobile service creation—third-parties and access to OMA service enablers—and a reduced time-to-market potential, with the use of interoperable and reusable interfaces, resulting in a reduction of the integration time for the adoption of new and innovative services. While these directions provided a closer alignment with the customer demands, in terms of mobile multimedia service capabilities, innovation and evolution in the market revealed the significance of the social nature of communications, built on the intrinsic needs of all human relationships: personal and professional. These shifts appeared in the form of Web 2.0 and its continuing enhancements, in terms of immersive communications toward Web 3.0, which has been referred to as the Transcendent Web.* The four significant attributes of the evolutionary Web include: (1) *Social Web*—augmentation of the experience of social networking in terms of search, location, and interest-driven recommendations, (2) *Semantic Web*—augmentation of experience through the use contextual attributes that provide information relationships in information search and knowledge advancement, (3) *Internet of Things*—augmentation of experience through the flow and presentation of information across machine-oriented and human-oriented interfaces distributed over the fabric of a mobile and fixed World Wide Web, (4) *Artificial intelligence*—augmentation of experience through the shaping, filtering, aggregation, analysis, and presentation of multimedia/multimodal—machine or human centric—information, for a variety of personalized applications and services. The significant role of the World Wide Web in the experiential plane underscores the need for an integration of these capabilities into the mobile network infrastructure. The blending of the universality of the World Wide Web within the mobile network infrastructure, together with its accommodation in the telecommunication business models ad strategies, unveils the boundless realm of imagination and innovation, in the shift from a network-centric paradigm to a human-centric and experiential paradigm.

---

* Sabbagh, K., Acker, O., Karam, D., and Rahbani, J., *Designing the Transcendent Web: The Power of Web 3.0*. Booz & Company, June 27, 2011.

On the other hand, the proprietary nature of many of the attractive multimedia services available over the World Wide Web imposes barriers to convergence and interoperability with services crafted for delivery over the mobile network infrastructure. These barriers compound the challenges to meet the demands of real-time telecommunication services—voice, text, and video— relevant for a variety of mobile device enabled communications across personal and professional endeavors. The WebRTC* (Web Real Time Communications) framework provides a pathway to convergence between the World Wide Web and the multimedia subsystem capabilities, within the mobile network infrastructure. This convergence is facilitated through the two prongs of interoperability: protocol and API specifications. The evolutionary directions in the delivery of browser embedded applications, with the advent of HTML5, unveiled a departure from the traditional requirement of the browser needing a plug-in to provide the necessary interfaces to support a browser embedded application. The RTCWeb† provides a set of enabling building blocks for interoperability among browsers, over the Internet, through a JavaScript API. As a complement, the WebRTC‡ specifies a suite of APIs to enable real-time communications between web browsers and for them to interact with mobile devices, including its audio-visual interfaces, such as speakers, microphone, webcams, the transport capabilities for mobile multimedia, and the media content encoding/decoding capabilities. Among the significant benefits of the WebRTC is in the interoperability that it offers, as result of the decoupling from the underlying platform-specific protocols, which in some cases may be proprietary or not open sourced. This is a departure from the plug-in model (e.g., Adobe Cirrus, which needs flash player plug-in for peer-to-peer communications, and requires the underlying RTMFP [Real-Time Media Flow Protocol]) for browsers. On the other hand, WebRTC utilizes the video element of HTML5 for media streams, while not requiring any specific underlying communication protocol, as long as that protocol complies with the WebRTC API requirements.

The WebRTC capability allows OTT (Over The Top) applications, based on HTML5, to integrate real-time media streams, voice and/or video, within native web browsers. This capability removes the need for an application to be used without a browser-specific plug-in or without manual intervention, such as the download, install, and configure cycle of an application. While this enhances the QoE for the user, it also enables the ease with which the functions of mixed-media stream management, codecs, and API development are integrated by browser developers. It provides the tools for web development, such as JavaScript and HTML, to integrate a web-oriented real-time communication services. The use of these ubiquitous web development tools enables the ease of OTT application development, which allows a convenient integration of the multiscreen capabilities of WebRTC into third-party applications and websites. This translates into an attractive potential in the building of innovative real-time communications frameworks that interoperate across a variety of web browsers that operate over different runtime environments and OSs.

The WebRTC architecture utilizes the principle of a separation between the signaling and the information (media) planes. It deftly allows both interoperability and implementation flexibility, by allowing the signaling plane to be left to the implementation choices of application

---

* WebRTC. http://www.webrtc.org.
† Holmberg, C., Hakansson, S., Eriksson, G., and Ericsson, "Web real-time communication use-cases and requirements," draft-ietf-rtcweb-use-cases-and-requirements-14.txt, RTCWeb Working Group, February 12, 2014, IETF.
‡ WebRTC (Web Real Time Communications). http://www.w3.org/TR/webrtc/, An initiative of the W3C (World Wide Consortium).

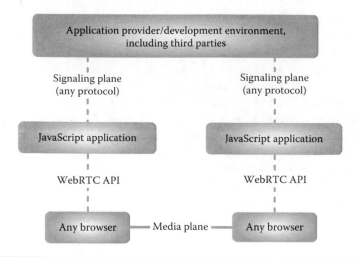

Figure 4.22   WebRTC architectural model.

development communities. The WebRTC architectural model is depicted in Figure 4.22. The signaling protocols that are widely adopted include both SIP* and XMPP.† The primary objective of these signaling protocols is to facilitate the negotiation of information transport–related parameters and the information media format, between any two endpoints of an information communication stream. This is accomplished through the use of SDP,‡ which provides a description of the signaling-related information. This description is utilized by the signaling protocols, for exchanging the signaling information, between the corresponding endpoints, with a sequence consisting of an offer description followed by an answer response. The JavaScript session establishment procedure§ describes the signaling process, based on this principle of signaling information exchange. The principle of allowing choices in the implementation of the signaling plane, while completely specifying the operation of the media plane, is an experiential augmentation for both the application development communities and mobile multimedia service users through an expansion and ease of customizable choices, over a universal communication landscape. In this paradigm, the significant aspect is the exchange of the multimedia session description, between the participating endpoints, while the style of signaling is choice driven, from an application development perspective, which in turn enables innovation and evolution.

Along these directions, the JSEP (JavaScript Session Establishment Protocol) allows a decoupling of the signaling state machine embedded within any browser, from a JSEP-controlled signaling state machine. This decoupling effectively removes a browser-specific signaling state machine from being the main control mechanism. The only requirement is for an interface for the application to negotiate the local session descriptions and the remote session descriptions using any signaling mechanism that is compatible between the communicating entities. This decoupling is especially useful, where a refresh or a reload of the browser results in a loss of the state information,

* RFC 3261, SIP: Session Initiation Protocol, IETF, June 2002.
† RFC 3920, Extensible Messaging and Presence Protocol (XMPP): Core, IETF, October 2004.
‡ RFC 4566, SDP: Session Description Protocol, IETF, July 2006.
§ Uberti, J. and Jennings, C., JavaScript Session Establishment Protocol, Internet Draft, draft-uberti-rtcweb-jsep-06, IETF, February 2014.

which can be recovered through an automatic push to the browser, from the corresponding state stored at a remote server, with a minimum interruption. Such a capability is aligned with the optimization of mobile multimedia service quality. The fabric of a WebRTC capable browser is exposed through an API,* which hinges on the complementary notions of enabling media streams and peer connections. In the WebRTC protocol stack, media stream represents the information media plane, and consists of two types, which are used to distinguish between local and remote media stream behaviors. Examples of local media streams include the capture of information through sensors (e.g., camera, microphone, biometric, etc.) that are local to the mobile device browser, where the information rendering occurs through the HTML5 video tag descriptor. In the case of remote media streams, the information capture occurs via sensors and a browser associated with the remote device, while the rendering occurs through the HTML5 video tag descriptor. The remote information stream traverses intervening networks between the communicating device browsers. Since the WebRTC specifications allow the negotiations for any codec, compliant capabilities may be incorporated into any browser implementation, in a mobile device. Peer connection represents the other component of the WebRTC protocol stack that manages the communication of information streams between two peers. A JavaScript object abstracts the capabilities of this component to hide the potentially complex and arduous underlying details that are associated with audio and video characteristics, including bandwidth adaptation, automatic gain control, echo cancelation, packet loss concealment—minimization of the impact of packet losses in a VoIP† service, such as VoLTE,‡ through and extrapolation or interpolation of detected gaps in the voice information stream—the use of ICE§ (Interactive Connectivity Establishment) for NAT/Firewall traversal, and any other peer communication specific nuances. The use of ICE typically obviates the need for the insertion of media stream relays, such as TURN (Traversal Using Relays around NAT), and avoids impediments in the media forwarding path to avoid delays or additional failure points, especially for delay-sensitive services such as VoLTE. The principle of avoiding intermediaries along the media stream path between browsers is intrinsic to the design of the WebRTC framework, while security is facilitated through media plane encryption. On the other hand, information-steering intermediaries, such as SBCs (session border controllers), across administrative domains, permit flexible configurations for NAT/Firewall traversals, private IP address pool management, and IPv4/IPv6 interworking. The style of media encryption is different between the Internet-homed OTT service providers and the mobile service providers. In the former case, the master encryption key associated with SRTP (Secure Real-Time Protocol) is exchanged within an end-to-end secure transport channel, facilitated by the DTLS (Datagram Transport Layer Security), where the master encryption key is not exchanged in the clear; in other words, it is opaque. This requires the use of a media relay server that decrypts and encrypts media streams to meet the demands for intercepting media streams that correspond to user identities, marked for legislative information collection. On the other hand, mobile service providers typically utilize SRTP and SDPSD (Session Description Protocol Security Descriptions¶) for media streams encryption, where the exchange of the SRTP master encryption is in the clear, using SIP signaling. SIP is the established signaling protocol, within the IMS framework, for IP multimedia

---

* Bergkvist, A., Burnett, D.C., Jennings, C., and Narayanan, A., WebRTC 1.0: Real-time communications between browsers, W3C Editor's Draft 21, March 2014.
† VoIP—Voice over IP.
‡ VoLTE—Voice over LTE.
§ RFC 5245, Interactive Connectivity Establishment (ICE): A Protocol for Network Address Translator (NAT) Traversal for Offer/Answer Protocols, IETF, April 2010.
¶ RFC 4568, Session Description Protocol (SDP) Security Descriptions for Media Stream, IETF, July 2006.

services, such as VoLTE and RCS (Rich Communication Suite). In the latter scenario, a separate decryption/encryption media relay server would not be required for intercepting users marked for legislative information collection.

The accommodation of QoS requirements, associated with the WebRTC media streams, is allowed to suit the network condition. The leveraging of DSCP (Differentiated Services Code Point) markings for the flow of real-time media streams through the core network of a mobile service provider, from a WebRTC framework perspective, is reasonable. The preferential marking of packets, within the flow of traffic, may not always be beneficial from the perspective of the QoS demands of any given service, which is dependent on the nature of the related media stream (real-time, mixed-media such as audio, video and data, interactive, etc.). The convergence of real-time multimedia web services, while avoiding the interoperability, integration, and cost challenges associated with proprietary clients and plug-ins, is facilitated through the use of an adaptation* of the IMS framework with the WebRTC framework. This adaptation is realized through an interworking solution approach illustrated in Figure 4.23. On the other hand, if a majority of media streams that are marked for treatment are either audio or video for a given network condition, then it may either result in a resource blockage for other media types or it may result in a capacity limitation for media streams that require a preferential QoS treatment. A selective usage of DSCP markings for packets associated with specific media streams, to support browser communications, within

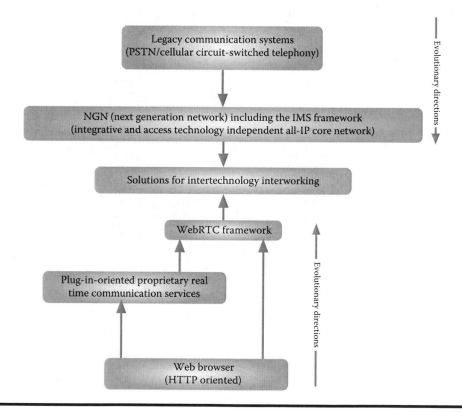

Figure 4.23　WebRTC–IMS interworking for web service convergence.

---

* 3GPP (Third Generation Partnership Project), Study on Web Real Time Communication (WebRTC) Access to IP Multimedia Subsystem (IMS), v12.0.0, TR 23.701.

the WebRTC framework, facilitates consistent default real-time communication behaviors, based on priorities and media types. At the same time, a configuration of packet flow markings, based on network policies and dynamic resource conditions at the edges—mobile connectivity network—promotes flexibility and choices for the mobile service provider, from an operational perspective.

The PCC* (Policy and Charging Control) framework is utilized to configure the appropriate QoS for the WebRTC media streams, using the Internet APN (Access Point Name), which is accessed via the mobile connectivity network. This strategy in the mobile connectivity network at the edges complements the use of DSCP markings for QoS-oriented priority configurations, for real-time media streams between browsers over the Internet. The adaptation of the IMS framework to the requirements of the WebRTC environment provides architectural directions for the orchestration of real-time communication services across web browsers in a variety of mobile devices and those elsewhere in the Internet. From a strategic and business model evolution perspective, these architectural directions provide opportunities for mobile services providers to render the richness and the ubiquity of web services on mobile devices. At the same time, such architectural directions promote opportunities to expand the experience of real-time communication services, including VoLTE, beyond the mobile connectivity network environment. The potential for a universal availability of real-time communication services is the recipe that the WebRTC framework provides for IMS-oriented services that are intrinsically confined to a mobile connectivity environment. This potential is provided independent of mobile device types, smartphones, tablets, etc., and independent of access technologies. A universal availability of WebRTC mediated real-time communication services that are created in the IMS domain is afforded as a result of a decoupling of knowledge of the nuances and complexities of the mobile connectivity technologies, from the application development process for rendering services over the web. This implies a wider participation from the web application development community to rapidly expand, the availability of the IMS domain type of real-time communication services, over the universal fabric of the web. It also enables the provision of IMS domain type of services, independent of IMS framework–oriented credentials, through the use of third-party WebRTC clients on any mobile device. It allows a bridging of the web and the IMS domains, enabling innovative directions for mobile service creation, new revenue opportunities/business models, and for enriching the mobile service experience across the universal fabric of the web. Interoperability across these inherently disparate domains—web-oriented browser and IMS-oriented mobile device—is facilitated through an interworking solution, which embodies a translation of the signaling and media components, consisting of SIP signaling, ICE, and the coding of media streams, as depicted in Figure 4.24. Interoperability, in its

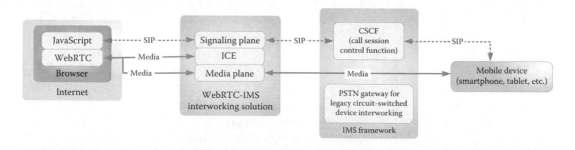

Figure 4.24　Interworking solution—WebRTC and IMS frameworks.

---

* 3GPP (Third Generation Partnership Project), Policy and charging control architecture, v12.4.0, TS 23.203.

ultimate interpretation and application, implies a departure from the barriers imposed by boundaries for a behaviorally uniform availability of mobile multimedia services. Web-oriented services are intrinsically composed for universal uniformity, in contrast with the constraints of market boundaries, business model boundaries, and access technology boundaries that are inherent in the composition of services within the mobile connectivity environment.

An interoperable adaptation of services across the web and the mobile connectivity environment allows new paradigms, where a diversity of participant actors are attracted to engage in service development, for innovative revenue opportunities and market expansion. This leads to concordant avenues for a trusted framework that inspires the establishment diverse service development communities for a bridging of service operation over both the web and the mobile connectivity environments.

In the convergence of the web and the IMS framework–oriented mobile connectivity environments, the enrichment of services is enabled through user-centric and intuitive perspectives. These perspectives harness the simplicity of ontology*-based models, conceptual models, that describe and represent information, in terms semantics and knowledge, which is shared across different environments. The description of knowledge as concepts and relationships allows flexibility and relevance that enable the sharing of knowledge across disparate environments. An ontology-based publication and discovery of knowledge promotes a widespread participation of consumers in the service composition and development process. The availability of intuitive tools for service creation, or for new service composition, is facilitated through the use of simplified building-blocks, such as REST-oriented APIs to access service enablers, in the mobile connectivity environment. These directions enable consumers to easily compose and publish innovative, and market-aligned mobile multimedia services that suit different business and revenue-sharing models, resulting in market expansion, for mobile service providers. The extension of the web to include semantics, for information flows over the fabric of the Internet, including the mobile connectivity environment, promotes a direction toward an enhanced interpretation, interaction, and cooperation across human and machine interfaces. The availability of Open APIs† fosters a collaborative landscape to discover, compose, and publish services across the entire Internet fabric independent of whether connectivity is mobility oriented or fixed. Since the intrinsic attribute of an API is to expose the attributes of an interface to a variety of application development communities, with varying degrees of understanding associated with the nuances of the underlying technologies, the use of abstracted concepts in the specification of APIs promotes widespread ease and collaboration—for example, in a Web 2.0 and beyond landscape of semantically enriched applications—in the process of mobile service innovation, adaptation, and adoption—a shared, and user-centric marketplace for service value creation. This leads to the shaping of strategically poised vistas of innovation and evolution, through coordination and collaboration. A convergence of the mobile services and web services is orchestrated through a global participation in the offering of the capabilities of services in the mobile service domain to web service domain and vice versa. In the mobile service domain, the IMS serves as the entity that offers the mobile services capabilities, derived from the IMS framework, and for example, the OMA service enablers, to the web domain for the creation of web-oriented services and mashups. In the web service domain, the web-oriented capabilities (Web 2.0 and beyond, etc.) can be harnessed for embellishing the mobile service domain with the richness and ubiquity of web services. Common and open APIs promote interoperability and cooperation across heterogeneity

---

* Berners-Lee, T., Hendler, J., and Lassila, O., "The semantic web," *Scientific American*, vol. 284, no. 5, pp. 34–43, May 2001.

† Mulligan, C., *The Communications Industries in the Era of Convergence*. Abingdon, U.K.: Routledge, 2011.

Figure 4.25   API operational model—Web-IMS service convergence.

of domains that include the mobile service and web service domains, together with the corresponding cloud* and virtualized renditions of the underlying networks. The IMS framework capabilities that are intrinsically session oriented are exposed through Open APIs that utilize a RESTful[†] style of specifications. The IMS capabilities, session oriented in nature, in the mobile service domain are exposed through REST-oriented APIs, which provide simple and scalable transactions with the web domain, using easily definable information referred to as resources, which are identifiable using URIs. The RESTful style of APIs permits a standardized utilization of HTTP methods, such as GET, POST, PUT, and DELETE, which allow a client to easily monitor or modify the state of a resource. These methods are reflective of a client-server communication paradigm, which is stateless.[‡] This allows the development of APIs that can be customized, while at the same providing a common and attractive platform for a variety of web-oriented application development communities. A conceptual operation of an API client is depicted in Figure 4.25. The operation of the API client hinges on a translation of the actions taken on the resources and their attributes, into SIP procedures on the IMS framework side of the IMS interface exposure layer. Conversely, the SIP procedures are mapped into actions taken on the resources and their attributes on the web domain side of the IMS interface exposure layer. The translation process is initiated, when the API client selects the collection of resources referenced by the URI that corresponds to the invocation of a mobile service that utilizes the IMS framework, such as the launch of a voice or a multimedia call.

Service specific policies in the mobile service domain and the related QoS constraints, for example, negotiations, via SDP,[§] of a given multimedia service in the IS domain govern the latencies associated with the setup and updates that pertain to the associated service. From a mobile device perspective, latencies and battery life are traded off, while from a network perspective, the prioritization of radio and core network resources are traded off with the latency bounds associated with a multimedia service. The core concept in this RESTful and common information format approach to convergence is the adoption of a web-oriented paradigm for the exposure of the IMS capabilities that energizes a widespread adoption of IMS APIs.

In an experiential world there two pivotal notions: (1) user-centric nature mobile service, and (2) Internet as the universal fabric of service creation and consumption. Services may be broadly categorized as customizable combinations of features rendered as mashups, and multimedia content syndication from a broad and emerging landscape of content providers. Mobile devices enabled with web-oriented API clients serve as interfaces for the distribution of user-generated content over an emerging semantically oriented web, in the realm of service innovation and evolution: presence, content feed readers, music distribution, social networking, located-oriented information, video content (YouTube, etc.), contextual information associated with communicating users, etc.

---

* Mell, P. and Grance, T., "The NIST definition of cloud computing," September, 2011. http://csrc.nist.gov/publications/nistpubs/800-145/SP800-145.pdf.
† Fielding, R.T. and Taylor, R.N., "Principled design of the modern web architecture," in *Proceedings of the 22nd International Conference on Software Engineering*, Limerik, Ireland, pp. 407–416, June 4–11, 2000.
‡ Fielding, R., Gettys, J., Mogul, J., Frystyk, H., Masinter, L., Leach, P., and Berners-Lee, T., "Hypertext transfer protocol—HTTP/1.1," June 1999. http://www.ietf.org/rfc/rfc2616.txt.
§ Session Description Protocol.

The Semantic Web reveals a pathway toward experientially enriched creation and consumption of mobile multimedia services.

4.8.4.1 OSE Framework

Service-oriented architectural elements embody the capabilities to support a variety of business logic rules and service provider policies derived from service components and provided by other ingredients, such as applications. For instance, billing models for services may be derived or abstracted from the characteristic of an application. The business logic rules and service provider policies are also capable of abstracting the behavior of OSS/BSS systems. A given application may be offered on a subscription basis or may be offered as a part of advertising revenue generation. This characteristic of an SOA framework, where a given application may be utilized across different usage models, is significant in that it allows an abstraction of network technologies and capabilities, including OSS/BSS systems. This characteristic of the SOA framework lends itself to an integration of OSS/BSS systems with runtime applications. The OSE is reflective of the nature of the SOA blueprint, and exhibits an alignment with the foundational tenets of the SOA framework, through an identification and specification of the I1, I2, and the I3 interfaces associated with the service layer, within the OSE. This service rendition fabric serves as an SDF realization. The OSE utilizes IMS as an intermediary for the service layer to harness the capabilities and the resources of the mobile connectivity networks such as LTE, Wi-Fi, etc. and fixed networks. The requirements for utilizing IMS are adopted within the common OMA enablers, such as XDM (XML Document Management), Presence and Messaging, which utilize the SIP (Session Initiation Protocol) suite of capabilities. The service execution environment is defined through notions derived from the Parlay Group, web services, and APIs.

The elegance of the SDF model is embedded in the idea that the service components that shape the SDF are interoperable and independent of the nature of the physical resources—heterogeneous networks, implementation platforms, etc. The interoperable interfaces of the SDF service components are available as functional exposures for other SDF service components, across administrative domains, while being aligned with the principles of loose coupling and information hiding. This promotes simplicity in service orchestration and life cycle management. The service component functional interfaces that are exposed allow access to a variety of capabilities that are extensible to meet the requirements of a service. For example, life cycle management capabilities could be envisioned to include capabilities such as service configuration, provisioning, fault management, upgrades, billing, charging, campaign management, etc.

Since the SDF model utilizes SOA guiding principles, the service-related policies and business logic can be implemented in other components, outside of the service components, to enable flexibility and adaptability for the service components. These other components could, for instance, abstract OSS/BSS systems, such as the derivation of billing models from an application, where a given application may either be subscription revenue oriented or advertising revenue oriented. Conceptually, this illustrates the abstraction, independent of the underlying physical resources and network technologies, provided by the SDF model represents a significant benefit, since the same application can be switched between two different billing models. The abstraction of the underlying functionality allows common and reusable approaches in a dynamic fashion, for service orchestration, over assorted runtime environments. The SDF model is essentially the glue that binds the disparate environments of the underlying network, and application developer resources. The inclusion of service-related context—presence, location, social networking and related attributes—management in the SDF model is vital along a trajectory of experiential embellishments for the user. These enhancements are enabled through the specification and utilization of APIs that represent network layer and service layer abstractions. This represents a service innovation potential, through open API frameworks that augment the QoE. The OSE is aligned with the SDF model, through the use of the I0, I1, and the I2 interfaces.

The guiding principles of the OSE include the following*:

- The service enablers embody the foundational aspects of SOA, where their exposure to third-party applications is policy oriented—management and enforcement—within a mobile connectivity network or at its edge.
- Web service and REST techniques and SOA concepts are leveraged by the service enablers to expose interfaces or capabilities for use by applications and application development communities (e.g., well-defined REST-oriented APIs between the mobile service provider and the content provider).
- The service enablers are characterized by their Lego system nature and the corresponding interfaces, without affecting or dictating the shape or color of an envisioned service.
- The exposure of resource interfaces allows implementations that leverage the underlying network resources of assorted technologies that may be implemented on different platforms, administered by different service providers or other SDFs. In other words, the exposed interfaces represent the underlying network abstractions, such as the IMS reference points linking the service layer, such as ISC, Sh, etc. and REST APIs.
- The SDF services are composed of service components through the functional interfaces of the service components. The use of functional interfaces allows the rendering of a function, a completion of a task, or information transformation. Any requestor may utilize the set of functions associated with a functional interface. Each service component is provided such that it is independent of other interfaces it may utilize for the realization of a service.
- The interfaces to lifecycle management provide the capabilities for suitable management and operational capabilities of service components, where the lifecycle management includes functions such as billing, usage charging, provisioning, configuration, administration, upgrades, problem resolution, etc.

These guiding principles serve as the fabric of the context and components of the OSE framework layers depicted in Figure 4.26. The framework allows the use of a generic interface (I0 + P) for third-party application developer communities to access the service enabler building blocks, using the "I0" interface for consistent service behaviors, using service provider–specific policies that are established using the "P" interface. The policy enforcement shapes the interaction between the application layer and the service enablers, whenever policies are in place. The service enablers serve as building-blocks of intrinsic functions, which are able to interact with other service enablers, within the service enabler layer architecture of the OSE, and with the underlying networks as needed and with the IMS layer, through the I2 interface. The service execution environment consists of service management, load balancing, infrastructure, and O&M (operations and management). The guiding principles behind the motivation for lego-like nature of service enablers hinge on the notion of reusability of functionality, while allowing creative implementations of mobile services. This implies that the functionality of the service enablers does not intrinsically dictate or imply its usage with other enablers or its usage in the creation of new mobile services.

---

* TR139, Service Delivery Framework (SDF) Overview, Release 2.0.

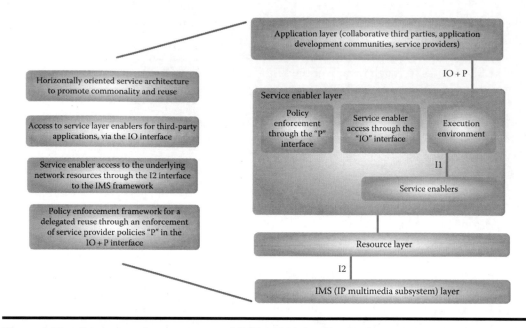

Figure 4.26 Context and components of OSE (OMA Service Environment).

These attributes of service enablers allow and promote its applicability potential to a virtually unbounded landscape of forward-looking and innovative mobile services.

The fabric of the IMS framework is composed from the NGN* architectural vision. The protocols, procedures, and extensions† execute over IP, and hinge on the SIP (Session Initiation Protocol) specified by the IETF. The principles of SOA are embodied in the constructs of IMS to enable service orchestration‡ schemes that facilitate a variety of telecommunication session-oriented services. The ISC (IMS Server Control) serves as a well-defined interface between an AS (Application Server) in the IMS framework and the S-CSCF (Serving-Call Session Control Function). The SIP message interactions with a mobile client are processed by the AS, via the S-CSCF. The service logic, within the IS framework, is shaped through the use of iFC (initial filter criteria) rule definitions, which are provisioned in the HSS (Home Subscriber Server) as a part of the mobile user profile, and the iFC is executed at the S-CSCF. The execution of the iFC rules allows the mobile device or any user equipment requests to be forwarded to the AS, via the ISC interface. The notion of an SiFC (shared iFC) consists of rule-set definitions at the S-CSCF, when the same iFC is applicable across a group of mobile users, where it is efficient to store the SiFC at the S-CSCF, using only an identifier pointer download from the HSS to the S-CSCF, for SiFC selection, instead of the entire SiFC download, a large XML file, for each mobile user registration. The SiFC pointer in the HSS is passed to the S-CSCF over the Cx interface. HSS simply stores the numeric identifier in this case and passes it to the SCSCF over the Cx interface. This reduces the traffic flow within the framework. The iFC or SiFC is a part of service profile, which is a part

---

* ITU-T Rec. Y.2011, "General Principles and General Reference Model for Next Generation Network."

† 3GPP—Third Generation Partnership Project.

‡ Dinsing, T., Eriksson, G., Fikouras, I., Gronowski, K., Levenshteyn, R., Pettersson, P., and Wiss, P., Service composition in IMS using Java EE SIP servlet containers, Ericsson Review 03/2007. Available http://www.ericsson.com/ericsson/corpinfo/publications/review/2007_03/04.shtml.

of the mobile user profile. The mobile user profile is downloaded from the HSS to the S-CSCF, during the mobile user registration with the IS framework. This is a basic service orchestration scheme that allows service combinations in an interoperable fashion.

The traditional service composition style categories, such as the web services protocol stack represented by WS-BPEL, Internet heritage, and the SIP services, Telecommunications heritage, protocol stack represented by iFCs, within the IMS framework, are intrinsically limited, since they were originally intended for disparate environments. These different service composition styles require coordination and cooperation, with respect to the demands of a blended service environment, to enable service composition within the foundational characteristics and architectures that represent the emergent mobile Internet. Convergence toward a rich multimedia service creation ecosystem is the objective of integration or blending of service composition styles, over heterogeneous mobile connectivity networks. The compelling direction of convergence is propelled via collaborative and widely relevant guiding principles that incorporate the service technology capabilities and strengths embodied within the Web 2.0 and beyond conceptual framework. Novel business models that embrace the Web 2.0 concepts hinge on its inherent building-blocks that promote agility in service evolution, while offering unique opportunities in the context of user experience both from an individual and a social perspective.

4.8.4.2 API Framework

The elements of experience in mobile communications could be broadly classified as the physical and nonphysical aspects of a mobile service. Both of these elements are distinct and complementary, with *mobility* as the cornerstone of services over the Internet of services. The physical aspects pertain to the physical characteristics of the mobile device, while the nonphysical aspects are associated with the perceived usage convenience and ease. Descriptors of the physical aspects refer to the ergonomics of the user interface, feature organization, etc. For instance, the ergonomic profile of the user interface rendered via touch, voice, etc., serves as vehicles of information input and output that approach human response and reaction. On the other hand, the nonphysical aspect of mobile communications extends well beyond the mechanistic aspects of the user-interface capabilities. The most significant distinction between these complementary elements of experience is that the nonphysical component of the experience profile is ill-suited to the notion of *one-size-fits-all*. The perceived usage convenience is as diverse as the human being that is uniquely molded through life and learning. While the value of a perceived usage convenience can be described and broadly recognized, the experience is unique to the individual. It is this understanding that colors the richness of user experience, in a pervasive mobile communication environment, characterized by rich multimedia services.

The various aspects of experience perceived via the physical conduit of a mobile devices and its interfaces are rendered via a combination of functions that are realized through the use of communication protocols and APIs. The communication protocols delineate vertical boundaries that separate communicating entities in a mobile connectivity network, and facilitate interoperability across the vertical boundaries. These entities—such as base stations and network gateways—are functionally partitioned, in alignment with the principles of the OSI* model of protocol layers. On the other hand, APIs provide a delineation of horizontal boundaries across different software stacks (e.g., a system of components consisting of an OS, a database, and a

---

* Zimmermann, H., "OSI reference model: The ISO model of architecture for open systems interconnection," *IEEE Transactions on Communications*, vol. 28, no. 4, pp. 425–432, April 1980.

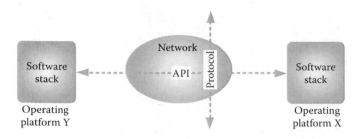

Figure 4.27   Distinction between the role of an API and a protocol.

programming language.). An API provides an application with a level independence from the detailed nuances of an underlying platform, which is a collection of software and hardware that renders a function or a set of functions. In other words, an API provides an implementation-independent abstraction of commands that an application can use to interact with an underlying platform. It then follows that an API serves as a horizontal bridge for applications to interact across different platforms. A logical model that depicts the distinction between the role of a protocol and an API is shown in Figure 4.27.

The challenges, at the physical level, are associated with preserving consistent behavior of the user interface, using the mechanics of interaction—touch, gesture, voice, etc.—and form factors. At the nonphysical level, the use of APIs enables the building-blocks for customized usage convenience that are motivated by individual preferences and choices, as shown in Figure 4.28. For example, a pervasive speech API could provide a customized level of convenience, for multimodal applications, within a mobile service aligned with a user preference. The objective is to allow a loose coupling between the physical and nonphysical elements of experience, where the nonphysical element is enabled to closely choreograph the choices preferred by the user.

From a mobile service perspective, OMA,* the API standardized specification strategy is to expose the mobile network infrastructure assets, independent of protocols and platforms to allow a widespread and expanding participation of application development communities.

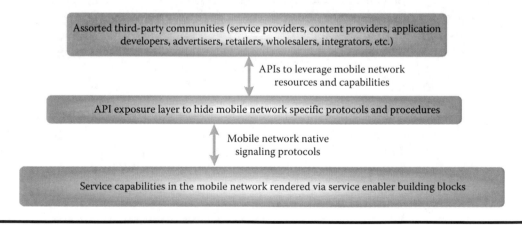

Figure 4.28   API framework in a service delivery environment.

---

* Open Mobile Alliance.

The standardization of APIs allows a uniformity of mobile service behaviors across diverse vendor platforms and application development communities. This avoids fragmentation and nonuniform service behaviors, where different APIs are used to provide the same intended functionality. The fabric of the web-oriented information flows and services is largely composed of web applications and APIs. The web APIs, which are built on REST-oriented styles, follow REST architectural principles* and it is their implicit simplicity that propels their adoption across diverse application development communities. The REST-oriented style[†] of API specifications utilizes the HTTP model of verbs, and URIs for both resource identification and interaction, which make them naturally aligned with the characteristics of the web fabric. The REST model hinges on client-server principles, where requests are stateless in nature. REST-oriented services are such that there is a decoupling between the resources and their representation, which allows the resources to be, accessed via a variety of different formats; in other words, format independence is exhibited. Web API specifications that are aligned with the REST-oriented style are characterized by their descriptions being resource-centric, while the access to information—the retrieval and processing of the information—utilizes HTTP procedures. The REST-oriented approach for Open APIs is followed by a variety of websites, such as LinkedIn, Twitter, Google, Flickr, etc., for inspiring a broad participation in the application development process, which is particularly significant in the mobile service ecosystem.

Open APIs inherently accelerate mobile service innovation and evolution in an underlying technology agnostic fashion, resulting in a simultaneous lowering of investments and entry barriers in the development of new services. The actors in a collaborative and cooperative Open API environment are depicted in Figure 4.29.

This is especially beneficial in the rollout of new mobile services, where the mobile connectivity network is harnessed for its communication capabilities, together with the creative endeavors of global and diverse application development communities.

These directions enable virtually unbounded possibilities in innovative combinations of interdependent information, acquired from a variety of different sources,[‡] such as those from which context-awareness (e.g., a combination of location, interest profile, mood, etc.) can be derived.

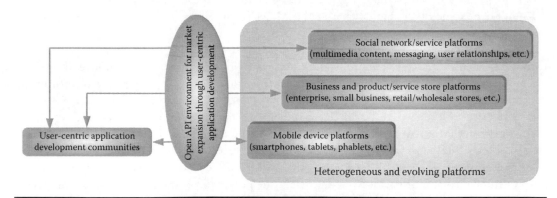

Figure 4.29 Open API framework in a service delivery environment.

---

* Fielding, R.T., "Architectural styles and the design of network- based software architectures," PhD thesis, University of California, 2000.
[†] Also referred to as the "RESTful style."
[‡] Mulligan, C., "Open API standardization for the NGN platform," *IEEE Communications Magazine*, vol. 47, no. 5, pp. 108–113, 2009.

Information collected and correlated across distributed sensors, using proximity access technologies such as NFC (Near Field Communications), unveils a plethora of user-centric mobile services, while lowering the barriers for new business models, partnerships, and customizable functionality. Open APIs serve as a malleable bridge for interactions across a network of assorted platforms. The flexibility enabled through Open APIs fosters an optimization of communication costs across different business entities, enhances the innovation potential, and allows contractual flexibility among the interacting business entities. Pathways to convergent* web fabric are indistinguishable and interdependent, across the traditional silos of the Internet and Telco worlds, where experiential mobile services transcend the functional boundaries. The emphasis is on user-centric uniqueness, within universal interests and pursuits (e.g., education, commerce, business, entertainment, health, well-being, etc.). These directions enhance the potential for market creation and expansion through imagination and innovation. The evolution and adoption of these direction hinge on a forward-looking specification of Open APIs through advocacy, cooperation, and collaboration across standards development communities, industry, and academia for bolstering the creation of innovative and interoperable mobile service enablers.

4.9 Internet of Things

The information age, in concert with the untethered connectivity afforded by wireless, fixed and mobile, technologies, ushers unprecedented multimedia and multimodal service possibilities, through information flows among things: humans and machines. The IoT is a fabric that represents a dynamic linkage of things that enable a realization of ubiquitous information transactions, among humans and machines. Machine-to-machine (M2M) communications is a dominant facet, in the landscape of IoT. M2M communications is associated with potentially unlimited, identifiable devices that operate in a symphony to provide services that attractively affect and advance lifestyles.

The distributed nature of applications and wireless devices in the M2M fabric unleashes enormous possibilities in advancing the experiential potential of information-driven services. A representation of some of the ingredients in the M2M fabric, within a broader context of IoT, is depicted in Figure 4.30.

The M2M landscape is part of a global infrastructure that links physical and virtual objects through the exploitation of data capture and communication capabilities. This fabric of diverse and interconnected objects—sensors, controllers, adapters, etc. —is characterized in terms of varying degrees of autonomous data capture, event transfer, network connectivity, and interoperability. The IoT serves as a bridge between the physical world and the information world. Machine-type communications in the form of assorted sensors collect, transform, and interact with information to create and impart lifestyle-enhancing value, directly or indirectly, as part of a contextual enrichment of mobile services. The scale of sensing devices of the order of nanometers—nanotechnology—enables embedded and distributed intelligence in the processing of information collection, transformation, and interaction with the environment within which they operate. Autonomous and coordinated decision making, together with self-organization, self-configuration, and self-healing, is among the value-added ingredients of distributed sensing schemes that form the fabric of

---

* Mulligan, C., "Arup, The Climate Group, Accenture. Information Marketplaces: The new economics of cities," 2011, available online at http://www.theclimategroup.org/publications/2011/11/29/information-marketplaces-the-new-economics-of-cities/.

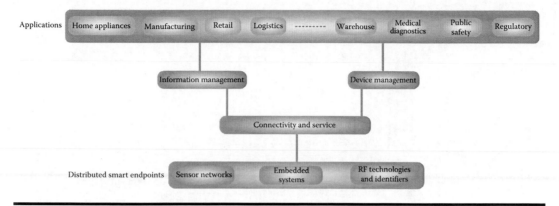

Figure 4.30   Representation of the M2M fabric.

assorted multimedia services, realized through wearable devices, smartphones, intrabody nanodevices, etc. The potential for mobile or fixed-wireless service innovation is virtually unbounded, where objects and distributed intelligence intersect in a variety of usage-specific scenarios. These emerging directions reflect a paradigm shift, where the distributed machine-to-machine processing of information over a mobile Internet surpasses the information flows initiated in human-to-machine interactions. Both connectivity and service are facets of the IoT. For example, the use of machine-to-machine communications—distributed sensors and intelligence, in the form of accelerometers—over the Bay Bridge that connects San Francisco to Oakland could have detected changes in the vibratory signature of the structure to predict impending structural failures that manifested as falling metal objects over a heavily travelled roadway.* The distribution of sensors and intelligence provides a generic strategy to enable detection and control capabilities, while implicitly embedding fault tolerance through redundancy, fault localization, self-organization, self-configuration, and self-healing in a scenario-independent fashion.

The most significant enabling attribute for simultaneously rendering both uniqueness and universality in a richly textured IoT is—interoperability. It then naturally follows that a global suite of standards that ensure interoperability across cultures and countries is pivotal. An architectural framework that embodies logical interoperability across the layers of protocol stack, in concert with open APIs across services, networks, and devices, provides enabling direction for innovation and implementation. Device management† is an implicit component, of such an architectural framework, for a customization of device behaviors, and remote configuration, through the use of standardized management objects. The considerations in such an interoperable framework include open interfaces, scalability, reliability, and modularity. Further the human-facing interfaces require appropriate levels of QoS-awareness mapping aligned with a given type of multimedia service for an adequate QoE. Interoperable interfaces with the prominent wide-area, local-area, and body-area network at the edges of an IP-oriented core network serve as wireless and mobile conduits of information, to a virtually unbounded diversity of sensors, actuators, or human interfaces, through a variety of mobile devices. A logical perspective of an interoperable framework for an IoT, with interfaces among machines and humans, is depicted in Figure 4.31.

---

* Conner, M., "Sensors empower the 'Internet of Things'," *EDN*, May 2010.
† Open Mobile Alliance, "Device management architecture," OMA-AD-DM-V2_0-20131210-C, December 2013.

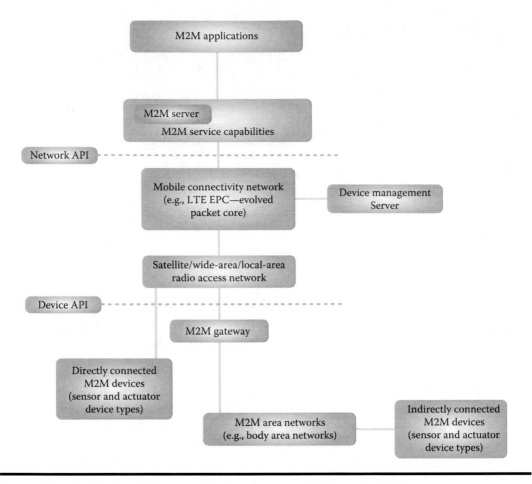

Figure 4.31   Logical and interoperable framework—IoT (Internet of Things).

The LTE system, with its all IP connectivity capabilities, offers a framework of M2M communications interconnectivity, across a variety of small and large coverage radio-access technologies. The interoperability and IP transport capabilities, inherent in the LTE and in the ongoing evolution of LTE-Advanced (4G) systems, promote a widespread heterogeneity of devices and their access to information connectivity on a global scale. Both macrocells and small cells, utilizing LTE-Advanced RATs, provide access to a variety of directly or indirectly connected sensors and actuators, within the fabric of IoT. With the widespread relevance and distributed nature of wireless sensors and actuators that engage in short-range or proximity-oriented interactions, the wide-area wireless coverage provided by the E-UTRAN,* within the LTE-A suite of radio-access technologies, is a portal into the Internet, while serving gateways to distributed sensing and actuating devices. For example, an E-UTRAN macrocell or a small cell base station may serve gateways, such as a smartphone gateway, a transportation gateway, or a home automation gateway. The smartphone gateway could serve a variety of wearable biomarker-sensing devices, such as for galvanic skin response, heart rate, blood pressure, blood glucose, and stress monitoring, for holistic health status indications, diagnostic and preventive information collection that triggers guidance feedback, from the

---

* E-UTRAN—Enhanced Universal Terrestrial Radio Access Network.

remote servers in the Internet, via the smartphone–human interface. The transportation gateway serves sensing and actuating devices for vehicular diagnostics and health monitoring, together with managing vehicular traffic flows, to avoid or to mitigate congestion conditions. The home automation gateway serves sensing and actuating devices for home climate control and management, optimization of energy management, remote audio-visual surveillance, etc. The processing of logistics, associated with the movement and storage of goods, is another arena of application for the sensing and communication of information such as temperature, humidity, location, duration of transit, type of goods, etc. Such capabilities in machine-to-machine and machine-to-human communications provide insights into an optimization of the flow, cost, and the quality of goods, through anytime–anyplace-oriented mobile multimedia services.

The predominant building-blocks in the realization of IoT include scenario-independent scalability, integration with IP as the universal glue, and interoperability. Distributed sensing and actuation capabilities provide and expansive IoT landscape of web services. The CoAP (Constrained Application Protocol)* provides session layer interoperability for distributed sensing and actuation devices that typically have low-energy consumption, variable level of channel reliability, assorted access technologies (NFC† schemes, such as RFID, ZigBee,‡ WLAN, etc.), and variable information packet size profiles. This protocol is specified based on a REST-oriented framework for applications and services that are associated with the management and control of sensors and actuators that are predominant in the fabric of IoT.

The E-UTRAN handles both machine-type and human interaction flows to and from the Internet, via the EPC, within the architectural framework of the EPS, which orchestrates the LTE—a wide-area connectivity. The proliferation of machine-type devices is in conjunction with the enormous diversity of content, bandwidth demand, and volume of information. For example, the information processed by these devices ranges from a few bytes associated with energy consumption metering, real-time sensitive alarm signaling, etc., to a large number of bytes associated with video monitoring applications in telemedicine, weather, etc. It then follows that the radio resource demands vary widely for the information flows associated with the enormous variety of usage scenarios inherent in machine-type communications. In the EPS§ the smallest unit of radio resource allocation for any mobile or fixed-wireless device is the PRB.¶ The dimension of a PRB may be expressed as 180 kHz in the frequency domain, and 1 ms in the time domain. Under good channel conditions, a single PRB can transport several kilobytes of information. This implies that a single PRB allocation to a machine-type device, which only transmits or receives a few bytes of information, would adversely impact the efficiency of spectrum utilization, within a serving EPS. For example, small distributed messages from sensors and to actuators, during fire or earthquake emergency situations, coupled with multiple emergency messages and alarms, with varying levels of QoS demands, have the potential to cause significant degradations in the EPS performance. This degradation caused by a deluge of machine-type communication messages, albeit short, in turn has the potential to adversely impact the experience of human-type communications, as a result of the potential constraints on resource availability. The breadth of usage of distributed and mobile sensing and actuating devices expands the volume of information flows from the point of

---

* Shelby, Z., Frank, B., and Sturek, D., "Constrained Application Protocol (CoAP)," IETF CoRE (Constrained REST Working Group), Internet Draft, draft-ietf-core-coap-18, 2013.
† NFC—Near Field Communications. Based on ISO/IEC standards for proximity communications that include the use of RFID (Radio Frequency Identification).
‡ ZigBee—Low power, short-range radio-access based on IEEE 802.15 standards.
§ EPS—Evolved Packet System within the LTE access technology domain.
¶ PRB—Physical Resource Block.

monitoring and control to remote management centers, such as in the case of telematics—remote monitoring, control, and storage of interdisciplinary information—applications and services.

Large-scale, or scale-free, behaviors within the fabric of IoT require universality of interoperability, through global advocacy, collaboration, and partnerships in the development of standardized specifications, with well-defined protocols and interfaces and interworking capabilities, across the prominent radio-access technologies. A harmonized approach to standardization that utilizes well-defined identification systems, protocols and procedures, is elemental in adoption and evolution of IoT. Differentiation, based on the breadth of radio-access connectivity (local and wide-area) provides flexibility and choice in terms of harnessing assorted types of spectrum, to optimize radio resource utilization that fosters an untethered flow of information, across machines and humans. SOA principles, together with service, network, and device context awareness, are among the enabling attributes of a generalized and logical architectural framework for machine-to-machine and machine-to-human communications, within the umbrella of IoT. This universally adaptable fabric has the potential to serve as a catalyst for synergies through cooperation and competition, aligned with a global advancement of knowledge, understanding, and trade. Human and environmental well-being is a prominent theme, for an IoT-mediated proliferation of innovative mobile services.

With this perspective IoT, represents an evolution of the existing interconnectivity provided by the Internet for information flows, associated with multimedia services that are rendered over human interfaces—a vast variety of mobile devices—such as smartphones, tablets, etc. In this existing model of the Internet, the services are launched and accessed in a silos, with none or limited interdependent context. This naturally implies that much of the context interpretation and management occurs via manual intervention. On the other hand, the evolutionary model of the IoT includes embedded intelligence, which is facilitated by a diverse fabric of machines—wireless and wired devices—that serve as sensors and actuators.

This fabric of interconnected devices autonomously interact with one another to provide a bridge, between the virtual and physical worlds, which is represented as human interfaces (e.g., smartphone, wearable, etc.) or as contextual information, associated with multimedia services. The interplay of the interdependent facets of personal, ambient, and social contexts that weave unique and universal profiles of experience, over the fabric of IoT, is shown in Figure 4.32. The intelligence embedded within the fabric of the IoT is derived and harnessed in the form of

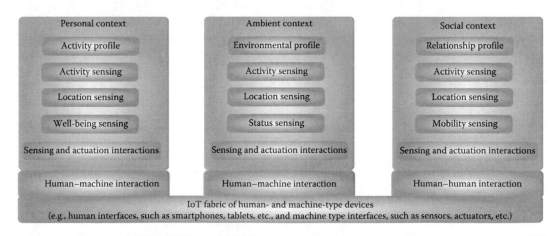

Figure 4.32 Interdependent contexts—IoT-rendered experiential service.

experiential contexts that are significantly enriching, along life's journey—in both routine and creative pursuits. Personal contexts are valuable in terms of services that enhance the experience of routine patterns of interest and behaviors, while also providing insights into an evolutionary potential: comfort and convenience that enhance well-being, optimize the efficiency of handling routine chores, etc. for lifestyle quality advancements. Ambient contexts serve to embellish the experiential world around the individual in terms of climate control, air quality, noise pollution, etc. Social contexts foster the establishment of relationships of common interest, such as research, learning, stress management through group activities, and meaningful interactions. Furthermore, they promote an evaluation of human mobility patterns for congestion avoidance, urban planning for cities and prominent locations, such as tourist attractions, museums, university campuses, enterprise buildings, large shopping centers, residences, etc., for virtually unbounded possibilities in IoT-mediated mobile services, aligned with experiential enhancements.

Information derived from distributed sensors provide experiential advances, via services for a variety of human activities ranging from routine interactions with things or objects in the environment—home, automobile, office, shopping, public transportation, navigation, etc.—to ideas and imagination related to education, health, well-being, social interactions, research, and creativity, to name a few. The information across these different facets of activities, when examined in terms of their interdependent aspects, provides insights into life's journey, from a user perspective. This in turn has a transformative and evolutionary potential to influence enhancements across societies, on a local and global scale. The embedding of distributed intelligence to sense and exploit multimodal[*] information via the IoT, in concert with information analytics, provides a scale-free approach to enriching the experiential landscape of human endeavor. Aligned with the conceptual framework, in Aristotle's *Categories*, the IoT bridges physical or virtual entities in the time-space domain that have identities and qualities or attributes, which characterize them. These entities, through their interdependent relationships, provide context and meaning that are choreographed as multimodal, multimedia services, via human interfaces. As Antoine de Saint-Exupery mused, "The meaning of things lies not in the things themselves, but in our attitude towards them." Our attitudes and behaviors are influenced and enriched through the value of contextual meaning provided by the fabric of interconnected things: physical and virtual. For example, using the concepts of swarm intelligence, a highly dynamic swarm of mobile smartphones can serve as a real-time, distributed information sensing and collection cluster, to get a snapshot of human mobility patterns, in cities or urban areas, to gain insights into managing congestion patterns, for experiential enhancements. The role of information sensing in urban areas, where about 50% of the world's population resides, is especially significant for enhancing the human experience, through IoT-mediated mobile services, with the anticipated doubling[†] of the urban population, by the middle of the third millennium. The Real Time Rome project[‡] aggregated information from mobile devices, in taxis, and buses to characterize the real-time profile of traffic patterns, spatial and temporal, in urban areas. This type of distributed sensing and customized information collection reveals insights, with respect to a dynamically changing environment—in this case, managing and mitigating congestion in urban areas—to enable intelligent decision-making, and to inspire innovation along experiential horizons.

---

[*] Different forms of information presentation, such as visual, audio, or haptic, rendered via different types of human interface form factors. (e.g., different video screen sizes, wearable devices etc.).

[†] Global Health Observatory, "*Urban Population Growth*," Geneva, Switzerland: WHO, 2014, United Nations.

[‡] Calabrese, F., Colonna, M., Lovisolo, P., Parata, D., and Ratti, C., "Real-time urban monitoring using cell phones: A case study in Rome," *IEEE Transactions on Intelligent Transportation Systems,* vol 12, no 1, pp. 141–151, 2011.

The self-organization, self-configuration, and self-healing tenets for autonomous and intelligent behaviors are significant characteristics in the realization of machine-to-machine communications, within a contextual fabric of IoT, since the intervention of sentient beings—humans—is absent. At the same time, the transformation of meaning and context, extracted from the interactions among interdependent machines, is pivotal in the enrichment of both unique and universal services, rendered over assorted machine-to-human interfaces. For instance, such interfaces would enable multimedia and multimodal presentation of information, where the services adapt and engage the user intuitively. Visualizations that are pervasive and adaptable, such as wearable eyeglass devices, Google Glass,* or those that are ambient, such as a display on an automobile windshield, promote experientially attractive mobile services that dissolve into life's journey of activities.

The information flows across a Web of Things imply a naturally pervasive distribution of an awareness of the environment, local and remote, where humans are enabled to engage in meaningful and purposeful interactions, through a variety of human interfaces. The information flows associated with the Web of Things, sensors, actuators, etc., have a variety of differences that are akin to the disparate sources, representations, and access methods of information over the World Wide Web, such as those managed by Google, Amazon.com, eBay, etc., and accessible via the associated proprietary web APIs. These differences in the types of information flows of the web are impediments, in terms of the creation of services that require the extraction of contextual meaning through an awareness of interdependence, across the disparate ingredients of any composite service. The principles of linked data† of concepts, terms, and relationships, Semantic Web, enable an integration of disparate types of information flows, embellished with contextual meaning, in a global and discoverable information space of humans and machines. Such directions are enabled through an interlinking of the machine-type device information ontologies, with the information over the World Wide Web for discovery, decoding, encoding, and global access for a variety of multimodal and multimedia services. The machine-type communication platforms enable these directions through interoperable interfaces that expose their capabilities and information to an implementation-independent space of applications, sensors, and actuators.

An integrated machine-type communication gateway, for example, served by the EPS‡ framework that provides wireless mobile access to varying sizes of coverage footprints, such as macrocell or small cell coverage, requires an awareness of the network of machine-type devices that the gateway serves. This awareness consists of, for example, the resource constraints, such as memory, processing bandwidth, etc., associated with each of the machine-type devices, power availability limitations (e.g., battery lifetime), latency tolerance of the information sensed or used for actuation. Knowledge of these profiles, associated with the machine-type devices that are served by the machine-type communication gateway, facilitates intelligent decisions, in terms of the scheduling, prioritization, routing, and caching of information flows, associated with each of the machine-type devices, for an optimization of performance and resource utilization. The integration of SDR (software-defined radio) technologies within a machine-type communication gateway enhances the ease of configuration of machine-type gateways for connectivity to short-distance wireless (RFID, Zigbee, WLAN, etc.) machine-type devices. The use of SDR-enabled machine-type communication gateways is naturally attractive as a result of the implicit ease of configuration.

---

* Starner, T., "Project glass: An extension of the self," *Pervasive Computing, IEEE*, vol 12, no 2, pp. 14–16, April–June 2013.
† Bizer, C., Heath, T., and Berners-Lee, T., "Linked data: The story so far," *International Journal on Semantic Web and Information Systems (IJSWIS)*, vol 5, no 3, pp. 1–22, 2009.
‡ EPS: Evolved Packet System that utilizes the LTE suite of radio-access technologies.

The flexibility and ease of configuration of SDR is particularly suitable for IoT-oriented services in home environments, or in access constrained environments, such as in hazardous areas, such as nuclear power plants, or in remote geographic topologies or sites, such as mountains, satellites, volcanic, underwater, etc. The characteristics of SDR are defined by the Wireless Innovation Forum* as "a collection of hardware and software technologies, where some or all of the radio's operating functions (also referred to as physical layer processing) are implemented through modifiable software or firmware operating on programmable processing technologies. The use of these technologies allows new wireless features and capabilities to be added to existing radio systems without requiring new hardware." The adoption of SDR technologies facilitates the evolution and adoption of new features and capabilities, through software/firmware updates, while allowing backward compatible solutions, by avoiding requirements for hardware modifications.

Architectural frameworks equipped with SDR technologies allow flexibility, adaptability, and interworking, in terms of enabling the usage of multiple protocols, protocol revisions, software enhancements, and computational resource sharing. Machine-type gateways equipped with SDR technologies pave frontiers for serving machine-type devices, over multiple radio-frequency carriers, multiple frequency bands, and assorted radio-access technologies. Examples of mobile vehicular machine-type gateways that are SDR enabled have the potential to access a variety of sensors, distributed within the vehicle for diverse diagnostic information collection, engine performance, climate control system, audio-visual system, usage patterns for performance improvement–related system tuning, navigation system, etc., over short-distance wireless communication technologies (RFID, Bluetooth, WLAN technologies in the ISM [industrial, scientific, and medical] band, etc.)

The vision of a universal fabric of IoT, for connectivity and information flows that provides a contextually enriching experience, requires scenario-independent interoperability and scale-free or scalable design implementations. Interoperable specifications are a cornerstone for wide applicability and adoption, across different ranges of radio-access coverage among machine-type and human-type interfaces. The management of a potentially enormous volume of mixed-media types of information across distributed things is formidable, despite the design concepts for simplicity and support for high bandwidth information flows, embedded within the LTE suite of packet-oriented technologies. The flexibility afforded by interoperability, within the logical architecture of the EPS, provides topological choices, for the off-load of machine-type information flows to short-range radio-access technologies to alleviate the challenges of handling the scale of distributed machine-type devices. The confluence of information flows across machine-type and human-type interfaces marks the advent of an IoE (Internet of Everything), where experience embellished with individualized context awareness becomes a dominantly attractive theme in multimedia and multimodal mobile services. These directions unveil an era of unprecedented potential for experiential advancements, across a milieu of personal, educational, cultural, economic, sociological, and philosophical evolution, mediated through service mobility.

4.10 Virtualization and Cloud Computing

The inflection point in the evolution of the information age is influenced by the cycles of technology innovation and human behavior. Beyond the mechanistic ramifications of technology, business, and monetization, both innovation and behavior are humanistic in nature. The evolutionary

---

* Wireless Innovation Forum, "What is software defined radio?" http://www.wirelessinnovation.org, SDR Version 2.0, 2014.

tools enabled through technological advancement are primarily motivated by the changing and evolving needs of society. They serve human demands and desires toward understanding, and fulfillment of diverse endeavors through the journey of life. Purposeful and meaningful interactions are the crucible of creative thought, innovation, and social well-being. From this perspective, mobile phones, just viewed as gadgets, are less significant relative to their ability to provide an anticipated experience of human communications. This is revealed in the universality of the desire to communicate, independent of the mobile device vintage, which is governed by regional cost, and availability conditions. Satisfaction and well-being are rooted in the flexibility of technology to allow communication modalities that promote good feelings and emotions, beyond the mechanistic and transactional interactions that substitute quantity for quality. An enabling of the absence of these communication modalities is just as vital in the incubation of creative thought and well-being in all human endeavors. This is aligned with the notion that while digital or virtual is everything; everything is not digital or virtual, such as human experience that translates into thought, imagination, innovation, and physiological, psychological, and biological well-being. These gems of human experience are essential ingredients in the orchestration and virtualization of mobile multimedia services and mobile connectivity services.

Virtualization reveals the landscape of digital transformation, which is not limited to technology and unveils a vision that integrates all segments of society across personal interests and professions, in the fabric of experience and well-being. It promotes directions, where cooperation and competition coexist, with collaboration serving as a catalyst for shared value, and market expansion. Innovation, exploration of new ideas, insights, and best practices are some of the significant benefits of digital transformation. It ushers a departure from the perspective of value being limited to "bright shiny objects" and embraces the paradigm of value embedded in the QoE across all arenas of human endeavor. From a human perspective, within which technology is one facet, empathy and understanding are elemental aspects to harness the exciting and evolutionary potential of digital transformation. In terms of mobile services, virtualization opens new vistas for widespread collaboration across people, processes, and technologies to enable stellar directions in unique and universal QoE, while simultaneously optimizing capital and operational investments.

The ubiquity of computing platforms has enabled a proliferation of the creation and consumption of information. Mobile communication continues to accelerate the universality and the uniqueness of information through the unbounded nature and modalities of service. In this sense, the advancement of knowledge through the proliferation of information has effectively introduced the notion of virtualization from a social perspective. The social dimension, with its emergent behaviors of an anytime, anyplace connected life paradigm, serves as a catalyst for virtualization across a multitude of endeavors, as a result of the virtual nature of information and knowledge. With the emerging demands of an information society, a postindustrial paradigm, the notion of virtualization in the arena of mobile connectivity and services is a natural consequence.

The dramatic evolution of computing augmented by mobility and wireless communications is foundational in the shift toward a global information society: a postindustrial society. The virtualization of mobile connectivity and service serves as an evolving conduit in the advancement of knowledge that is at the core of value proposition, across all human endeavors. Peter Drucker notes the significance of knowledge advancement, which underscores the role of virtualization, along the interdependent frontiers of mobile communications.

The basic economic resource—"the means of production," to use the economist's term—is no longer capital, or natural resources (the economist's "land"), or "labor." *It is and will be knowledge.* The central wealth-creating activities will be neither the allocation of capital to productive uses, nor "labor"—the two poles of nineteenth- and twentieth-century economic theory, whether classical, Marxist, Keynesian, or neoclassical. Value is now created by *productivity* and *innovation*, both applications of knowledge to work. The leading social groups of the knowledge society will be *knowledge workers*—knowledge executives know how to allocate knowledge for productive use, just as capitalists knew how to allocate capital for productive use; knowledge professionals and knowledge employees research knowledge, create knowledge strategies, and apply knowledge for implementing innovative products and services. Practically all these knowledge people will be employed in organizations. Yet, unlike the employees under capitalism, they will own both the "means of production" and the "tools of production": the former through their pension funds, which in all developed countries serves as a value resource, the latter because knowledge workers own their knowledge and can take it with them wherever they go. The economic challenge of the postcapitalist society will therefore be the productivity of knowledge work and the knowledge worker.*

With the merging mobility and communications capability, the virtualization of the creation and consumption information is on a path of continuous expansion in terms of the volume of traffic, the modalities of content, and connectivity—human- and machine-oriented—service innovation, and the impact on global social behaviors. In this regard, the emergent global information society is characterized in terms of a shift from an industrial or a capital-centric to an information-centric landscape, and the dominance of the information-centric fabric over an economic one, where the value proposition of information and imagination surpasses commoditization of value in the economic sphere.† Virtualization of mobile connectivity and service not only satisfies these aspects of mobile communications, but also enhances the potential for creativity and well-being across a global society, through optimization and automation that enable a reduction in infrastructure costs and a minimization of manual intervention. The symbolism of multimedia and multimodal mobile communications enhances the richness of experience in the human context, which is fundamentally social in nature. It is an arena, where the hardware assets of computing in the network serve as the invisible and distributed scaffolding for the orchestration of virtualized functions. Virtualization in all aspects of mobile communications services represents the evolving horizons of digital transformation.

The evolution of computing continues to enable possibilities for a separation between hardware processing engines and the software functions that operate over them, for the delivery of multimedia services. Historically, in the world of telecommunications, there is a tight coupling between the hardware platforms and an associated software functions. The motivation for tight integration has been shaped by the need to satisfy both high-availability and high-reliability. These are inherent performance requirements characterized by the nature of telecommunications, where technology mediation particularly serves the real-time and immediacy—voice and messaging—nuances of human communications. On the other hand, the rigidity of tight-coupled and customized frameworks—hardware and software—of circuit-switched technologies, necessary to serve a narrow suite of voice and messaging services, are inadequate to suit the enormous diversity and flexibility demands of IP multimedia services, rendered over a mobile Internet. Loosely coupled, highly cohesive, and interoperable architectural frameworks of packet-switched technologies are elemental consideration aligned with a rendering of multimodal and multimedia-rich contextual experience. While these directions are largely embedded within the guiding principles of the EPS, further decoupling of the functions realized from the underlying software/hardware processing platform is critical for two primary reasons: protection of mobile infrastructure platforms and ease of adoption of technology advancement. A simultaneous protection of investments and ease

---

* Drucker, P.F., *Post-Capitalist Society*. New York: Routledge, 2011.
† Toffler, A., *The Third Wave*. New York: Bantam Book, 1970.

of exploitation of technology advancements are feasible, as a result of a decoupling of the realized functions from the serving software/hardware fabric. In a nutshell, such a functional realization is an artifact of virtualization.

The two prominent characteristics of the EPS ecosystem are the architectural flatness and the separation, between the bearer plane information, and the control plane information. While this shift bodes well in terms of resource sharing leading to lower capital expenditures, relative to the legacy of mobile access networks, the effects of commoditization tend to diminish the return on investments. The exploration of avenues to reduce the operational and capital expenses, through the use of virtual functions that overlay the underlying infrastructure platforms, is attractive from a revenue perspective. With this direction toward virtualization, interoperable enablers are pivotal in the establishment of functional overlays realized in the form of SDNs (software-defined networks). The rich diversity and openness of the Internet that continues to shape the tenor and texture of a global information society has been the motivation for a convergence of the Internet principles and mobile communications. This convergence is reflected in the architectural model of the EPS, and in the IMS framework.* Logical interfaces, open and standardized protocols, serve as interoperable enablers for innovative implementations and deployment scenarios at the access and service layers. These attributes promote the extensibility and enhancements of features and capabilities, in terms of mobile connectivity and services.

In the case of mobile connectivity, the value proposition is characterized by access to information independent of location and time. This value proposition is limited in terms of connectivity differentiation across providers. The availability and reliability of mobile connectivity are fundamental attributes, where the scope for differentiation across providers is limited, as access technologies continue to mature and are subject to a declining revenue potential. On the other hand, service differentiation, over mobile connectivity, in an experiential landscape, is virtually unbounded. To mine the enormous potential of the mobile service landscape, a functional virtualization, over a loosely coupled, distributed, and interoperable framework of information, decoupled bearer and control planes, processing platforms, is foundational. The potential for service innovation thrives, where architectural frameworks contain astutely designed open interfaces, across high cohesive and loosely coupled information processing platforms, where there is a separation of the information control and forwarding logic. Such directions promote an agile adaptability to dynamic market conditions, as well as the demand for end-user driven experientially attractive services. It is testimony of the continuing evolution of mobile broadband connectivity, on a global scale, which has spurred market demands for *anytime, anyplace* service availability and reliability expectations. These evolving demand profiles have motivated considerations for architectural frameworks that are malleable for functional virtualization, which facilitate cost-effective solutions together with the ease of service creation, configuration, integration, and delivery.

Open interfaces between the control and forwarding logic are foundational in the framework of an SDN,† which enables a virtualization of the mobile network infrastructure.

A conceptual model of the SDN framework is shown in Figure 4.33.

The strategy to separate the control logic, from the data or traffic forwarding logic, is well-suited to the inherently shared resource, and distributed nature of packet-switched routing of control and data packets of information. This enables mobile service providers to program—control—the behavior of the mobile access network, from well-positioned centralized or distributed locations in

---

* Camarillo, G. and Garcia-Martin, M.-A., *The 3G IP Multimedia Subsystem (IMS): Merging the Internet and the Cellular Worlds*. Chichester, U.K.: John Wiley & Sons, 2006.
† Software-Defined Networking (SDN), *The New Norm for Networks*. Open Networking Foundation.

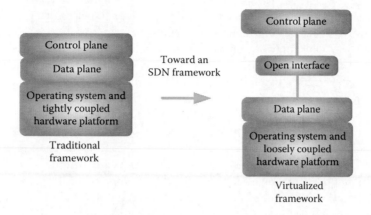

Figure 4.33 Transition toward a virtualized model of functions.

the network topology, using well-defined software functions. These directions mitigate cumbersome provisioning and configuration methods that are required in the traditional, tightly coupled hardware-software network elements. On the other hand, in the SDN model, such control procedures can be distributed across the selected sections of a mobile access network, using, for instance, the OpenFlow* protocol. The control instructions utilize the associated APIs to map the mobile access network into a virtualized environment that is represented as a SDN. The separation between the control logic and the information-forwarding logic holds the conceptual key in the information routing element paradigm, which enables a virtualization of functions that are no longer bound within the same physical entity. The OpenFlow protocol represents a standardized structure for message transactions, over an open interface between the control and forwarding logic. The openness of this interface is a natural catalyst for the promotion of a multivendor environment. In the SDN framework, the information-forwarding logic is in the data plane, while the control logic is allowed to physically reside elsewhere, in separate controller entities that may be distributed to serve the demands of mobile connectivity. Resource sharing† the network is an elemental attribute of virtualization. Agility and adaptability are at the core of the SDN guiding principles, which makes the SDN framework attractive for performance adjustments, optimizations with respect to QoS and service context, together with other aspects, such as border control, load balancing, energy savings, and security.

A fundamental advantage of the SDN framework is that it can be introduced incrementally into sections of the EPS, specifically in the EPC. The prominent attributes of the SDN framework are network programmability, virtualization, and orchestration. The orchestration aspect of SDN allows the control logic to manage the relevant applications, requirements, and policies. These aspects are characterized in the model of the OpenFlow SDN framework, where network programmability is enabled through intuitive high-level language constructs, which is attractive for controlling, managing, and automating virtualized network profiles and behaviors. Some of these notions are reflected in their early stages in the shift toward a relatively *flat* and packet-switched *shared* resource model of the EPS. The TCO (total cost of ownership) is inversely proportional to

---

* OpenFlow Switch Specification. http://www.openflow.org/documents/openflow-spec-v1.1.0.pdf, February, 2011.
† 3GPP TS 23.251, "Network sharing; architecture and functional description," v. 12.1.0, 2014.

the rate of innovation. These directions are intended to adopt the ideas of *Lego-like* interoperable networking platforms, with customizable control and management software. Passive network element sharing, in the mobile network infrastructure, such as antennas and site locations, provides some alleviation of costs incurred, while not affecting service differentiation. Much like IP packet–switched technologies at the network layer, which inherently allow a loosely coupling and resource utilization efficiencies through sharing, active network sharing in the RAN (Radio Access Network) provides concomitant benefits. These directions provide flexible deployment choices, cost optimization, and innovative opportunities for mobile service delivery, where ownership is distributed across the core network and the radio network layer of the EPS. Resource sharing in the RAN spans elements such as antennas, base stations, and spectrum, across a multitude of mobile connectivity providers: virtual and those that own the physical radio network assets. The tenets of virtualization in the core network, which combine loose coupling among functional entities together with a partitioning of the control and the data-forwarding logic, are also significant in the radio network. A realization of these new horizons demands collaborative recipes that bridge diverse and innovative business models, to effectively create a uniform and consistent fabric of connectivity for mobile services. Interoperability across the functional entities in concert with virtualization, through resource sharing in the radio network,[*] is pivotal in enabling the realization of forward-looking and innovative business models, for expanding value creation in the marketplace. The virtualization of the base station resources is foundational for enabling new frontiers, in terms of collaborative arrangements, where both autonomous and interdependent behaviors of shared resource utilization coexist, based on a variety of bilateral or multilateral agreements among the participants.

The sharing of capacity is among the suite of benefits that accrue as a result of virtualization, where participating mobile connectivity providers are enabled to acquire available capacity from one another, based on unique capacity demands. Dynamic load balancing and distribution among services related to human-type interfaces and machine-type interfaces, based on time-of-day scheduling preferences, or QoS-oriented priorities—QCI[†] that specifies[‡] a compatible treatment of information flows—are facilitated with relative ease, with widespread network function virtualization. These directions set the stage for both forward-looking business models that hinge on advanced operational and service delivery horizons, through the use of SDN-oriented mobile network infrastructures. Spectrum sharing is a complementary ingredient along the directions of virtualization, which require the appropriate regulatory compliance, interference mitigation strategies, and the management of shared resource access demands. While dynamic spectrum access and cognitive radio techniques facilitate spectrum sharing, barriers from a practical perspective include the negotiation of shared resource access, and coordination across different administrative domains.

Cloud computing technologies facilitate virtualization, in terms of connectivity, processing, and storage resources through distributed systems of hardware and software, where these systems are remote—the cloud—relative to a mobile service access location. A mobile connectivity event to a network edge radio link enables the potential for access to a variety of cloud-oriented mobile services. This model extends the notion of virtualization beyond the mobile network to include mobile devices as well, where the software and hardware resources to render a mobile service are resident in the cloud, instead of the mobile device. Various combinations of local and cloud-oriented processing capabilities are feasible, allowing a variety of user-centric choices. For machine-type devices,

---

[*] 3GPP TR 22.852, "Study on Radio Access Network (RAN) sharing enhancements," v13.0.0, 2014.
[†] QCI—QoS Class Identifier.
[‡] 3GPP TS 23.203, "Policy and charging control architecture," v13.0.1, 2014.

cloud-oriented processing capabilities allow a proliferation of low-cost sensors for the collection and aggregation of ambient intelligence and context, to embellish the mobile service experience. A preservation of configuration profiles and preferences is particularly attractive for nomadic users, where consistent service behaviors are critical, independent of access device type or location.

In contrast with a traditional OS, resident within a mobile device, a web-oriented OS resident in the cloud offers a variety of inherent benefits: service redundancy, reliability, and availability. Longer battery life and data security, as a result of remote processing and storage, are among the virtualized and distributed mobile service environment of the cloud. Consistent service behaviors are sustainable, with minimized dependency on the nuances and changes related to the hardware/software processing platform, local to a mobile device. Variations in the design or updates of OSs or clients that are locally resident, in a mobile device, have the potential to impact the familiar metaphors of service behaviors, sometimes with an adverse experiential impact. Specialized reconfiguration or optimization may be required to restore preferred service behaviors, which from a user-centric perspective may be undesirable or unappealing. On the other hand, cloud-rendered service designs have the potential to minimize behavioral variations and reconfiguration inconveniences, amidst a continuing proliferation of device types and capabilities in the marketplace. Since the cloud processing server farms are finite and managed, there is an implicit reduction in the vulnerability of the relatively large and diverse scale of mobile devices, to potential security threat scenarios. The cloud computing paradigm offers ubiquitous mobile service discovery and access, while allowing user-centric configuration choices and experiential preferences. A cloud-oriented management of devices—human- and machine-type—is critical for scale-free software and firmware downloads and updates, with the enormous continuing proliferation of devices: fixed wireless or mobile. With Moore's law driving both technological and social changes around the globe, both size and cost of devices continue to diminish, which serves as a catalyst for the realization of a virtually unbounded space of sensor-enabled environments and services.

The synergy among the capabilities offered through a functional virtualization of the mobile network, and cloud computing, is exhibited in a multifaceted suite of beneficial advancements across the entire mobile ecosystem: cooperative and collaborative business models, service innovation, market expansion, experience-centric context, and customization, in a user-centric participating evolution of mobile services. Smart environments are enabled, using combinations of virtualized and cloud computing frameworks, where mobile services are rendered via collaborative arrangements, across multiple stakeholders to serve a diverse variety of users, in a scale-independent fashion, with high levels of availability and reliability over a distributed fabric of mobile connectivity. Sophisticated, contextual, and customizable services, aligned with a cloud-oriented vision propelled through mobile network virtualization, are elemental for a realization of IoE: a vision where the mobile multimodal and multimedia services shape lifestyle-enhancing experiences for a global society, through an interdependent symphony of human and machine-type interfaces.

4.11 Pathways to Well-Being

A blending of technology mediation aspects together with the virtually unbounded facets of the experiential realm contains the enriching elements across a plethora of human communication modalities. In concert with personal and social aspects, innovation in the blending of these elements lends itself naturally to the affective components of mobile evolution, rooted in well-being.

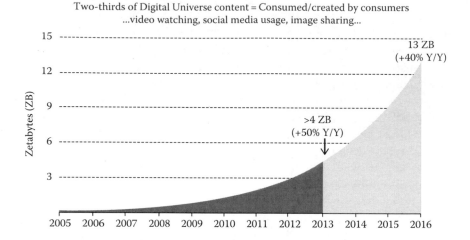

Figure 4.34 User-generated information proliferation. *Note*: 1 petabyte = 1MM gigabytes and 1 zetabyte = 1MM petabytes. (Data from IDC Digital Universe.)

It is ONE, orchestrated next experience, which combines and renders information without intrusion, intuitively resolves uncertainty to orchestrate a physical presence experience. The personalized nature of the smartphone category of mobile devices establishes itself as a conduit for the distribution of multimedia content and services. Figure 4.34 illustrates the evolutionary trends toward and expanding landscape of user-generation information creation and consumption.

In the context of ONE, user-generated content creation and consumption becomes a natural path in the evolution of technology mediation, in a one-to-many, many-to-one, and a one-to-one distribution of experience. It promotes universality—oneness of experience, across a shared reality—while simultaneously allowing the uniqueness of a customized experience at the individual level. This direction aligns with the digital transformation of a physical presence experience that simultaneously promotes a space-time-independent modality of human communications, which is naturally mobile. These attributes of ONE are enabled through the mediation of pervasive computing technologies, where mobile devices and services provide enabling capabilities in a variety of usage scenarios. Among these scenarios, health and well-being are significant arenas in promoting the quality of life. The use of smartphones that serve as conduits of health and well-being information provides transformational opportunities in the evolution of human well-being—one where flexibility and convenience thrive, while at the same time optimizing the flow, application, and leveraging of relevant preventive or treatment information.

For example, the use of smartphone mediation for ECG (electrocardiogram) information monitoring cardiovascular health, for preventive measures, as well as for treatment of less than optimal conditions, paves new frontiers for the promotion of well-being. A transducer that doubles as an ear-bud, for music or audio programs, also serves as a nonintrusive reflective photosensor* that provides ECG information to processing in the cloud, via a smartphone interface.

---

* Ebling, M. and Kannry, J., "Healthcare," *Pervasive Computing*, vol. 11, no. 4, pp. 14–17, 2012.

The ear-bud could also serve as a sensor for providing biofeedback, for relaxation and stress management based on information derived from the impact of behaviors and lifestyles. The mobile device and service opens a variety of innovative possibilities, as a personalized platform of technology mediation that is nonintrusive, while enabling ubiquitous connectivity to resources and repositories of information.

Personalized technology mediation resonates with the attractiveness of a virtually unbounded landscape of experiential themes. These themes that are staged over the ubiquitous fabric of IoE, where all interactions and services may be broadly classified into two categories: M2M (machine-to-machine) and H2M (human-to-machine) communication modalities. While technology mediation provides convenience enhancing automation, a concurrent growth in complexity is detrimental from a practical usability perspective. Mitigation of complexity is realizable through a contextual awareness of individual behaviors. These behaviors are governed by experiential factors, such as how one feels, which affects the quality of life. With contextual awareness, technology mediation vanishes unobtrusively into an experience-sensitive mobile service. The ideas on the subject of human–computer interaction are intended to soften the intrusiveness of interactions between people and machines. However, the impact of technology in terms of the manner in which it affects experience, while the intended service is effectively delivered, requires careful examination and enhancement. The extent of unintended effects of technology mediation on service experience depends on the nature of personal preferences and context.* Contextual augmentation of personalization promotes an experiential awareness within a mobile service, while the mechanistic components of the technology mediation process provide a desired function.

An experiential landscape hinges on intrinsic human attributes such as imagination, creativity, moods, reasoning, privacy, and well-being: it is within this landscape, where technology mediation has virtually unbounded potential in harnessing the interdependence between personalized context and service functionality, for a dynamic alignment and compatibility with human behaviors. Wearable devices that comfortably map the ambient and personal context are illustrative of nonintrusive technology mediation: reflective of a shift toward a rendering of mobile services, where empathy is exhibited to enrich the human experience. Such capabilities, within technology mediation, are unbounded and have enormous potential for service innovation. They exist in the gaps between the subjective nature of personalized experience and the mechanistic nature of technology-mediated functionality. Technology mediation, which is empathy aware, is broadly classified as empathic computing. This type of technology mediation is a natural pathway, for the enrichment of mobile services, which are intrinsically personal. It opens new vistas for uniquely personal choices, while simultaneously embracing a universally attractive human characteristic of empathy. On the subject of empathy, George Orwell mused in his writings, "If you see somebody begging under a bridge you might feel sorry for them or toss them a coin, but that's not empathy, its sympathy or pity. Empathy is when you have a conversation with them, try to understand how they feel about life, what it's like sleeping outside on a cold winter's night, try to make a real human connection and see their individuality."

A few examples of the adoption of empathy into technology mediation include self-surveillance: monitoring of a variety of biomarkers, such as sleep quality, heart rate, blood pressure, temperature, calories burned, etc. A realization of self-surveillance is manifested in a mobile wearable application

---

* Pedersen, I., *Ready to Wear: A Rhetoric of Wearable Computers and Reality-Shifting Media*. Anderson, SC: Parlor Press, 2013.

called Health,* which tracks biomarkers and serves personalized guidance. The awareness of context from a mobile service perspective unveils uncharted trails in terms of modalities that promote human engagement through the universal and unique pathways of empathy. Kinetically† powered wearable mobile devices offer the convenience of automatic energy replenishment for a virtually continuous availability of mobile services that are mediated via the wearable device: natural and nonintrusive energy replenishment through a harvesting of the user's physical movements. Context-awareness at a personalized level, which contains markers of a user's emotional state of mind, opens new vistas for innovation for rendering effective and attractive recipes for mobile services. It is an arena where factual correctness or generalized policies—one-size-fits-all—while tenable, lack the empathic context that enriches the human experience, based on an alignment with intentions, nonverbal cues, and the emotional state of being. These aspects require the detection and processing of context, where technology mediation operates in the realm of affective computing. It suggests a confluence of ambient intelligence, IoT, human–machine interfaces, distributed sensor networks, wearable devices, psychology, and physiology, over a natural paradigm of mobility. According to the IWEC,‡ empathic computing is an interdependent and interdisciplinary form of technology mediation:

Empathic computing systems are software or physical context-aware computing systems capable of building user models and provide richer, naturalistic, system-initiated empathic responses with the objective of providing intelligent assistance and support. We view empathy as a cognitive act that involves the perception of the user's thought, affect (i.e., emotional feeling or mood, intention or goal, activity, and/or situation and a response due to this perception that is supportive of the user. An empathic computing system is ambient intelligent, that is, it consists of seamlessly integrated ubiquitous networked sensors, microprocessors and software for it to perceive the various user behavioral patterns from multimodal inputs.§

The benefits of these seminal directions suggest enormous potential in terms of lifestyle improvement, through the conduit of mobile service–oriented technology mediation. These subjective and innately personalized considerations usher an era of innovation in a global information society, propelled by the nature of human behaviors, which are simultaneously unique and universal over a fabric of mobility. Affective computing is intrinsically information driven, where interdisciplinary information together with contextual information is harnessed to derive and design mobile services, based on inferences from emotional markers to shape empathic responses. Some of the evolving aspects of empathic computing, aligned with the research as a part of IWEC, provide enormous insights for innovative mobile service designs, these are depicted in Figure 4.35.

Adaptations to individual preferences are implicit, in this type of personalization, while harnessing the power of specialized information, which is not constrained by the physical constraints of time and space (e.g., scheduling of visits or appointments with a well-being practitioner.). The widespread and real-time availability and access to health and well-being information of individuals and communities, across local and global dimensions, enables improvements in the processing and in the advancement of guidance for lifestyle enhancements. The implicit efficiencies

---

* Morris, A., "Why apple wants to help you track your health,". *MIT Technology Review*, June, 2014.
† Shirisha, K., "15 cook kinetic energy powered gadgets," Crookedbrains.net, Aug. 27, 2013. http://www.crookedbrains.net/2013/08/cool-kinetic-energy-powered-gadgets.html.
‡ IWEC, "International Workshop on Empathic Computing."
§ IWEC, "Overview," in *Fourth International Workshop on Empathic Computing (IWEC'13)*, Beijing, China, August 2013.

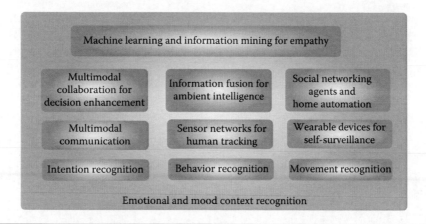

Figure 4.35   Model for mobile service enrichment.

in a mobile service–mediated collection and processing of diagnostic information translate into improvements in well-being on a global scale, while not compromising the uniqueness of every individual.

Mobility together with personalized diagnosis and guidance are the elemental pillars that shape the inherent attractiveness of mobile services, in the multifaceted and complex landscape of well-being.

Figure 4.36 exemplifies direction for a ubiquitous mediation over Internet of Everything for a delivery of services to adapt to lifestyles and to render choices for avenues of well-being and

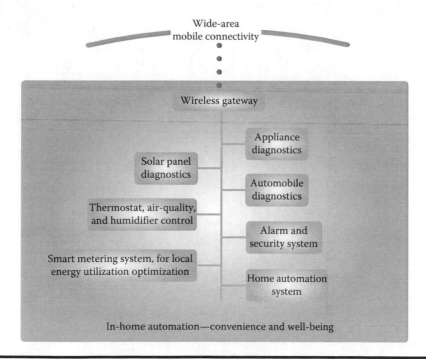

Figure 4.36   Internet of Everything—ubiquitous mediation for well-being.

health: a pivotal direction in the evolution of well-being, which in turn enables the experiential attractiveness, of human endeavor. An understanding of the situational aspects of human endeavor that precedes the shaping of a technology-mediated service enabler is naturally aligned with a direction to profoundly benefit advancements in lifestyles. Not the other way around. This perspective is a centerpiece that guides design and innovation, of a mobile service, along the experiential dimension.

Chapter 5

Interdependence
Renaissance of Multifaceted Convergence

Imagination is more important than knowledge.

—Albert Einstein

Commoditization to innovation: ushering a transition from tangibility to perceptibility, like shifting sands through the dawn of the information age.

The escalation of information spans all facets of human existence, from personal to professional. Entertainment, lifestyles, and business are among the arenas of information creation and dissemination. Information has been a central theme in human endeavor through the passages of history and civilization. It has shaped and molded the course of individuals, societies, and cultures around the world—a common vein pulsating through the colors and costumes of people and places separated in time and space. The spread of information on a global scale continues through the early conquests, trades, and wanderings that brought together disparate ideas and beliefs. Although history reveals both tranquility and turbulence during periods of information dissemination, in the aggregate, the distribution of information has largely been beneficial in the promotion of thought, creativity, and innovation. Through the cycles of conflict and calm, the widespread availability and adaptation of information continue to embellish lifestyles on an expanding scale, with global scope, diminishing a plethora of social, economic, geographic, political, cultural, linguistic, and religious barriers.

Scientific discovery and innovation, in the vortex of Industrial Revolution era, unveiled a collective consciousness toward the leveraging and application of intellectual capital to both understand and utilize the world around us through the power of creative thought and technology. Mankind's universal quest to communicate and connect has been a compelling arena for creative thought and innovation in science and technology. The widespread use of telephones, trains, airplanes, and automobiles is a testimony to the pervasive desire of all to communicate and connect. The tail end of the twentieth century saw the unbounded potential of the Internet on scale and richness, unimagined in the origins of voice telephony. This portal to new horizons in the dawn of the information age saw the beginnings of an untethered Internet—mobile Internet—conceived in the global vision of IMT-2000 (third generation [3G]).

From the early years of the twenty-first century, mobile communications continues to adopt, evolve, and reshape the rich potential of the Internet with the enormous attractiveness of untethered communication and connectivity. A technology-powered nomadic experience aligned with a natural state and nature of human communication. The inclusion of multiple dimensions of expression rendered through various combinations of voice, video, and data creates a virtual presence space, transcending the physical boundaries structured in time and space. The virtual presence space is experiential, and the richness of the experience depends on the nature and the effectiveness of the rendering—from the quality of connectivity to the quality of service (QoS).

The insights behind evolutionary and innovative designs, for synergistic and optimized mobile connectivity and service, naturally hinge on a convergence of technological and experiential creativity. A confluence paradigm of technology and art is depicted in Figure 5.1. Experiential creativity is in the realm of an ocean of human factors revealed throughout history in a rich fabric of archetypal stories. Stories build memories, which organically spread through traditions and cultures in various incarnations enabling the building of experiences—diverse yet with the common thread to imagine, explore, and discover. These experiences provide a holistic vantage point for new insights into design considerations. Since all experience results from communication, the insights from experience are not only significant but vital in the creative thinking that propels designs, for a compelling human experience—a dance of technology and art. The mobile Internet is a quintessential landscape, where a comfortable cooperation and collaboration of technology and art is imperative, since the mobile device continues to embed and fade into the fabric of our lives—a lifestyle accessory threading throughout the world.

"For knowledge is limited to all we now know and understand, while imagination embraces the entire world and all there ever will be to know and understand" Albert Einstein. Knowledge and imagination are pivotal components, in the evolution of ideas.

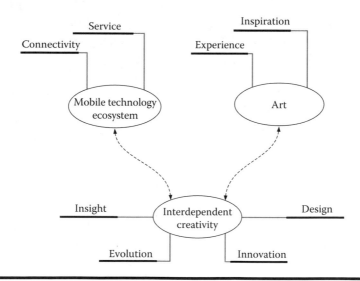

Figure 5.1 Confluence of technology and art—mobile ecosystem.

Why is imagination the fuel for all innovation? History and myth reveal and symbolize the vicissitudes of the human ethos through the ages. Imagination enables the proliferation of thought enriched through an understanding of human perspectives and experiences. This propels creativity—a catalyst for innovation.

5.1 Nature of Communications

Life is an expression of diverse and accumulated change, where expression is the motivation for communications. Throughout history and the universe, change continues to be pervasive. Ideas shape change and innovation. In turn, the progression of ideas continues to be shaped by change and thought—an interdependent cycle.

At the same time, notions of unifying aspects that form a common thread, across apparently different experiences, are abundant in the natural world. Existence in the natural world is identifiable through a dizzying array of observable classifications, such as color, texture, smell, taste, and sound that craft our perceptions. These perceptions carve out our experience, in the realm of imagination, memories, and emotions.

Information is transforming. It is an essential ingredient in the progress of understanding and value creation. Interpretation and creation of information is significant direction for the management of complex interlinkages, which engender widespread lifestyle conveniences.

Mobile broadband—packet-oriented—communications is a rich fabric for both applying and enabling the interpretation and creation of information. Accessibility and availability of information are the twin motivators in the mobile tapestry, where value and relevance are shaped by creative thought that encompasses the entire ecosystem as a whole. This tapestry transcends geographic boundaries, with an innate appeal and adaptability, across diverse social and cultural patterns on a global scale.

Thought: A vehicle of information is the engine that drives the evolution of the mobile ecosystem through standardization and innovation, elemental pillars that enable markets, through the ubiquitous realization of products and services.

Some points to ponder: What are the ingredients that make the mobile information age unique? Is it about the richness of data, malleability, and human factors? Is it interdependence across a multitude of existing and changing arenas of subject matter?

5.1.1 Xanadu

Linkages across imagination and logic are inherent in the information landscape, symbolized in a variety of inferences and interpretations throughout the progression of history. An insatiable quest for purpose and meaning in existence emerges as a consistent thread across a plethora of historical markers. Reflections that depict the human ethos present themselves in forms that shape our experience.

The travels of Marco Polo across the changing landscapes and cultures depict the interplay between imagination and experience that stimulate thought for exploration in a multitude of human endeavors—including research and innovation in mobile communications.

A state of mind observes, infers, and evokes holistic thinking for the incubation of new or different ways of design. The profoundly experiential nature of mobile communications demands a significant awareness of the need for a seamless blending of the enabling technologies and the human factors. The former is precise and predictable, while the latter is nonprecise and artfully evocative.

The artful vagueness is intrinsically human and unbounded yet malleable for personal customization—a lifestyle-enhancing potential. The seamless blending of these disparate aspects of innovative design is woven in the fabric of interdependence—between technology and human factors. Interdependence pervades in the world around us and is elemental in human endeavor. It is an artifact that must be embraced and leveraged, in the mobile information age.

The Gopher definition: Gopher: n. (1) Any of various short-tailed burrowing mammals, of the family Geomyidae, of North America. (2) (Amer. Colloq.) Native or inhabitant of Minnesota: the Gopher State. (3) (Amer. Colloq.) One who runs errands, does odd jobs, fetches or delivers documents for office staff. (4) (Computer tech) Software following a simple protocol for burrowing through a TCP/IP Internet – RFC1436.

The concept of the "Gopher" is profound: *search and retrieval* of information distributed over a *network of networks*—the Internet. The RFC1436 describes the enabling protocol: Gopher. The notion of a client and a server entity, where a client enables a user to access a server to perform a *search and retrieval* operation, is implicit in the realization of the "Gopher" protocol.

The "Gopher" concept is thematic in the hypertext concept, which was envisioned by Berners-Lee (circa 1989). An inspiration for his proposal was Ted Nelson's vision of interlinked documents realized in the *Xanadu model*, where users were enabled to access the contents of these interlinked documents. The hypertext concept extended the notion of the "Gopher" in an innovative quest to share, distribute, and contribute information across a community. The essential components included (1) hypertext markup language (HTML), (2) hypertext transfer protocol (HTTP), and (3) web browser. The significant underlying theme in this concept is that the components act in concert to promote convenience and ease of use, regardless of the intrinsic capabilities or characteristics of computing platforms or network technologies.

Images and stories serve as visual descriptors for the conception and incubation of ideas that shape design innovation. Ted Nelson's seminal idea conceived in the *Xanadu model* is reflective of *Xanadu*—a place in Mongolia, described by Coleridge, in the poem "Kublai Khan," which evokes visions of a magical, memorable domain with pervasive imagination and freedom. This historical epithet continues to be an underlying inspiration in the evolution of the World Wide Web: The *mobile cloud*—a user-centric, dynamic, interdependent, and evolving paradigm.

5.1.2 Broadband

Service ubiquity, while allowing service individualization (iconic services), is a fundamental note in the symphony of mobile broadband evolution. This theme allows the growth of individual choice, preferences, styles, and purpose—in the journey through life.

The notion of service ubiquity transcends the inherent boundaries of time and space, which also bounds our mortal beings that encapsulate a spirit—one that is both shaped by and shapes our experience. This innate formless framework is one where the experience of a service resides, beyond the bounds of wireless mobile access. The wireless mobile access provides a common transport framework for the service-related information, while the service experience is unique for every user. It is the collaboration between the transport technology and the service technology that provides the scaffolding for service ubiquity. The enabling requirements for service ubiquity are only a means not an end for service adaptability, where latter includes open and flexible capabilities for user-centric customization. Service adaptability allows a service to be tuned to a user's experiential preferences. Choices and selections are amenable to change with the desires, interests, and experiences of a user—with the same service, in combination with other services, or as new services appear on the horizon. It is an embodiment of function, elegance, and aesthetics—akin to the brilliance of nature around us, where an underlying theme of sameness and distinctiveness pervades through its creations.

It is to this distinctiveness that we must cater to explore opportunities for the shaping, nature, and expansion of services. Services that enable users to freely exploit the service benefits, adapted to distinctive behaviors, perceptions, and awareness, without external interference.

The service adaptability paradigm espouses the notion that human beings are not clones of one another but are distinctly manifested—externally and internally—and cannot be measured or categorized by rudimentary templates that are crafted by technological, commercial, cultural, religious, or political hues. It is the underlying experiential nature that carves the unfolding tributaries and streams in the river of life.

The value of service adaptability lies in its inherent amalgamation of the common and the distinctive, in virtually boundless manifestations. It hinges on iconization, where the user is the main actor and the service plays an adaptive supporting role in the journey of life in the mobile information age. This echoes of interdependent notes from the pages of nature, which masterfully blend a dizzying array of themes for a rich experience.

The subject communications, in the mobile information age, has intrinsic traits much like in the ancient times, where it included the study of rhetoric, which embodies the principles of memory, invention, delivery, style, and arrangement of content. These principal elements were woven persuasively by an orator to craft unique experiences for the audience collectively and individually for members of the audience. Much like the rhetoric pursuits, multimedia services in the world of mobile broadband weave unique experiences for the user. It is a propagation of service context and content, in the symbolic realms of thought, discovery, perception, emotion, and imagination.

The identification of the elements of value commoditization and value innovation, in mobile evolution, points to a collaborative interdependence between the creation and consumption of services. This interdependence—an expansive motif—promotes experiential inferences in the realm of thought and imagination, which invite potentially new directions independent of commoditization.

From success to significance—a change that is essential in the evolution of mobile broadband communication implies the becoming of a platform for the augmentation of communications in human existence. This potential for enormous significance is an attribute of the visceral essence of communications. The strategies, policies, governance, and regulations that shape the significance of mobile broadband communications have a profound impact across humanity, its behaviors and evolution, since it inspires the instinctive need to communicate universally, beyond the barriers of time and space. It is a modality of communication that is in stark contrast—technologically and socially—with a rigid command and control model, which allows mobile broadband communications to organically create, propagate, and sustain. It is a loosely defined communications framework, where free will and self-direction are elevated beyond the glitches of aberrant human behaviors that are sometimes intertwined with the notions of freedom and its benefits.

The loosely coupled framework of mobile broadband communications is inherently compatible with the self-directed, self-sustaining characteristics in the abundance of nature, which also reveals the inalienable truths that are also reflected within each of us as humans. These similarities between the building blocks that enable mobile broadband communications and the natural world elicit clues into the essence of the underlying loosely coupled framework. It is minimalistic and provides the tools for mobile communication pathways. In its ideal mode, it is nonintrusive and nonjudgmental with regard to the type of traffic that traverses the pathways that are set up by the framework.

An analogous example would be a framework of traffic lights to manage the flow of traffic—serving exactly one and only one purpose—not open for interpretation while not affecting choice or free will. Such a framework can be utilized to manage the limited availability of resources—roads or pathways—for effective sharing among users. The users are allowed to have a free will and choice, to drive from anywhere to anywhere. The Internet protocol (IP) exemplifies this notion in the managements of information flows over the highways in cyberspace. The framework intrinsically does not intrude or modulate the free will or choice—and so it is with the nature of a loosely coupled mobile broadband framework of access network, core network, and service layers, which opens the opportunities for a user-centric service experience.

5.1.3 Café

Patterns, intrinsic to human social and cultural behaviors, have no boundaries. As in dreams, so it is in ideas, where patterns are unbounded. Unbound and untethered patterns central to the possibilities are enabled by mobile communications. It embodies a virtual wanderlust—a departure from the early renditions of the telephone, which sets the stage for mankind to virtually transcend the traditional space—time hurdles. An appropriate metaphor is the enormously popular evolution and usage of the automobile representing motion and mobility—an invitation to the freedom of spaces and imagination spurring both improved lifestyles and innovation. Akin to this metaphor is the virtual world of mobile communications, a fertile pasture for new perceptions and conveniences, where technology provides the ingredients for a naturally untethered mobile service experience.

Howard Schultz's romantic pursuit transformed the mundane into a community experience—a feeling—much like the ingredients that have forged the intrinsic appeal of mobile communications. Against the beliefs projected by the incumbent experts concerning the viability and profitability of a café, Schultz's inspired vision—an intangible—was pivotal in the establishment of Starbucks, a universal attraction rooted in user experience. The elements of Schultz's vision were beyond the bounds of transactional business perspectives, where tangible goods and service are bought and sold for profit. It is in the foray into the boundless realms of experience, where commoditization and competition are traded for imagination and innovation.

Schultz's expeditions into the realms of experience elicit themes of memories—unforgettable recollections—as well as a universal longing for joyful moments. The impressionable experience evoked by a service is what makes the intangible immensely attractive. It is this nonspeculative tenet that comprises the heart and soul of innovative mobile broadband communication services. It is an anchor of possibilities in a complex and changing world, where the inexorable proliferation technology pervades with profound implications in the human journey.

The implicit boundaries etched into the perspectives of the post—Industrial Revolution era fade in a virtual continuum of form and function, elemental in the nature of the information age. In this continuum, the lessons and learning, for future endeavors in the evolving world of mobile communications, are supremely relevant when abstracted and applied selectively in the context of unpredictable change.

5.1.4 Human Factors

At a visceral level, mobile communications enables humans to use information in a plethora of experiential pursuits, while being connected to individuals or groups. The essential nature of communications is extended beyond barriers of separation defined in the dimensions of time and space. In this quest for experiential pursuits—personal or professional categories—the awareness of connectedness and interdependence is a fulcrum in the preservation of the natural modes of sensory-proximity-oriented communications. In collaborative and close relationships, awareness of the other is a catalyst in the direction of sensory-proximity-oriented interactions.

Although technological advances in mobile communications continue to bridge the time—space gap, the overly mechanistic nature of the tools of mobile technology—connectivity and service—demands procedural human intervention to suit the individual preferences of users. For example, a static default behavior may be embedded into a mobile device, which may not learn the context of contacts or a prioritized display of frequently accessed contacts, which are prone to change, subject to the dynamic nature of human preferences and behaviors. Interruptions or changes in the QoS may fluctuate in the event of mobility, changes in network conditions, or in roaming scenarios. Interruptions or changes in the QoS adversely affect the depth and range of emotional context inherent in a communication session.

The subtleties and the dynamic nature of contextual information, within a communication session, are vital ingredients in the fostering of personal and collaborative relationships. The awareness of the emotional states of the participants in the communication session adds to the richness, purpose, and the meaningfulness of the communication. At a personal level, contextual awareness promotes the level of connectedness, which leads in turn toward a productive and enjoyable exchange of information. Contextual information may include attributes such as habits and routines. Further, the contextual information serves as a catalyst for enhanced levels of consciousness through the tangible artifacts and attributes that characterize awareness. It is the enhanced levels of consciousness that lead to the forging of an attractive user experience.

The rhythms of communication are subject to change as the needs change, together with changes resulting from shifts between sensory-proximity-oriented communications and time—space separated communications. These changes suggest an adaptability of the underlying mobile communication technologies to adapt dynamically to user styles and choices for lifestyle-enriching communications.

Technology-mediated communications rely on the efficacy and the quality of both contextual and session-related information to approach the wholeness of sensory-proximity-oriented communications. The contextual components of information transactions surpass the experiential enhancements that can be achieved by simply augmenting the explicit quality of technology-mediated session-oriented content—voice, data, and video.

The traditional modes of communication that include physical media, such as letters, greeting cards, and packages, intrinsically provide elevated levels of connectedness as a result of the intrinsic multimodal sensorial experience. Technology-mediated communications utilize intrinsically artificial interfaces to transact information—virtual in nature—which do not intrinsically invite a multimodal sensorial experience. Although the technology-mediated communications—tablets, smartphones, and other data-oriented mobile devices—are pervasive and provide capabilities for fast, anytime, anyplace communications, they require not only seamless and high-quality information transactions but demand contextual awareness to promote approaches toward a natural experience. The contextual awareness encompasses presence information as well as ambient information, such as background activities and habitual rhythms.

The ambient background information, such as location spaces—living or working areas—serves as nonintrusive contextual content that serves to augment the interaction experience by way of approaching natural sensory-proximity-oriented communications. This type of augmentation avoids any additional user knowledge with regard to the artifacts of technology-mediated communications. It provides an ambient context—for example, audiovisual and tactile—for the communication session participants.

The underlying factors that influence connectedness among humans are a consequence of the nature of brain physiology, where the limbic component is not only a transfer function, between the environment and the body physiology, but also a processing element that shapes the emotional context. Communications—rich in emotional context—promote enhanced levels of connectedness through limbic resonance, which takes the technology-mediated information transactions closer to a sensory-proximity experience. The information exchanged takes on enhanced levels of purposefulness, relevance, and mutual benefit—personal and professional. The nature of the human nervous system is a control system lends itself to continual modulation and regulation, hinging on communication-driven contextual content—subjective—linked to the material and objective aspects of the information transacted, during a communication session. The subjective nature of contextual information that influences emotions and an ensuing experience are discernible through a variety of state transitions of this control system. For example, the limbic resonance between a mother and her child, through contextual information—subjective, such as an assuring glance and a warm smile—assures both that all is well in a seemingly unnerving situation, such as just before a test or surgery. An absence of such emotional context changes the states of consciousness that invite experiences of anxiety, fear, or depression.

Studies and experimental studies in psychobiology* reveal linkages between the internal states of awareness—consciousness that creates all experience—and the biorhythms. Biorhythms and circadian rhythms—sleeping and waking cycles—are profoundly affected by hormones, which in turn influences experience that modulate emotional context. In turn, experience—social, environmental—modulates hormonal profiles. For example, external cues, such as environmental—impact of natural light—are a universal time cue or a somatic rhythm known as zeitgebers.† Similarly, social cues, enabled through multimedia communications, since it is innate to human nature—whether or not technology mediated—universally influence human behavior and experience.

Social relationships have been well established as a marker of well-being and health,‡ where feelings of connectedness through relationships embellish both well-being and health. The mobile Internet—since it is rich in context and interaction—provides convenient and on-demand pathways for connectedness that effectively and synergistically supplements a face-to-face presence.§ The interactive nature of the mobile Internet—unlike broadcast communications—has a personalized impact that nurtures social connectedness. This is the most significant ingredient of mobile communications—an experiential realm, where opportunities for a confluence of technology and art thrive.

The attractiveness of mobile communications hinges on the extent to which it empowers users with choices to interact with others—remote or proximal—in intuitive and convenient ways. Traditionally, in telecommunications, "Perhaps we have been building crutches rather than

---

* Bentley, E., *Awareness: Biorhythms, Sleep and Dreaming.* New York: Routledge, 1999.
† Venter, R., "Rest, and specifically sleep, play a major role in athletic performance," *CME: Your SA Journal of CPD*, July 1, 2008.
‡ House, J.S., Landis, K.R., and Umberson, D., "Social relationships and health," *Science, New Series*, vol. 241, no. 4865, pp. 540–545, 1998.
§ Coget, J.-F., Yamauchi, Y., and Suman, M., "The internet, social networks and loneliness," *IT & Society*, vol. 1, no. 1, pp. 180–201, 2002.

shoes."* This is a metaphor for creative, humanistic design perspectives, where shoes—akin to mobile communication services—optionally complement form and function of the feet—akin to proximal communications—even when the feet function normally. Technology mediation must unobtrusively behave *like shoes* rather than *like crutches*. Presence, messaging, and e-mail features in mobile communications provide flexible, contextual, nonintrusive modalities for remote human interactions.

In other words, innovation in technology-mediated communications must facilitate flexibility, intuitiveness, and contextual richness to where it is attractive even when collocated communications—relatively shorter gaps of time—space separation between communicating users—are relatively feasible.

Innovation in technology-mediated communications allows new forms of social interactions, where the nature of contextual information engenders improvements in ties and relationships through emotional content. This is similar to collocated communications, with small groups—families, communities, etc.—where the members feel an implicit connectedness.† A harmonious balance in the social interactions—personal and professional—within the framework of mobile communications is of paramount significance through an imaginative inclusion of contextual information that is a sampling of the multifaceted elements that provide the richness of temporal and spatial dimensions. An imaginative mirroring of the elements of sociology and psychology is naturally aligned with the appeal of a harmonious balance, in the evolution of next-generation technologies that power mobile communications. These technology-mediated modalities will then behave like *shoes* in contrast with *crutches* that appear in rigid, silo-oriented technology design perspectives. The standardization and design frameworks must embody both sociological and psychological elements central to a human experience.

In the realm of technology-mediated untethered communications, the complexities and the ceaseless ongoing change in human behavior and cultures are an integral input into visions of new capabilities in the underlying technologies. The changing complexities portray both subtle and profound depictions of the social fabric, interpersonal relationships, beliefs, perceptions, and expectations that require an incisive exploration in the formulation of new ideas for spawning attractive mobile services.

The implications of a unification of technological and humanistic landscapes are powerful and profound, since the ideas that accrue influence and shape the virtually unbounded modalities in human connectedness. This is an experiential model—central to the theme of mobile services—emerging in a multitude of dimensions.

As in physical group activities—games and sports—the related exertion produces elevated levels of awareness and enjoyment, which make the activities memorable for all in unique ways. The excitement fosters existing relationships and promotes new ones. These activities conjure images of olden times, when cargo ships became icebound, during severe winters, and required smaller ships to *break the ice* to open paths for icebound ships to sail again. These metaphors suggest the appeal of exertion interfaces in mobile communications.‡ Emoticons, mood depictions, customized

---

* Hollan, J. and Stornetta, S., "Beyond being there," in *Proceedings of the SIGCHI Conference on Human Factors in Computing Systems*, Gaithersburg, MD. New York: ACM Press, pp. 119–125, 1992.
† Agamanolis, S., Westner, A., and Bove, V.M., Jr., "Reflection of presence: Toward more natural and responsive telecollaboration," *Proceedings SPIE Multimedia Networks*, vol. 3228, p. 8, 1997.
‡ Mueller, F., Agamanolis, S., and Picard, R., "Exertion interfaces: Sports over a distance for social bonding and fun," in *Proceedings of CHI 2003*. New York: ACM Press, pp. 561–568, 2003.

sounds, and haptic capabilities in user-centric mobile device interfaces are all ingredients that engender a memorable experience.

Across personal and professional endeavors, we experience connectedness, when we are a part of an interest group or a club. These engagements allow an expression of our unique selves, while being a part of different expressions and interpretations. The individual and integrated styles color the nature of the group as a whole. Technology-mediated communications—social networking—opens these endeavors to complement and nurture a face-to-face togetherness—an extended awareness.

Visual interfaces provide more than an objective rendition of images or videos depicted by the quality and clarity of the pixels. Other aspects such as aspect ratio, orientation, shape, viewing angle, context richness, intuitiveness—touch screen features—and adaptability to application requirements are among the significant considerations that shape the interface attractiveness. How does the input/output interaction features complement and enhance the display utility and ergonomics? The interaction features open possibilities to include common and unique preferences across different users—different in behaviors influenced by personal, societal, and cultural dimensions.

The level of interaction invited by a display depends on the various facets of display design and the intuitiveness of the arrangement/presentation of features and capabilities. This is beyond just the quality of the graphical display screen—type of surface, small form factor, clarity, depth, etc.—where the hardware and the software combine to create a certain ambience and mystique that is innately attractive, allowing both curiosity and revelations.

The mobile display content richness—color, presentation, and visual effects—and attributes convey both the individual and collective expressions across virtual communities. A unique facet of technology mediation that augments the content experience is a temporal compression of people, places, and events, related to a topic of interest, presented in a chronological order. This aspect is invaluable, since it vividly recreates a narrative—beyond the recollection of human memory. In turn, it offers purposefulness and inspiration in the context of collaborative ideas. Allowing interaction and uniqueness* of styles, through rich media,† is a profound aspect of mobile communications that promotes connectedness and meaningfulness in human endeavors. From a design perspective, openness, flexibility, and customizability are elemental in the integration of the mobile device display and the presentation of the user interface (UI) for interaction and connectedness in the experiential realm. Figure 5.2 depicts a human paradigm in technology-mediated communications, embellished with mobility. This paradigm is explored in the ensuing sections.

Boundaries must yield to enable value and uniqueness, in the domain of the long tail‡—a proliferation of relatively few customized services to many. The particulars must be sufficiently abstracted for a generalization, necessary to effectively mine the gaps of an emerging interdependence.

Max Planck, the quantum physicist and Nobel laureate, postulated: "As a man who has devoted his whole life to the most clear headed science, to the study of matter, I can tell you as a result of my research about atoms this much: There is no matter as such. All matter originates and exists only by virtue of a force which brings the particle of an atom to vibration and holds this

---

* Dourish, P., Adler, A., Bellotti, V., and Henderson, A., "Your place or mine? Learning from long-term use of audio-video communication," *Computer-Supported Collaborative Work*, vol. 1, no. 5, pp. 33–62, 1996.
† Bly, S.A., Harrison, S.R., and Irwin, S., "Media spaces: Bringing people together in a video, audio, and computing environment," *Communications of the ACM*, vol. 36, no. 1, pp. 28–47, 1993.
‡ Anderson, C., *The Long Tail: Why the Future of Business Is Selling Less of More*. New York: Hyperion, Corp., 2006.

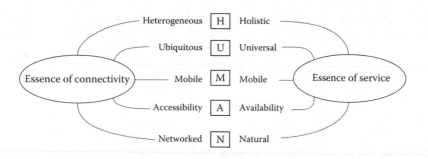

Figure 5.2 Mobile services—a human paradigm.

most minute solar system of the atom together. We must assume behind this force the existence of a conscious and intelligent mind. This mind is the matrix of all matter."

This unifying power is present in the natural world, of which we humans are an integral part. It is an organizing intelligence that allows unique experiences, while simultaneously allowing us to share our experiences. The dawn of ubiquitous mobile communications complements this innate nature, beyond form and boundaries.

5.2 Essence of Connectivity

The insatiable desire to connect to among human beings as individuals, groups, communities, and across the world at large continues to be assisted and shaped through the evolution of information and communications technologies (ICT). The evolutionary directions in connectivity are captured in the spirit of the Internet—where connectivity thrives in a loose framework of decentralized and distributed networks. These directions suggest autonomy of connections that weave a global fabric, without the impediments of a centralized command and control. Distribution of connectivity, in the exchange of information, heralds a transformation in experience that is allowed to manifest at different levels of interest and participation—individual, group, community, and global. The ICT-mediated transformation in information connectivity ushers new models for exploration and experience in the human social context.* Mobile connectivity—a natural communication modality—is a cornerstone in the ICT-mediated information landscape.

Mirroring the natural modalities of human communications and interactions, technology-mediated mobile information symbolizes the advent of new horizons of awareness and experience. Mobile information transcends the time—space gap and adds the richness of mobility—a personal companion and portal to an information universe. Information creation and consumption in the form of mobile data continues to evolve through multimedia applications, such as social networking, TV, Web 2.0, ambient awareness, augmented reality (AR), and content mash-ups—the possibilities are virtually endless and hold enormous potential for innovation to enrich and enhance lifestyles in a user-centric paradigm.

It is these nascent possibilities that have inspired the evolution of the mobile connectivity capabilities, through long-term evolution (LTE)—a global partnership for the development of

---

* Castells, M., "Information technology, globalization and social development," UNRISD discussion paper No. 114, September 1999.

forward-looking interoperable capabilities that provide the foundations for broadband mobile connectivity design and deployment. With the architectural model of LTE-oriented mobile connectivity, steeped in the concepts of decentralization and distribution, it hinges on IP transport through the edges of the mobile ecosystem. At the same time, the information transport objectives that improved end-user throughputs; reduced user-plane latency; enhanced capacity, QoS management, and self-organization; and reduced operational cost are anticipated to serve the potential for an unparalleled mobile user experience. While the LTE architectural model, with strategically envisioned interfaces, between the radio segment and the wired segments, has flexible building block capabilities, it maintains a high level of flexibility to interwork with other radio access technologies. This flexibility gracefully accommodates mobile connectivity migration strategies and different business models, to enable an inexorable service rendering potential, where information proliferation is an innate ingredient that virtually permeates all human endeavor.

5.2.1 Next

The ubiquity and enormous potential for the creation and dissemination of information over shared transport—exemplified in the Internet and realized through IP—set the stage for a transition of the mobile technologies toward a similar paradigm. The circuit-switching techniques that originated in the 1G and 2G era were adopted from the perennial models of fixed telephony and were an artifact of the nature and objectives of voice communications.

With the compelling richness of multimedia content—voice, video, data—rampant in the fixed Internet, the adoption of similar models in the mobile ecosystem is a natural step in the evolution of mobile systems—3G, 4G, and beyond—where shared transport is not only essential but vital from both a cost and resource utilization perspective. In this emerging paradigm, mobile communications is subsumed in the term Internet, where a distinction between wired and wireless fades in an ocean of information flows shaping unique and ubiquitous patterns in user experience.

The 3G connectivity, established under the auspices of the IMT-2000 initiative, identified and adopted the benefits of a packet-switched model, hinging on IP as the unifying glue across assorted access layer protocols. This marked the first step in a transition toward a logically distributed architecture, realized through IP transport just beyond the first-hop wireless access segment, between the mobile device and the mobile infrastructure. The next-generation evolution directions ushered in an affirmation of packet-switching techniques in the form of architectures and protocols centered on IP and adopted in the framework of LTE system architecture evolution (SAE). The wireless interface evolution for 4G connectivity is a vision promulgated by the IMT-Advanced vision.

The evolving vintages of packet-oriented connectivity have a common objective—to enable the creation, implementation, and operation of a viable transport network for rich suite of mobile multimedia services that hold enormous potential for imagination and innovation. Categories of informational types range from conversations to content, with temporal and nontemporal constraints. The emphasis and focus of these emerging forms of connectivity is to maximize the opportunities for mobile service customization—an unlocking of the experiential domain. Together with the identification and specification of pivotal service enabling application programming interfaces (APIs), the decoupling of the transport and service planes fosters an independence of mutual constraints, while strengthening the service enhancement potential through collaborative interdependence. The decoupling between the transport (connectivity) and the service promotes an independent provisioning of capabilities, in each of the domains without impacting the other.

The functional entities in the next-generation mobile evolution reflect a distributed nature in terms of managing policy, resources, media, sessions, security, and service delivery paving the path for creative and flexible implementation options. The distributed nature is crafted over open, interoperable interfaces, where the reference points are identified for the specification of interoperable procedures and protocols. The reference points may represent one or more open interfaces for flexibility in architecting operational choices that are driven by service delivery choices. These tenets constitute a fabric for generalized connectivity optimized for distributed service mobility, with the objective to sustain and optimize the service experience.

5.2.2 Elemental

The architectural tenets envisioned in the forward-looking packet-oriented mobile connectivity network contain elemental attributes that are essential. Among these are the attributes of distribution, autonomy, coordination, and flexibility. Each of these attributes holds the potential for innovation in design and implementation coupled with market expansion in an interdependent, collaborative, open landscape. These elements are embodied in the LTE SAE framework, which is a conglomeration of entities, interfaces, protocols, and procedures that orchestrate a symphony of mobile connectivity.

The LTE SAE framework provides mobile connectivity through the management of signaling and traffic between the Evolved Packet Core (EPC) and the Evolved Universal Terrestrial Radio Access Network (E-UTRAN), where the latter serves as point of connectivity for the mobile device. The prominent entities within the EPC are MME, SGW, PGW, PCRF, HSS, AAA, and PCRF, and the entity within the E-UTRAN is eNodeB. The connectivity is managed in terms of the initial attachment of the mobile device as well as in terms of the handover of the mobile device to a new attachment point, in the event of mobility. The entities within the EPC and the E-UTRAN are distributed and have nonoverlapping, orthogonal functions, which enable them to be naturally autonomous. The separation of these entities, over selected reference points that are open and interoperable, embellishes flexibility, in terms of deployment choices. The attributes of distribution, autonomy, and flexibility demand appropriate coordination capabilities across the entities in the EPC and the E-UTRAN.

A logical view of the elemental functions/entities in the LTE SAE is depicted in Figure 5.3.

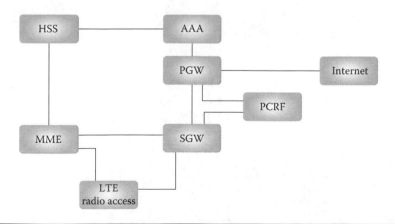

Figure 5.3   Functions/entities—LTE SAE.

The nonhierarchical nature of the connectivity framework in Figure 5.3 implies a natural distribution, which promotes a minimization of transport latencies, for delay-sensitive multimedia services, while retaining the potential for capacity and coverage management for assorted media content, over an all-IP network.

A conceptual view that is reflective of the attributes of distribution, autonomy, coordination, and flexibility, within the LTE SAE framework, is depicted in Figure 5.4.

The evolutionary attributes that are reflected in the nature of the LTE SAE framework complement the growing demand for resource sharing to optimize both CAPEX and OPEX as commoditization continues with technology improvements at the mobile access layer. Resource sharing is managed through a variety of functions, such as policy control, charging control, QoS, service awareness, and information flow profiling. These functions are among those provided by the PCRF to allow choices in terms of the nature and types of mobile services that are delivered to the user. Both online and off-line methods of charging for mobile services are enabled by the PCRF. The billing function and rating function act in concert with the PCRF to enable opportunities for a desirable user experience in the case of online (real-time) calculations of charging information. Selective crafting of policies can be utilized for a realization of fixed-mobile convergence of services using the shared resources of the EPC (Figure 5.5).

The management of QoS is vital over a shared resource connectivity network, since the wireless segment of resources is constrained in terms of availability and prone to link condition variability. This objective requires being in concert with the service-level expectations of the user delineated by subscription costs. The PCRF enables a differentiation and customization potential for mobile services.

The limited nature of wireless resources and the need to optimize the user experience, under mobile conditions, demand strategies to route information selectively to manage capacity. The local IP access (LIPA) feature enables access to mobile service resident within a local network served by a femtocell access point (FAP), a generic term for Home eNodeB (HeNodeB), in the LTE ecosystem. The selected IP traffic offload (SIPTO) feature enables access to mobile service resident in the Internet or in an enterprise network. The selective channel of information routes between the mobile device and a destination serves the twin objectives of a customizable service experience and capacity management. In the case of the LIPA feature, mobile access to user-preferred services

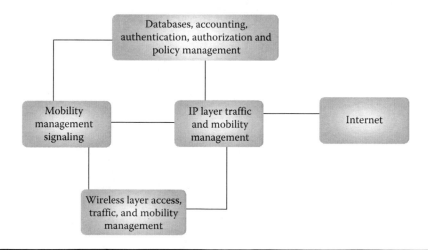

Figure 5.4 Conceptual view of functions/entities—LTE SAE.

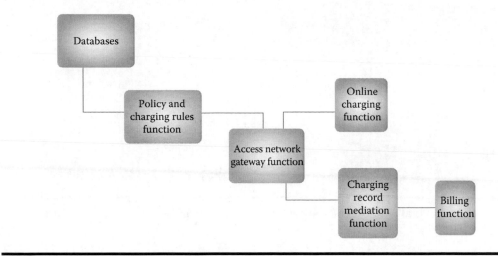

Figure 5.5 Policy and charging rules function context.

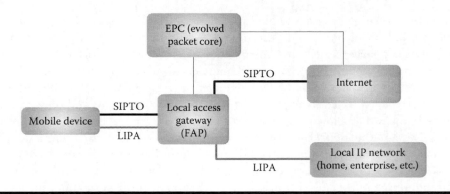

Figure 5.6 Elements of capacity management.

served via the FAP promotes a choice-driven appeal. A conceptual view of these two enabling features, both from a network and a user perspective, is shown in Figure 5.6.

In concert with capacity management, coverage diversity availability enhances the ability to optimize resources while sustaining an attractive user experience. To enhance the radio-link performance limit, the LTE-Advanced framework includes capabilities to enhance the spectral efficiencies per unit of coverage area. This implies coordination across a diversity of wireless coverage techniques—heterogeneous networks. These networks, which are combination of macro, pico, femto, and relay transceivers, provide increasing levels of coverage granularity, which facilitates flexible deployment options. The additional granularity of capacity distribution through coverage topology choices provides more malleability in tuning the balance between available capacity and the mobile service experience.

The harnessing of coverage diversity through heterogeneous networks is particularly relevant in the context of a boundless growth potential of mobile multimedia services. This is an artifact of the shift in growth from circuit-switched voice services to packet-switched multimedia services. The latter type of service, being multimodal and more widely used—regardless of whether the

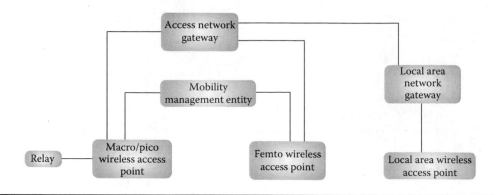

Figure 5.7 Contextual view of a heterogeneous network environment.

user is indoors, outdoors, or under mobile conditions—demands higher link budgets and coverage augmentation, relative to the relatively monolithic circuit-switched voice services. Figure 5.7 illustrates a contextual view of the elements of a heterogeneous network with the framework of LTE SAE.

A web of information pervades around the human context of routine, recreation, rejuvenation, and research in mobile world. The information is embedded in things around us pertaining to virtually every aspect of human endeavor. These things—such as food, clothing, automobiles, planes, trains, buses, exercise equipment, health supplements, health monitors, medical treatments, cameras, and toys, to name a few in a sea of objects—have embedded intelligence that can be harnessed to augment the human communications and lifestyle experience through techniques that allow these objects to interact with one another in a complementary fashion. This is a foundational motivator in the Internet of Things that bridges the islands of objects through machine-to-machine (M2M) communications that provides a contextual fabric in the evolution of mobile communications. In this ambient context of things that have the potential to interact and supply either real-time or non-real-time information, the information presentation requirements and related complexity typical in direct human interaction scenarios are minimized.

The edges of heterogeneous connectivity to an object or sensor are bridged to the Internet of Things through the LTE EPC. The heterogeneous connectivity enables a diversity of coverage footprints and access technologies to allow a proliferation of objects of sensors based on the proximity to the related network attachment points. In the case of objects that connect to a network over very short wireless distances, such as in the case of those enabled through near-field communication (NFC) technologies, the volumes of such objects are very high as a result of the very large usage scenarios. A few NFC examples include mobile payments, file sharing, gaming, and identity cards, among numerous other possibilities. Similarly, the other access technologies, such as the E-UTRAN, pico, femto, Wi-Fi, and Bluetooth, provide remote interaction capabilities across machines—a large category that includes objects, sensors, and devices that have no direct human interface. Figure 5.8 illustrates a generalized connectivity framework for heterogeneous access and coverage for an Internet of Things.

A multitude of object may interact for information synthesis, before the synthesized information is required to be processed for human intervention or consumption. For example, the climate control system settings in a home could interact with the external weather sensors for an optimized setting of the home temperature and humidity, before presenting the information on

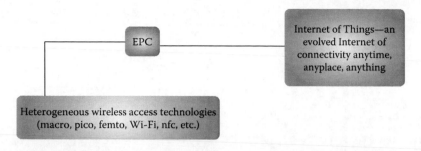

Figure 5.8   Generalized connectivity for Internet of Things—EPC.

a user's smartphone or tablet. Optimizations of traffic and signaling are critical ingredients to manage the nature of communications—packet size, frequency of transactions, delay sensitivity, real-time, non-real-time—among the machines. These optimization enhancements provide the flexibility and scalability of the EPC to effectively enable the Internet of Things, while providing a generalized heterogeneous connectivity, in the face of an unbounded information transaction potential across machines.

The proliferation of human-type devices—smartphones, tablets, etc.—and machine-type devices demands enhancements to match the corresponding capacity and coverage demands. As alluded earlier, it is the shift toward the multimodal nature of packet-switched multimedia communications that imposes new challenges in terms of increasing coverage and capacity demands that were absent in the homogeneous world of voice-centric communications. The increasing capacity and coverage demands also imply an increasing operational costs and burdens to manage the user experience for mobile services.

Self-organizing capabilities within network complement the distributed nature of the LTE SAE architectural vision, which is essential for autonomous behaviors necessary to combat the complexity challenges—resulting from the capacity and coverage demands associated with packet-switched multimedia communications. Among the enabling features of self-organizing networks (SON) are automatic network configuration, optimization, healing, and plug-and-play capabilities. The distributed nature of connectivity networks through diverse levels of heterogeneity requires SON capabilities to adapt to the user-experience demands, while reducing operational costs and enhancing the utilization of resources in shared all-IP mobile connectivity framework. Heterogeneity in coverage sizes and technologies includes a variety of small cells of connectivity coexisting in a macro connectivity environment. The elements of SON are characterized by self-configuration, self-optimization, and self-healing. The guiding principles of the SON framework are depicted in Figure 5.9.

With the linkages in information, being a significant factor in information complexity, a distributed model has distinct advantages: simplification as well as robustness—through a minimization of problem replication, while promoting value replication. It exemplifies an enhanced management of localized information through decentralization and autonomy.

The guiding principles of the LTE SAE framework embrace the attributes of simplicity for operation and deployment, through distribution, autonomy, and self-organization. Together with these attributes, the LTE-Advanced radio interface offers a high level of granularity for scaling bandwidths from lower than 5 to 20 MHz, with concatenation capabilities for wider mobile transport, which promotes flexible deployment choices, hinging on spectrum holdings and business models. Overall, the LTE SAE connectivity framework optimizes operational costs

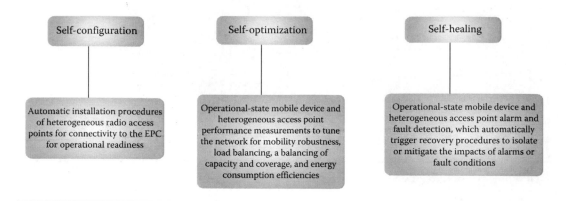

Figure 5.9   Elements of the SON framework—autonomy in distribution.

and efficiencies through a reduction of processing nodes and interfaces for flexible configuration choices that reflect the spirit of a rich mobile service experience, while enabling robust service availability conditions.

5.2.3 Shift

The heritage of mobile communications is rooted in conversational transactions over wireless transport. The dominant themes were relatively narrow and well-defined, such as the quality of voice, availability of transport, reliability of transport, and latency. The small packet sizes and the ability of the human brain to correctly interpret conversations, even in the presence of packet losses, allowed the formulation of an archetypal architectural tenet that defined the cellular voice communications system. Circuit switching and the use of dedicated resources to guarantee acceptable latencies influenced the shaping of the system-level architectural tenets. In a nutshell, the type of technology mediation—circuit-switching—together with the nature of voice communications implied that centralization would be a compatible architectural tenet in the origins of mobile voice communications. The supporting protocols, procedures, and specifications followed suit to match the objectives of this hierarchical and centralized architectural framework.

On the other hand, the advent of the Internet and its unifying protocol (IP) embraced the opposite—nonhierarchical, decentralized, archetypal, architectural tenet. This signified a profound shift in the nature of technology mediation in mobile communications. The architectural framework, in the untethered communication modalities, continues to recognize the distributive nature of IP-oriented procedures and protocols, in the identification of the logical entities and interfaces. It represents a loosely arranged collection of logical entities and interfaces that enable mobility, compatible with the nature of human communications—user-centric and distributed. The diversity of mobile devices, mobile applications, and always-connected mobility paradigm ushers multifaceted motivators for enhancements—virtually boundless—a reflection of human imagination, preferences, and lifestyles.

The LTE architectural framework fosters an always-on connectivity, which is crafted to a variety of information traffic profiles, over the wireless, mobile segment of connectivity. The local entities and interfaces allow implementations to suit business models for compatible trade-offs, such as user experience, system performance, signaling efficiencies, self-configuration, self-optimization, self-healing, and mobile battery-life longevity. The radio access network (E-UTRAN) and the mobile device (UE) are enabled to cooperate over different duplexing schemes: time division duplex (TDD) or frequency division duplex (FDD). The mobility paradigm hinges on the efficacy of energy management for usage longevity in the mobile device to preserve and promote the communications experience, over untethered connectivity, for a variety of multimedia applications. This implies that mobile connectivity preservation is realized, while energy consumption approaches levels that are as close as possible to an idle, but connected mobile device.

The enhancements for mobile connectivity demand cooperative and cross-protocol-layer information transactions to adapt to changing connectivity and mobility conditions, which are generated or sensed by the access network or the mobile device. These event triggers serve as the signaling intelligence that evoke the appropriate procedures for corresponding state changes, in either the access network or the mobile device. These interoperable procedures and protocols cooperate to select the suitable trade-offs between the system performance and battery longevity, while being responsive to a multitude of applications that are served over mobile connectivity to meet the user expectations. The management of wireless and wired access network resources provides the necessary levels of configuration and control to balance responsiveness and resource utilization for an optimized service experience. At the radio level, radio resource management (RRM) and discontinuous reception (DRX) are examples of procedures that are configurable to manage the trade-offs between responsiveness and resource utilization efficiencies. The DRX concept enables a mobile device to turn off its transceiver periodically for energy conservation, where the periodic wake-up intervals to listen to the network for any newly information, destined for the mobile device, can be configured dynamically. The configurations are driven by intelligence governing the mobile device activity patterns, with interval resolutions of the order of subframes (1 ms). While the periodic switching off of the transceiver enhances the resource utilization, it reduces the responsiveness, translating into latencies that could potentially adversely impact the user experience of latency-sensitive applications.

The rising sophistication of smartphones, tablets, netbooks, laptops, and embedded modems for the creation and delivery of a cornucopia of multimedia services, with complex transportation demands, requires a matching flexibility and cooperation, from the radio access and core networks. The LTE evolved packet system (EPS) is an embodiment of open interfaces and functions that are adaptable for various deployment scenarios to satisfy the information transportation demands associated with a plethora of multimedia services. The information traffic profiles, with a variety of demands, are handled in the EPS through QoS and policy management techniques. The categories of trade-off considerations, to satisfy the application demands, for an appealing or at least tolerable user experience span signaling overhead, transport latency, network resource utilization, and mobile device power consumption. The optimization of trade-offs for information transport depends on the nature of the applications, and a QoS classification of applications is driven by the corresponding system and application deployment scenarios. The handling of trade-off optimizations for mobile connectivity over data-oriented, decentralized architectures invites a continual adaptation to an ongoing changing and evolving mobile communication market landscape.

Beyond the vistas of human-oriented mobile communication, machine-type communications (MTC) modalities represent M2M communications adapted over the LTE connectivity ecosystem. This is a fabric for the Internet of Things and suggests an enormous addition to the range and the type of transport demands on the transportation of information, between and among machines—an ambience of things.

5.2.4 Impact

The elemental shifts embodied in the nature of the LTE SAE framework for mobile connectivity has far-reaching implications in the context of mobile evolution. The characteristics of this evolutionary mobile connectivity network embrace the next-generation mobile network vision to establish a foundation for an attractive user experience, seamless access to mobile services, blending of assorted content types, coexistence and interoperability in a heterogeneous access technology ecosystem, and the adoption of a convergent all-IP connectivity fabric.

These attributes are reminiscent of Peter Drucker's vision of a hybrid organization, where decentralization is a dominant theme, where coordination among the disparate and autonomous entities mimics the benefits of centralization, without the ailments of rigidity and cost that discourage creative thinking and evolution. This was a profound insight that enabled Toyota to find the sweet spot between decentralization and sufficient coordination for enabling innovation and preserving a high quality of user experience. This balance crafted in a coordinated decentralization is the profound impact that the LTE SAE framework ushers into the realm of mobile connectivity evolution. It exemplifies a loosely coupled set of entities and functions that provide coordinated connectivity for promoting a compelling and memorable mobile service experience.

The attributes that guide the direction in mobile evolution are represented in Figure 5.10. The attribute symbolizes the impact of the LTE SAE connectivity framework. Why are these directions natural and appealing in the unfolding horizons of mobile multimedia service? It is the essential nature of information—distributed and universal—that demands a shared and diverse connectivity combined with operational efficiencies to provide value-added and customized mobile services. This in turn dictates the adoption of new technologies, strategies, business models, and operational paradigms that are aligned with the notions that these underlying attributes symbolize. It reveals the potential to leverage the potential of partnerships that allow innovative business models, where incremental value is leveraged.

The impact of these imaginative and practical insights is to invite new and different perspective in thought and innovation across the various facets of business and personal pursuits. The implications demand compatible changes in streamlining policies and regulations that promote the mining of the enormous intrinsic potential within the virtually unbounded vistas of user experience. The attributes shape the nature of evolutionary mobile connectivity models that serve as blueprint for a harmonious evolution of the human ethos on a universal scale. They represent harbingers for new value creation in the realms of mobile service creation that augurs a return to simplicity in a plethora of technologically powered complexities.

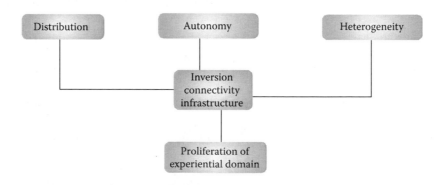

Figure 5.10   Mobile connectivity attributes—directions in evolution.

5.3 Essence of Service

The changes in the nature of the mobile connectivity framework, realized in the LTE SAE guiding principles, foster directions toward broadband transport, which is an integral attribute of the wired Internet. These changes usher a transformation in service creation and delivery. The transformation encompasses a variety of arenas in the service domain for new business models, collaboration, cooperation, and interdependence. The robustness of mobile connectivity, enriched with heterogeneous connectivity topologies of coverage, over an all-IP fabric enables the evolution of mobile services that leverage the tenets of service-oriented architecture (SOA). The convergence of assorted mobile connectivity via the generic and flexible nature of the EPC architecture promotes a convergence of services in terms of mash-up-assorted media that offer the potential for a rich experiential domain. Content infused with assorted media that convey sensorial information—sight, sound, and touch—highlights the motivational objective of SOA in a contextual world rendered in virtual and AR descriptors.

Contextual information enriches the mobile service experience through an awareness of the conditions—geolocation, usage profiles, connectivity conditions, interest profiles, etc.—associated with a mobile device and the mobile user. The distributed nature of all-IP connectivity networks enables an integration of multiple sources of contextual information in a scalable and extensible fashion, for harnessing a multitude of applications over SOA that are created by third-party application providers over open interfaces and protocols.

5.3.1 Horizons

In the changing horizons, what are the revelations and what do they imply? How do these revelations influence and complement ideas that inspire designs aligned with change? Commoditization to innovation—from tangibility to perceptibility—portrays shifting sands that usher new horizons, in the dawn of the mobile information era.

Along with the technology-driven changes, business models must shift in the wake of the mobile information proliferation. It demands a shift where solutions grounded in one-size-fits-all must yield to flexible, diverse, and open thought and application. Unsustainable Maginot lines and moats must vanish into the creative vistas, where spatial and temporal barriers neither promote nor inhibit the potential for a compelling user experience.

After the end of World War I, the French government—entrenched in siloed thought patterns—invested tremendous resources to construct the *Maginot* line, extending across the boundaries of Germany and Italy, to protect itself from a potential invasion. History suggests that the German air force simply flew over it, and the German army entered through Belgium. The Maginot line serves as a metaphor for walled gardens of services, where wide-area wireless access is guarded with respect to third-party services. The ubiquity, openness, and interoperability of IP allow these third parties to potentially circumvent the wireless access layer—Maginot line—thereby allowing users to access third-party applications. With the tremendous and growing popularity of smartphones, with multimode wireless access—Wi-Fi, LTE, CDMA—and multimedia applications supported by Android and iPhone operating systems, third parties have emerged as a vibrant component of the mobile user-centric landscape. For example, on the voice/video communication front, Skype provides a viable and convenient option—independent of the access technology.

These revelations identify shifts in the business models to accommodate, sustain, and promote the changing trends embodied in the distribution of players in the value chain. The shifts demand a departure from the rigid to the fluid, where the storms of commoditization are avoided through partnerships and technology innovation.

The widespread and mobile-powered dissemination of information paves uncharted paths in the formless realms of thought ripe for harvesting in a changing world of conversations and communications. These uncharted paths invite collaborative visions toward interdependence—not only between business and technology—across all imaginable endeavors for value and service creation. In such a narrative, the following archetypes are relevant in the exploration of insights in the waves of change that shape and influence the information age.

Conversations paint a tapestry of us and others, through the unique and the common in sight and sound. Voice is that distinguishing signature that describes our interpretations and ideas motivated by thought and emotion. It is a palette that renders a multitude of compositions that describe our personal and public hues of expression. The web of the Internet-orchestrated services extends and complements this most compelling mode of interaction in a fabric of multimedia. Mobile communications—a burgeoning facet of the Internet—embellishes this narrative through one-to-one and one-to-many expressions. Mass media, such as broadcast media—TV, radio— lack the levels of fluidity that untethered communications bring with their natural modes of interaction. It is the openness and the noncontrived aspects of conversations that enrich through spontaneity. The hyperlinked nature of the web avoids a hierarchical stage, which inspires and encourages new conversations and experience. In this ocean of information, content and context are perceived through the lens of expression and experience as depicted in Figure 5.11.

While procedures and protocols provide the technological structures for end-to-end mobile communications, it is the freedom of thought within and beyond this structural framework that provides the richness of experience in mobile services. Christopher Alexander noted*: "At the core, is the idea that people should design for themselves their own houses, streets and communities. This idea… comes simply from the observation that most of the wonderful places of the world were not made by architects but by the people." This stemmed from an observation of the natural beauty, attractiveness, and character of cities and monuments from antiquity, where architects and designers adapted their creative holistic vision to match the needs of the people being served,

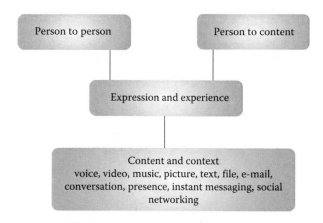

Figure 5.11 Lens of expression and experience.

---

* Alexander, C., Ishikawa, S., and Silverstein, M., *A Pattern Language: Towns, Building*, Construction. Oxford, U.K.: Oxford University Press, 1977.

within the constraints of building regulations. A resonating metaphor, where building regulations are akin to standardized enablers—procedures and protocols—where a creative holistic vision is vital, be it for the creation of buildings with timeless beauty or mobile services that invite a compelling user experience.

Context is embodied in the background as well as in individual expressions—molded in unique styles—through the narratives of language, gestures, and motion. Mobile augmented reality (MAR)* promotes a vision, where expression encapsulated in content—audio, video, picture, and text—is streamed on cue, into a smartphone or tablet from an ambient locality. An MAR smartphone browser, known as Argon,† enables access to ambient information, selected through spatial orientation sensors—GPS, accelerometers, and vision. Eye contact, as in face-to-face communications, such as gaze awareness,‡ provides seamlessness in technology-mediated communication. The context provides a palette for individual expression that seeks resonance in one-to-one, one-to-many, and many-to-many conversations.

Global locality awareness harnesses and augments the vibrancy of expression, across the chasms of distance and time. Mobile devices enabled with geolocation, orientation, and tactile sensors provide interfaces to this rich expression. A sampling such as—Dopplr, Google Maps, Fire Eagle, Loopt, to name a few—from among a dizzying and evolving array of applications rendering expression.

Expression in the form of mash-ups—where web service APIs enable the mixing of information from different sources to create new services—is an artifact of Web 2.0. The Web 2.0 framework is a framework of design patterns that provides new perspectives for architecture, content, supply chains, and business models, associated with information sprawling over the global information pathways in a mobile-powered Internet. The main actor is the mobile device—smartphones and tablets—ubiquitously connected to the Internet while simultaneously allowing expression among people for practical and memorable experiences. Examples of Web 2.0–like services include blogs, Wikipedia, Google AdSense, syndication, Flickr, Napster, and search engines.

Awareness of a global locality—global availability of local information—becomes a corridor into an enrichment of expression conveyed through a mash-up of information. It then serves as a window to the world around us, enabling natural modalities of expression that transcend the separation of space and time.

It is the experience that inspires new ideas. From an architectural vision and design standpoint, a profound shift in perspective is demanded—one where the innovation in the underlying technology must simultaneously and seamlessly consider the stage and context where the scenes of service are played out. The *how* of technology must include the nuances of the *why* and the *where* of mobile services, while their roles are enacted in the experiential realm.

The innate beauty and attractiveness of expression, forged through contextual awareness, hinge on the labyrinths of mobile information flows. These flows—inward and outward—enable expression through a plethora of human-facing and nonhuman-facing devices and sensors. The fabric of mobile devices and sensors becomes an ambience of interaction, where the labyrinths fade indistinguishably into an experience of expression.

In these horizons of mobile information, expression enables discourse, which in turn propels the emergence of new and different markets, with technology innovation serving as the

---

* KHARMA , KML/HTML augmented reality mobile architecture. https://research.cc.gatech.edu/kharma/.
† Georgia Institute of Technology, 2014. What is Argon? http://ael.gatech.edu/argon/what-is-argon/.
‡ Ishii, H. and Kobayashi, M., "ClearBoard: A seamless media for shared drawing and conversation with eye-contact," in *Proceedings of Conference on Human Factors in Computing Systems (CHI '92)*, ACM SIGCHI, Monterey, CA, pp. 525–532, May 3–7, 1992.

unobtrusive bedrock of incubation of experiential mobile services. From the origins rooted in windows through technology, encumbered within the constraints of boundaries and monolithic, mechanistic features of a keyboard and display hardware, an adoption of the contextual, multi-mode interaction and ambient awareness must be at the forefront of technology architectures and designs. An ambient architecture—where the sensorial environment is described in the language of gestures, tactile feedback, color, sound, gaze, location, and mobility—is elemental. For example, contextual content, such as color, sound, and tactile feedback, could synergistically augment the expression in a text or an e-mail message. The environmental information content significantly enhances the experience, relative to a text or e-mail message in a silo or in isolation.

Information ubiquity continues to spawn participation on a global scale, in the creation, interpretation, learning, and dissemination of knowledge and expertise unparalleled in the history of civilization. Organic in nature, the universal conversations propelled by the power of mobile communications encourage new possibilities through an interest and convenience-driven motivators for both personal and professional enrichment.

Wikipedia, among other types of mash-ups, inherently promotes the sourcing of information embedded in populations throughout the world. *The Wisdom of Crowds*, envisioned by James Surowiecki, exemplifies the significance of collective knowledge, which expertise tends to morph from silos to boundaryless repositories. This shift reveals the common themes and insights, in the information age, with mobility at the epicenter, where rigidity fades into a symphony of ideas, be it technologies, business, marketing, or personal endeavors—from the hierarchical to a teleocratic distribution. These considerations prompt a shift from the notion of offering products and services to actively gauging, interpreting, and dynamically responding to the needs of the crowds—a rich tapestry of consumers. This implies a subtle transformation of decision making from a centralized to a collaborative one, between the producer and the consumer. Expertise or decision making hinges on a blending of collective information, instead of being vested solely in the producer of mobile services. This movement, in the evolution of mobile service offerings, aligns naturally with providing few—niche or customized—services for many. It is one that reflects the role of the *long-tail* model, as information evolves and proliferates.

With the virtualization of information availability, in the mobile landscape, innovation possibilities continue to expand on a global scale in concert with commoditization, which in turn propels further participation in technology-mediated mobile communications. In these changing and evolving ICT dynamics, the role of the consumer shifts from a monolithic consumption-oriented role to an active collaborative participant, together with service providers, in the crafting of compelling mobile services. Examples include Web 2.0–like applications, promoting a variety of content mash-ups, such as LinkedIn, Facebook, YouTube, Skype, and Second Life, which reveal growing levels of interdependence—departure from the classically disparate and categorized silos of value and market creation.

Customization of user profiles, identity management, NFC, SON, peer-to-peer (P2P) communications, and artificial intelligence (AI) are significant considerations in the research and standardization of new mobile technologies and applications in the quest for lifestyle-enhancing capabilities—pathways to inspire innovation.

In these changing vistas, policy and regulatory aspects require reform to both adapt and serve as a catalyst in the evolving paradigm of information mobility. The inherent movement toward decentralization of networks and applications is driven by the natural propensity of data that allows distribution. In federation of individually operated hot spots that serve a coverage region, the traditional view of an operator changes, in that a user is responsible for managing their connectivity with minimal or no assistance from a serving network. P2P communications naturally demand the use of distributed resources, such as in gaming and voice over IP (VoIP) applications such as Skype. A natural decentralization of the mobile communications is both necessary and opportunistic. The decentralization leverages the enormous potential of the mobile Internet for a reformation of policies and regulations that are compatible with the flat, distributed architectural models, such as in the case of LTE. The resolution of the related policies and regulations must be applicable in the spirit of open access, privacy, security, and service usage at the consumer level. This shift promotes new opportunities in terms of nonintrusively examining trends at the user level in the enhancements of policy and regulation models—for example, in detection of congestion to manage spare capacity availability for communicating emergency situations.

5.3.2 Enablement

The primitive, elemental building blocks are the invisible agents that orchestrate the crafting of innovative multimedia applications and services. The fabric of standardized building blocks serves as the agents of interoperability in a field of innovative potential, which is crucial for both market expansion and attractive horizons in the experience of mobile services.

From visual to sonic to video tapestries, these building blocks of multimedia mobile services are crafted with a view to enable, without constraining imagination leading to design innovation. Services—virtually boundless—rendered in terms of mash-up possibilities meet potential usage scenarios as well as create new trails in lifestyle enhancement features. Why are these building blocks significant? What are the insights in the realm of enablers? Some of the forward-looking enablers are explored in terms of concepts to reveal insights into the nature of the types of patterns and primitives that characterize emerging trends, in an ocean of mobile services.

A combination of contextualization and localization provides a template for capturing the descriptive, temporal, and spatial dimensions associated with interactive media. This type of blending of multimedia information is foundational in the enhancement of perception. It is an approach that lends itself to an augmentation of perceived reality. The mobile infrastructure advancements toward higher information throughputs and bandwidths have created a viable environment for the execution of AR applications. A broad range of services that hinge on AR include tourism, entertainment, gaming, advertising, and mobile search. The hardware entities in a mobile device—smartphone—that facilitate a rendering of AR applications include sensors, accelerometers, GPS, display, and camera.

The conceptual primitives that act in concert to enrich the mobile service experience are represented in Figure 5.12.

For a realization of MAR, the mobile device requires to register for a context information service from a home macro network that serves as a repository of the user identity and related profiles that modulate connectivity, service, interests, privacy, security, customization, etc. The user launch of a context-aware service triggers the mobile device interactions with the appropriate repositories of third-party applications that acquire any relevant mobile device ambient context information for augmenting the service experience. The actors and functions are enabling entities that are conceptually relevant independent of implementation scenarios.

The perceptibility enabled through spatial and temporal dimensions synthesizes a tangible experience through the use of AR. Haptic* and visual content descriptions are contextualized

---

* Azuma, R.T., "Survey of augmented reality," *Presence: Teleoperators and Virtual Environments*, vol. 6, pp. 355–385, 1997.

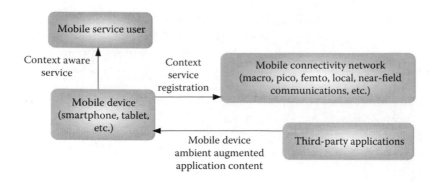

Figure 5.12   Context service registration for MAR.

to augment the experience of mobile information communications. The overlay of geolocation information together with content merged, via an AR browser in smartphones, provides a multi-modal user experience. The orientation of the camera in the smartphone serves as a trigger for an intuitive invocation of related content. This is a graceful departure from a user-mediated paging of browser content. Geolocalization enabled through GPS location and device camera orientation allow a user selection of point of interest (POI) triggers characterized by content types—localized information, such as tourist attractions, city guides, cultural themes, historical monuments, and museums. The augmentation of experience is promoted through a mash-up of different types of information, where the context is user selectable.

The user of markers provides links to information associated with user-selectable POI, which may indicate further opportunities for an augmentation—such as premium content rendering, for example, 3D viewing, or more detailed information—if available. Interaction with the information source, for navigating through the detailed information, serves to further embellish an intuitive experience. Such interactions may also be invoked on repositories of stored information for subsequent viewing. Either direct or content mash-up augmented viewing may be dynamically selectable from a user perspective on the same device, in different sessions.

Interoperability is a significant ingredient in the preservation of experience across different AR applications and mobile device platforms. Universality and consistency of experience are essential for the sustainability and expansion of AR through the use of enabling, building block frameworks.

The creation of content must allow the transport, storage, and viewing across different types of access technologies and rendering platforms. The primitives in the building block framework must serve as the foundational set of capabilities to ensure interoperability, in the design of lightweight AR applications suitable for low-power mobile devices. User-selectable filtering and customizations of AR contents allow users to carve out that only the desired content is streamed for used.

In the spirit of the experience-oriented realms in mobile service, a framework of API serves to provide intelligent linkages between the rich communication service (RCS) features and a globally distributed communities of Web 2.0 application developers—third parties. The universal exposure of RCS features encourages creative and participative contributions toward the development and integration of compelling services. The Open Mobile Alliance (OMA) Network API is referenced in the creation of functional APIs fabricated in alignment with the requirement of user–network interface (UNI)/long tail and the security framework for authentication and authorization.

Unlike fixed usage scenarios, tethered to the Internet with larger displays, interfaces, and power availability, mobile usage scenarios require adaptability in terms of wireless access technologies, latencies, bandwidth, device capabilities, small UIs, and displays. A direct mapping of ideas from the fixed to the mobile space is fraught with incompatibilities, in terms of usage efficiencies, such as viewing, browsing, and UI navigation capabilities. Apart from the form factor and UI aspects, the appropriate application of the relevant API functions is essential in the crafting and adaptation of applications for mobile usage through well-defined UNIs. The network–network interfaces (NNIs) allow interoperability across intervening networks between servers and mobile clients.

The shift toward packet-oriented services, over LTE-enabled radio access system, requires the use of IP-centric, real-time, and non-real-time applications. The packet session control of real-time voice and video calls is managed by the IP Multimedia Subsystem (IMS) layer—control layer—between the packet access and the service layers. A framework of concepts and generalized techniques is embedded in the SOA model. This model operates over the LTE access network layer to enable service creation and delivery. It allows for an implementation of loosely coupled service interactions, policies, contracts, service description, service discovery, execution, etc. The SOA model may also be adapted to process information semantics and ontology logic for transformations that discover synergies within or across domains of information. According to Tim Berners-Lee, "The Semantic Web looks at applications that enable transformations, by being able to take large amounts of data and be able to run models on the fly—whether these are financial models for oil futures, discovering the synergies between biology and chemistry researchers in the Life Sciences, or getting the best price and service on a new pair of hiking boots."

Figure 5.13 highlights the prominent entities in a Web 2.0 context, over device connectivity, where the content creator and consumer are enabled to collaborate in a flexible and interdependent service environment. Among the desirable characteristics for the functional definition of the API—for exposure toward a universal community of Web 2.0 developers—the following are significant:

■ Generic and simplified protocol bindings (e.g., HTTP like representational state transfer [REST]).
■ High levels of abstraction.
■ Enable thin clients, such as web-oriented clients.
■ Hide protocols while representing user/client behaviors.

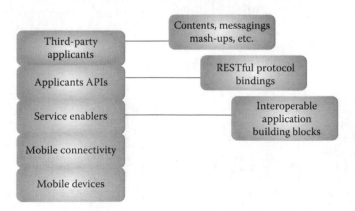

Figure 5.13   Entities in a Web 2.0 context.

- Align with the user agent (UA) model, where the user/client behaviors are represented through the UNI, native to RCS.
- Perform operations, with the supporting resources for the service layer, such as the network layer.
- The functionality provided may be selective—partial or full set.
- Preferably, the server exposing the API functionality handles the context and state information, instead of the client accessing the API functionality.
- Access to data for API functionality is limited to user accounts, subjected to authentication.

Some of the prominent features enabled for RCS through the use of REST-oriented APIs include SIMPLE (SIP for Instant Messaging and Presence Leveraging Extensions)-oriented chat, SMS (Short Messaging Service)/MMS (Multimedia Messaging Service), network address book, image/video sharing, file transfer, presence, and service capability indication.

The logical relationship between an API and a protocol binding is illustrated in the Figure 5.14.

As in the case of all enablers, both policy and privacy aspects are considered, for a third-party rendition of RCS enablers. In alignment with user privacy choices, credentials are not required to be presented to an application for its usability. The user authorizes the scope of accessible data that may be delegated to an application, and this delegation is revocable by the user.

The envisioning and creation of web-oriented network API functional descriptions are pivotal ingredients for both mobile service providers to expose the necessary information and capabilities for application development communities. This exposure promotes opportunities for innovation and expansion in the arena of application and content design. The use of well-defined, open, and REST-oriented APIs, for the RCS enabler, also reduces the time to market through reduced development cycle times, resulting from the API-enabled hiding of the mobile network information complexities, over which the applications execute.

Games are an expression of the creative and entertainment pursuits—a flavor of communications in an experiential costume. An experience creates memories—database for an evolution of new ones in a human context. User-friendly interfaces—displays, sensors, and touch—in nomadic world serve as a personal portal into game-rendered experiences. Smartphones and information transport over mobility-powered broad-bandwidth access continue to evolve in appeal and creativity, leveraging ongoing technology advancements. A catalyst in these directions is the definition of UI functions, for mobile devices, to enable the development of gaming applications, such as the OMA Gaming Service API (GSAPI).

Openness in the envisioning and in the definition of API—through collaborative and global standardization endeavors—paves new frontiers in the design of mobility-oriented applications

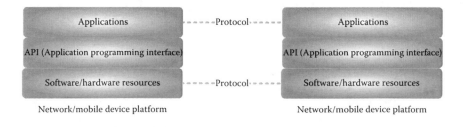

Figure 5.14 API and protocol interface orientation.

and services to both adapt to and influence a variety of service provider business models. The choices for users, in turn, allow an exploration of assorted mash-up or widget incorporation possibilities—a participative landscape that incubates a continuous innovation potential. Since the APIs hide the details of access technologies—LTE EPS, 1xEV-DO, WiMAX, Wi-Fi, etc.—portability is inherent. Further, the widespread usage of well-defined APIs ensures robustness, availability of updates, adaptability, consistent behaviors, availability, and developer support. This in turn allows service providers to sustain service quality as well as to conveniently explore new market potential through API-enabled service creations, through a harnessing of broad and diverse application development communities. Broader participation of development communities—enabled through non-access technology-specific expertise—facilitates a realization of user-specific service demands in the realm of the long tail. The organic evolution of user-specific and customizable services, in the long tail, is fundamentally a market expansion paradigm, in a distributed and decentralized direction.

5.3.3 Ties

The flexibility of the LTE SAE framework for broadband mobility connectivity accommodates access layer independence from a mobile service perspective. In turn, this decoupling of the service layer from the access layer provides for interdependence at the service layer, in terms of enabling a variety of ties that have the potential to augment the mobile service experience. These ties are manifested in a variety ways through a mining of the contextual information available to the mobile devices in temporal/spatial dimensions.

The IMS layer offers a common transport and control plane for devices or sensors that have connectivity over heterogeneous mobile access technologies. The confluence of sensors from a health and well-being perspective and mobile devices enables a variety of monitoring and remote diagnostic capabilities. These components of information may harnessed as metadata in a social context for research, education, and other virtually unlimited usage scenarios, driven by user-interest and participation profiles in various social networking groups. Other scenarios are relevant as well, such as in the case of the Internet of Things, entertainment, and public safety. Service mash-ups are enabled through SOA models using RESTful APIs and Web 2.0 design constructs.

The IMS Web 2.0 gateway, shown in Figure 5.15, serves as an interworking function for embellishing the mobile user experience through context augmentation for a variety of mobile services.

5.3.4 Ambience

E.M. Forster postulated: "Surely the only sound foundation for a civilization is a sound state of mind." A foundation that is built on experience in turn shapes the foundation.

Mobility and information together weave patterns of ambient experience. Mobile services shape the ambience of expression influencing experiential paths in connectedness and emotional well-being to serve as a significant cog in an evolving wheel of lifestyle enhancements and possibilities. Inherent in these directions is a localized ambience, where the global is bridged in a dizzying array of communication pathways, with mobility at the epicenter. Twitter—a Jack Dorsey vision—exemplifies the interdependence between the global and the local through a localized ambient context—an evocation of feelings, which is at the heart of human communications.

Technology and information evolution in this realm is a participative venture, where the serendipitous insights appear and thrive, promoting a culture of innovation. This spontaneity

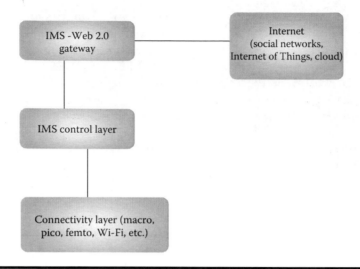

Figure 5.15   Context awareness—IMS Web 2.0 gateway mediation.

is paramount in the orbits of inspired innovation and transformation of thought toward paths less trodden—open to shift propelled via widespread collaboration across traditionally disparate domains of knowledge and expertise. Implicitly, these trends demand a balance between scale and uniqueness at the user level—one size or one solution does not fit all, much like in life.

We now know from clinical case studies that music can affect—in a variety of ways—human neurological, psychological, and physical functioning in areas such as learning, processing language, expressing emotion, memory, and physiological and motor responses. How the brain perceives and processes music also differs depending on whether or not one is a musically oriented. The effects of music raise intriguing questions about both early brain development and brain plasticity later in life.

Service composition and creation promote a quest for excellence in expression and experience through the collective wisdom of the crowds, who serve as both producers and consumers of information. Each cycle of innovation is a plateau for exploration and discovery in the thought landscape. It symbolizes a changing knowledge space—a fusion of ideas crafting form and function in mobile multimedia services. It encapsulates an abundant variety of ambient information through an articulation of elemental concepts, within SOA—a stage for individual and collective experiences that bear the potential for simplifying and enhancing life's journey.

Innovative thinking, in conjunction with the ubiquity enabling capabilities of standardization, is vital for a realization of the products and services in the mobile information age. The complexities of information creation and delivery are increasingly a function of the linkages across information components. The strategies and the architectural frameworks that are necessary to manage and leverage information complexity may be broadly categorized in terms of centralized and distributed approaches. In the mobile information context, the intelligence and information are resident at the points of creation and consumption. At the creation point, a vision guided by lateral thinking is a paramount tenet, for the composition of value, in information complexity. It is an art form—where technologies and human aspects blend. At the consumption point, a rich variety of attributes such as appeal, experience, and memory shape the value significance. These attributes have the potential to be mined in the form of ambient information that may be available

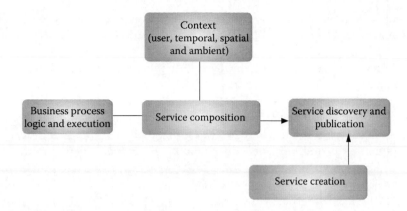

Figure 5.16   Ambient context for mobile service composition and creation.

for mobile services to harness, in the vicinity of the mobile device through temporal and spatial dimensions. The nature of service composition is shaped in the procedures that reflect the business requirements for a formalized description of information handling. A model to harness this information, for service composition and creation, through discovery and publication is shown in Figure 5.16.

5.4 Transformation

Events and states described in the language of information contain the potential for shaping a multitude of experiences. The mobile Internet is a fabric for purveyors of these experiences through the twin components of connectivity and service. These experiences in the human dimension are reflective of an archetypal transformation of the connectivity and service that defines an evolutionary mobile Internet.

The transformational aspects in connectivity and service mirror an interdependent paradigm, pervasive in the natural phenomena in the world around us. It is returned to the memory of the way of nature—a distribution of information, in an interdependent dance. The mobile Internet, along with its predecessor—the fixed Internet—projects a nonhierarchical paradigm of interlinked information—stories, documentaries, research, news, entertainment, education, health, leisure, shopping—available anytime, anyplace. It is a tapestry of a global locality—a virtual marketplace—that transcends the microcosms of boundaries that have existed through centuries of human civilization. Yet the evolutionary mobile Internet model embraces the timeless pursuits of human beings—tales, trade, and travel—which are inherently interdependent. These innate motivators serve as the pillars of transformation that shape a vision toward new archetypes for technologies that form the building blocks of mobile evolution.

In his book, *The Third Wave*, Alvin Toffler emphasized the notion that the culture inherited from the progress of industrialization created an artificial dichotomy between ourselves as producers on the one hand and as consumers on the other hand. This polarization created an artificial rift in "the underlying unity of society, creating a way of life, filled with economic tension, social conflict, and psychological malaise." The artificial divisions or silos are incongruent with the nature and potential of information, which is inherently malleable and must be allowed to flow openly for exploration and innovation. It is the open flow of information, coupled with purposeful motivators, that promotes cooperation and collaboration toward innovation in the experiential domain of mobile communications.

Openness directs the sails of evolution toward a nonhierarchical paradigm, which implies both distribution and autonomy, in the next-generation mobile architectures. This is a departure from the centralized command and control models that were pervasive in technology evolution since the Industrial Revolution. These models worked well for products and services that were relevant in geographically disparate localities, shaped by the dominant local choices and preferences. As market boundaries and understanding expanded, together with technological advances, boundaries inevitably faded. These shifts were especially relevant in the realm of information, which by nature is virtual and rooted in the realm of imagination, which has no boundaries. The fixed Internet and its universal appeal established the significance of its underlying technology models to delve into the related insights and enormous possibilities in an untethered world, at the dawn of the third millennium. The advent of third-generation mobile systems marked this dawn of transformation of architectural models, innovation, and standardization.

Interoperability is the attribute that serves as the bridge across disparate islands of innovative and evolving technologies. It transcends the artificial separateness of services, where the attraction of familiar metaphors associated with a silo of service, such as iTunes, Facebook, Skype, among others across numerous arenas, including customer loyalty programs, while beneficial hinders the exploration of virtually unlimited possibilities across these artificial boundaries. Possibilities could spawn a plethora of innovative benefits through collaboration and cooperation. *Jingle* is an exemplification of an interoperable media (voice and data) signaling protocol standardized by the XMPP Standards Foundation (XSF), where XMPP (Extensible Messaging and Presence Protocol, RFC3920) is standardized by the IETF. *Jingle* is standardized as an XMPP extension, developed by Google and others within the XSF as an open VoIP standard.

Since XMPP embraces interoperability and openness, it is capable of bridging different silos of capabilities through the definition and specification of extensions to XMPP to serve new requirements. For example, Jingle allows widespread collaboration and cooperation through its open nature, which is a significant distinction from the widely popular Skype that uses proprietary technology models—protocols and procedures. While the latter is popular, it is inherently limited in terms of broad collaboration and cooperation, which are an imperative for evolution and expansion. The benefit is that the new extensions to an open and interoperable protocol inherit the same attributes as the foundational open protocol. The result is a preservation and proliferation of a consistent user experience in an evolving mobile communications ecosystem. It is a model that invites a variety of value-chain relationships, which are broad and incremental, rather than being confined to a walled garden. In the experiential realm, a shift from silos of products and services targeted for captive users to open and interoperable products and services, augmented with user autonomy through interdependent relationships, is a vibrant pathway for a sustainable mobile evolution.

The LTE SAE is an archetypal template that embodies the spirit of interdependence through open interfaces that are nonhierarchical. This architectural model enables flexibility in terms of decentralization and autonomy to promote new vistas for implementation and deployment in the pursuit of ubiquitous, broadband mobile connectivity and service. These tenets provide a rich fabric that evokes innovation, ideas, and imagination for new horizons in mobile technology–mediated human communications. The novelist William Gibson mused, "The future is already here—it's just not evenly distributed."

5.4.1 *Symphony*

The experiential nature of mobile communication demands a symbiotic relationship across the different actors of mobile access, mobile service, and business models. It is a relationship of inter-dependent and customizable ingredients. This implies a derivation of the conceptual nuggets that create a fabric that espouses assorted and evolving aspects of interdependence. Context provides purpose and richness to the consumption of information.

Content and transport—distinct in essence—collaborate to simulate a virtual experience for mobile users. The content in its various forms represents the context, together with a presentation of the information, which is either being produced or consumed. Software and hardware innovation continues to lower the thresholds for both mobile access and mobile service. In this evolving paradigm, users continue to become an integral part of the changing fabric of mobile communications, in the inexorable human heritage of ideas and perspectives. It is a cycle that blends, promotes, and organically advances the nature of mobile communications.

The traditionally disparate layers—where specific protocols operate in the processing of information—while necessary for structure and logic, and their participation across their designated boundaries have increased in relevance. For example, from Web 2.0 service mash-up paradigms, such as Wikipedia to the heterogeneous mobile access system architectures, hierarchy continues to dissolve into nonhierarchy and autonomy. This shift is applicable and is beyond access and service, into the realms of business and art, to foster a cross-domain, collaborative model.

The levels of cross-domain collaboration, in the next-generation mobile communication system, are essential for not only the advancement of experience but also for a sustainable progress of creative ideas that fuel innovation. In this sense, the collaborative model facilitates an organic environment, fertile for leveraging ideas in a decentralized manner—not limited to established research and development institutions—with the user being a vital actor in the wave of creative ideas.

From an access technology perspective, small wireless coverage zones enabled through the use of Wi-Fi and Bluetooth have enabled the both the expansion and the role of mobile communications, beyond the rigid boundaries of hierarchically established and accepted models espoused for decades, in the voice-centric world of mobile communications. These models of yesteryear, while significant, have served their purpose as a stepping stone toward the inherent nature of human communications—inherently experiential, multimodal, contextual, and distributed.

In Leiden, Netherlands, there is Wireless Leiden,[*] which is a wireless community network. It embodies the idea of a volunteer-based, distributed access network. The access network is implemented and managed with support from volunteers and sponsors, using open source and standardized technologies. This encourages further contributions from individual in such communities through an experience of the benefits, in the quest for nomadic and seamless communications. Users experience a location independent—home, school, library, business, café, etc.—access to information and services, in a distributed array of wireless coverage areas.

The participation of users in the creation and consumption of information is an element of the experiential fabric of communications. P2P transaction of information is an example of such transactions over an assortment of wireless access technologies (e.g., E-UTRAN, EV-DO, and Wi-Fi). This type of virtualization[†] allows innovative and decentralized possibilities beyond the spatially confined segments of the underlying physical infrastructures. Examples of virtualization

---

[*] Verhaegh, S., "From simple customer to warm user: Who cares about the maintenance, of community innovations?" *Observatorio*, vol. 1, no. 3, pp. 155–184, 2007.

[†] Stephanos, A.-T. and Spinellis, D., "A survey of peer-to-peer content distribution technologies," *ACM Computing Surveys*, vol. 36, no. 4, pp. 335–371, 2004.

include content transactions (e.g., Tribler, BitTorrent, and Skype), content search (e.g., Google and Bing), and content mash-ups (e.g., Wikipedia), which represent some of the prominent and evolving indicators of user-level participation.

The user-level participation in technology-mediated communications is an archetypal and necessary shift toward a global symphony in the nature of mobile ICT. Decentralization suggests complementary and aligned changes to the policy, regulatory, and business models to foster innovation and market expansion in the proliferation of information orchestrated through mobile communications. The convenient and evolving nature of user-level participation, in an inherently mobile landscape of human endeavor, in turn influences the evolution of policy, regulatory, and business models.

While the fabric of technology-mediated elements—access and service enablers—is tangible, it is the patterns of experience that shape the DNA (deoxyribonucleic acid) of experience—both unique and universal. The fabric comes to life in these patterns that inspire and create experience. Experience is a symphony of multiplicity that transcends the fabric. It is incubated in the realm of thought (unbounded), while the fabric (bounded) is subject to commoditization. The patterns—the building blocks of experience—infuse the soul and attractiveness of the fabric, the mobile device pathway to mobile services. Experience is unbounded and subject to migration over the fabric, while the fabric itself is bounded—spatially and temporally—in the form of structured artifacts, such as devices and networks. It breathes life, character, and uniqueness in the consumption and creation of mobile services. The inflections of change in the evolving horizons of decentralization invite interdependence across the actors and roles in the mobile ecosystem for augmentations in experience. The rigid boundaries disappear in a symphony of interdependence carving new frontiers of experience in the age of information.

The role and participation of users in the mobile information paradigm hinge in decentralization, where the sources of information are distributed through self-organization principles. It portends a sea change—a departure from centralized and command—control notions of mobile network infrastructures and business models. It symbolizes an organic, user-inspired *inverse infrastructure* epitome. Examples include Web 2.0 service mash-up, P2P, and local area access, such as Wi-Fi. The ingredients of an *inverse infrastructure* motivated evolution of mobile communications include

- Self-organizing principles across mobile devices and systems
- User-influenced design innovation
- Incremental investments for innovation
- Leveraging of interdependence for widespread collaboration
- Distributed mobile access networks
- Promotion of service-oriented directions for market expansion
- Coexistence of heterogeneous of wireless access technologies—macro and local access network
- Incubation of a multitude of service providers to harvest the long-tail opportunities
- Shift from the burdens of the silos of mobile service to partnerships that optimize user experience

The traditional hierarchical and top-down mobile communication architectures must yield to autonomous and distributed architectural models. This reflects a shift in the standardization approaches, accepted as the norm in the centralized circuit-switched mobile networks to one that embraces both autonomous and distributed architectural styles that fit well in the evolving

IP-centric mobile ecosystem. A natural consequence of autonomy and distribution is that it inspires both user-level participation and imagination on a global scale. Principles of self-organizing mobile networks bolster and expand the proliferation of decentralized, nonhierarchical architectures for mobility and transport. Policies and regulations, in the light of these impending trends, and the enormous market potential must serve as a catalyst that fosters an organic evolution of the mobile Internet, in the ageless saga human communications.

The inference of the nuggets that hold the confluence of connectivity, service, and human factors in the changing horizons of the mobile ecosystem requires an awareness of the interdependent nature of the creation and consumption of information. The computational nature of the underlying technologies provides clues into the nature of thought that inspires innovation through imagination and inquiry. Computational thinking—ensconced in a rational and logical process—is geared toward problem solving, regardless of the subject matter. It seeks to understand the requirements, context, and the environment within which a solution is sought. This type of thinking reveals opportunities to explore the bedrock of nuggets that appear as patterns across seemingly different categories of either intradomain or interdomain classes of subject matter–related information—technology, arts, medicine, business, marketing, etc. Thought processes that reveal the invisible and intangible underpinnings of subject matter are significant in the imagination space that is at the core of the information age. Some of the aspects that are explored in the computational arena—central in the realm of mobile communications—include processing resources, algorithmic techniques, power consumption, processing latencies, user-to-device interfaces, not just as individual considerations but as interdependent entities that dance together to provide an ocean of user-oriented experiences.

The analytical nature of technology-oriented thinking and the artful elegance and fuzziness of experience must blend in the identification of the unchanging patterns applicable in the changing and innovative designs that are pivotal in the propulsion of mobile evolution. It is an evolution that is shaped by natural usage metaphors for mobile services, where the enabling technologies disappear in the elegance of an intuitive and lifestyle-enhancing experience. This is particularly relevant in a sea of diverse user choices and preferences, where connectivity and service innovation chart new courses through the power of possibilities embedded in interdependence.

The mingling and rearrangement of expected patterns across contextual boundaries creates frizzante and syncopation—a foray into leveraging interdependence—which like the enrichment of music enhances the service experience through an evocation of different emotions. The human brain is impressionable with respect to the elements of surprise and unpredictability. These observations provide clues into the nature and benefits of harnessing interdependence across the structured layers of connectivity and services at the edges of mobile evolution.

The extraction of the unchanging and repeating patterns/characteristics is elemental within the standardized building blocks of LTE SAE vision of connectivity for access and service experience. A fabric of these patterns that motivates the foundational architectural tenets includes the following:

- Implications of a shift from circuit-switched islands of hierarchy to packet-switched oceans of distribution.
- Aggregation of wireless highways through enhanced resource allocation and scheduling techniques to suit connectivity and service demands.
- Enhanced modulation techniques to preserve link robustness and spectrum utilization efficiencies and to adapt to dynamic mobility conditions for an establishment of the necessary wireless transport QoS targets for a seamless and consistent service experience.

- Voice is just another instantiation of data, with specific attributes, among a virtually unbounded data representation of information types, with customizable attributes—voice, video, mash-ups of content.
- Oceans of distribution are applicable in directions toward a cloud model for connectivity and service.
- Potential for a blurring of artificial silos for evolutionary business models through cooperation and collaboration.

Within the embodiment of interdependence, in the evolution of mobile communications, the notion of decentralization is a dominant one and has profound ramifications in terms of compatible and relevant policy models. The policy models must adapt to the nature of decentralization, which involves the arenas of regulation, governance, and usage. The user's role in the enactment and in the enforcement of policies relative to those of a serving network becomes more collaborative in a decentralized mobile ecosystem. Cooperative partnerships across all the segments in the mobile connectivity and service space—business, communities, and individuals—are vital in the effective leveraging of efficiencies and market expansion, where there is an alignment across technology innovation and the associated policies that enable diverse adoption scenarios.

The liberalization of access scenarios, to a virtually boundless array of services, requires innovative interdependent behaviors across intellectual property assets—content such as music, video, and files—connectivity technologies, subscription profiles, policy profiles, and user profiles. For example, content from providers, which are subject to Digital Rights Management (DRM), may be available DRM-free, based on interdependent relationships across connectivity providers, subscription profiles, policy profiles, and user profiles. These interdependent relationships could be modeled in virtually unlimited flavors, in the mobile paradigm, hinging on the partnership agreements in the value chain. This interdependence is crucial in the evolution of mobile ecosystem, where disintermediation in a simplistic bilateral sense—between end users and an application provider—will be inadequate, as a result of the nature of the mobile wireless link, in addition to a multitude of other factors, such as managed customer care, service differentiation, and service customization at the user level.

The traditional notions of service are fraught with limitations, driven by the belief that a rigid template for the rendering of services is adequate. As the nature of information—its extreme malleability, usage scenarios as distinct as individual human experiences, dynamic and unpredictable—continues to evolve and be adopted in the mobile landscape, the rigid and top-down strategies for rendering services will continue to diminish in value, since such strategies are highly prone to commoditization. The progression of mobile evolution, innate to its untethered and decentralized usage nature, demands a consideration of interpretations and ideas from complex adaptive systems (CAS), which has parallels in quantum phenomena.*

Mass market dimension—What does it mean? How does it play a role in evolution? Why? This section is about the ways in which mobility serves as a catalyst for social and business change and evolution, in a nomadic, mobile world.

From success to significance—a change that is essential in evolution of mobile broadband communications, a platform for change in human existence, in the theme of communications, which is central to the human ethos. The strategies, policies, governance, regulations, etc., that shape this landscape must have a profound effect across humanity, its behaviors, and evolution

---

* Gell-Mann, M., *The Quark and the Jaguar: Adventures in the Simple and the Complex.* New York: Freeman & Co., 1994.

since it is related to the instinctive need to communicate—inform, educate, evangelize, disseminate, etc.

Command and control models cannot create, propagate, or sustain. It is a loosely defined framework, where choices for innovation are elevated beyond artificial constraints that are unnecessary in the information creation and consumption archetypes. Decentralization and distributed architectural models and implementation provide the autonomy that is reflective of mobile communications—human communications.

An example would be the framework of traffic lights to manage the flow of traffic—serving exactly one and only one purpose—not open for interpretation and is not applicable to choice or free will. Only a few such simple frameworks essentially manage and utilize a resource that may have limited availability, such as a road for transport—which cannot physically transport an infinite number of vehicles. But people have a free will and a choice to drive anywhere. The framework does not intrude or modulate the free will or choice—so it is with services that are indeed user centric—iconizing.

Information is transforming. It is a requisite ingredient toward understanding and value creation. It imbues a direction, for the management of complex interlinkages, which engenders widespread lifestyle conveniences. Mobile communications provides the fabric, for leveraging the creation, accessibility, and the availability of value propelled through innovative thought. Transcending geographic boundaries, it is a fabric with an innate appeal and adaptability, across diverse social and cultural patterns—a global fabric.

Suggestive of *inverse infrastructures*, which are spontaneous in nature, the mobile paradigm, with the user as the primary actor, interacting in a sea of information, thrives where the services rendered adapt dynamically to user-oriented behaviors, untainted by rigid or preordained procedures assumed by a service provider. In the vast and changing seas of information, the user is empowered to craft choices and participate in the usage and contribution of information in ways that cannot be predicted via the rigid methodologies that worked well with voice-centric services, where value was solely measured in terms of the availability and the quality of connectivity. The shift to a generalized model, where data of any vintage (voice, video, files, or any combination of these) represent information, has resulted in a much richer and broader perspective of mobile communications that is reflective of the human dimension—intrinsically a multifaceted, interdependent information processing model.

5.4.2 Aesthetics

Among the various elements of mobile evolution, the decentralization effect—an artifact of the creation and consumption of information via technology-mediated mobile devices—is essential in the shaping of the evolutionary trends. In these new vistas, deeply rooted in the human journey enveloped in the natural world, the role of form takes on an unprecedented significance in addition to the technology-mediated elements of function. It is here that the lessons and findings in the natural world color and shape innovative thinking. Aesthetics not only plays the role of attractive and appealing forms but also in the intuitive adaptations of familiar and natural metaphors inherent in human interactions.

Benoit Mandelbrot's observation of nature's replication of self-similarity in the works of nature gave birth to the mathematical notion of fractals—from the Latin word "fractus" implying fractured. It is nature's enabler of beauty and attractiveness, in a symphony of aesthetics, where each building block can be replicated to create a whole. Jagged edges repeat themselves to form shapes of leaves, coastlines, and mountain ridges to cite a few creations. The model reflects spontaneity and randomness at the experiential level to evoke aesthetic appeal uniquely and commonly compelling to the observer. These ideas from the pages of nature's masterful creations provide a wellspring of guiding principles in innovative design styles that combine standardized building blocks to promote aesthetic appeal in the rendition of mobile device, connectivity, and service.

The heritage of large and centralized systems prevalent require yielding to the natural transformation toward smaller, decentralized systems with increasing levels of autonomy. Such a shift appeals to the evolution—creation and consumption—of information, which lends itself to enormous possibilities resident at the user level, where ownership is localized with a global reach. Examples of these shifts are illustrated in the attractiveness of Wikipedia—distributed repositories of information service and distributed networks of wireless connectivity to name a few among a virtually unbounded space of innovative possibilities. The localization and autonomy are naturally suited for customization and increasing levels of quality through the fabric of convenience, aesthetics, and function. The aspects of form are exemplified in a vast array of renditions, such as the personalization and presentation of a user home page, where form meets function in user-customizable layouts aligned with familiar metaphors and styles that appeal in different ways at the resolution of the individual. Among the popular renditions and application of such possibilities, iGoogle* is an illustration of an observable in the customizable and intuitive widget-driven aesthetic layouts of a personalized web page. An example of localization is an inversion of infrastructure—the user leverages aesthetics, convenience, and function, comfortably adapting to the changing and evolving needs and understanding in the human journey.

Self-organization is intrinsic in the decentralized paradigm, where policies enable autonomous behaviors across the different segments of a system. The intuitive and adaptive nature of self-organization lends itself to appealing renditions applicable in a wide range of scenarios such as in the design of user interfaces, user-profile-driven service suggestions, connectivity choices, self-provisioning, and self-healing, among others. The implicit elegance of decentralization lies in its natural propensity to be small—low overhead, teleocratic—unencumbered by a complex and stifling centralized hierarchy of rules, policies, and regulations that are implicitly at odds with the objectives of adaptive and unbounded levels of service customization. It is indeed a localized paradigm naturally designed to serve locally, regionally, or globally leveraging a sprawling fabric of the ICT landscape, with information and user mobility at its epicenter. The aesthetic possibilities in this confluence of technology and art are virtually limitless in the imaginative creation of services that hone in on catering to the multifaceted, multimodal nature of human communications—an experiential dimension.

In the web of information, ebbing and flowing through the veins of mobile connectivity and service, seamlessness is a foundational ingredient in the shaping of experience. Within seamlessness, the ability to interact and consume information requires aesthetics as a centerpiece in the shaping of experience. The foundational aspects of mobile connectivity and service—crafted in the adaptive and flexible architectures and protocols of LTE-EPS, in concert with the service enabler and APIs—are both essential and catalytic to foster value creation while averting disintermediation of the segments between a client and a server. Disintermediation, just like any other capability in isolation, has limited knowledge and flexibility to groom and customize a service for an optimal experience. The probabilistic and dynamic nature of the wireless link requires cross-layer cooperation for enhancing the aesthetic possibilities that render a memorable and impressionable experience.

---

* http://www.google.com/ig.

The mobile service expansion portends a vast landscape unbounded possibilities shaped in the realms of imagination and incubated in the cradle of usage behaviors and related experience. The nomadic aspect of communications, which adds a transcending of the space dimension in addition to the time dimension, complements the nature of usage behaviors,* which elucidates the uniqueness of mobility in human communications. With the usage behaviors, the mobile device serves as a portal for access to connectivity and service, which naturally then invokes a personal attachment to the look, feel, and familiar metaphors associated with the mobile device, from an experiential perspective. The mobile device in its role as a personal accessory takes on the role of a *shoe* as compared to a *crutch* that must be endured for connectivity and service. Here, function blends into form, in the innovative evolution of smartphones and tablets. It is an interdependent dance in the making of a mobile *shoe* that fits comfortably to the desires and needs at the individual level, the connectivity, and the service. The range of innovative thought spans technology designs, business models, and marketing campaigns to shape mobile connectivity and service with distinctive differentiation. Through the portals to connectivity and service integration, the aesthetic attributes of a mobile device are a complementary facet, in arena of social and fashion† preferences, which are inherently of personal nature. The stylistic‡ customization choices, available to the user, reveal an enhancement of personal engagement and attachment to the mobile accessories that serve as sensory portals to the experiential realm of information in its various incantations.

The intimate association between the mobile device and its role as a portal to connectivity and service places mobile evolution at the heart of sociological and psychological aspects, which are central to the theme of human communications. This association is unique and personal unlike other forms of technology-mediated communications, such as desktops and, to lesser extent, laptops, which are space bounded, via wires or local area wireless coverage without any seamless handover for wide-area connectivity. The availability and use of connectivity and service—anytime, anyplace—renders a personalized relationship between the user and the mobile device—a personal accessory. The perception of this accessory as a sensory portal to the unbounded field of communications and interaction, motivated through the expression of assorted information modalities—sounds, pictures, touch, and words—establishes the smartphones, tablets, and the like as customized personal possession. The aesthetics—appeal and intuitiveness—of these mobile device categories influences the sensory experience, which in turn reveals vestiges of new value creation promoting opportunities for innovation and market expansion, with the user as the leading actor on the stage of mobile evolution. The adoption of mobile devices in their various emerging incantations depends in their adaptability to the profoundly distinct user-level styles and preferences.§ A metaphor that resonates with the personal nature of an *experiential distribution* of mobile connectivity and service is "Kännykät"—a term in Finnish for a mobile device and literally implies "handy." It is a compelling symbolic representation in the human dimension.

The *handy* nature of the mobile portal for professional and personal interaction coupled with access to abundant vistas of information is suggestive of a strong correlation between a perception of the mobile device, where both the object representation and its usage implications are

---

* Van der Heijden, H. and Junglas, I., "Introduction to the special issue on mobile user behaviour," *European Journal of Information Systems*, vol. 15, no. 3, pp. 249–251, 2006.
† Katz, J.E. and Sugiyama, S., "Mobile phones as fashion statements: Evidence from student surveys in the US and Japan," *New Media Society*, vol. 8, no. 2, pp. 321–337, 2006.
‡ Blom, J.O. and Monk, A.F., "Theory of personalization of appearance: Why users personalize their PCs and mobile phones," *Human–Computer Interaction*, vol. 18, no. 3, pp. 193–228, 2003.
§ Belk, R.W., "Possessions and the extended self," *Journal of Consumer Research*, vol. 15, no. 2, pp. 139–168, 1998.

synthesized cognitively by the user. The notion of aesthetics is pivotal in human–computer inter-action (HCI)* as well as in the realm of user experience† conceived and manifested in sensory and emotional fabric through a variety of information interfaces—mobile visual, audio, and haptic pathways.

As in the cognitive dimension, customization of the look and feel of the mobile information interfaces, to suit choices and desires at the individual level, is a prominent facet of aesthetics that fosters a favorable perception of the mobile device and its *handiness* as a personal accessory to the rich portals of multimedia information transactions. In this context, an individualized perception of beauty affects the HCI modality in mobile communications. The flexibility to customize these interfaces serves as a measure of attachment to the mobile device as personal accessory.

The degree to which a specific rendition of a mobile device—smartphone, tablet, and the like—blossoms as a personal accessory depends on a perceived blending of the look and feel of the mobile device with the richness of ubiquitous connectivity and service that is available for lifestyle-enhancing experiences—personal and professional. The elements of aesthetics play a dominant role in a convergent communications experience, where the mobile device is both a personal acces-sory and an information portal simultaneously.

The aesthetics that propels a convergent—device, connectivity, and service—mobile commu-nications experience hinges on the appeal of the HCIs where the graceful and naturally intuitive interfaces augment the experience. The user's aesthetic reaction (UAR)‡ is a measure of both the form and the design style of the mobile device HCIs.

The significance of the mobile device as a personal accessory is deeply tied to lifestyles, a sense of handiness, self-expression, beauty, intuitive appeal, ubiquitous connectivity, and service, which represent a symphony in aesthetics that is vital to a compelling *experiential distribution*. The elements of aesthetics lead to an emotional attachment to the mobile device, where its absence equates to "cannot do without it."

The opportunities for a variety of segmentations are virtually unlimited in the rendering mobile device designs, connectivity, and service, for aesthetic appeal, as well as in the crafting of new business model that hinges on shared value, through a distribution of revenue across the value chain. Aesthetics then becomes not only a unique component in an attractive *experiential distribution* but also fundamental in the symphony of ingredients that are elemental in mobile evolution.

Mass communication research in uses and gratification (UG)§ theory, with its inceptions in radio and television, has revealed a variety of motivators at the user level in the virtual corridors of mobile communications. The motivations for user attachment to mobile device, connectivity, and service may be classified into functional, hedonic, and social categories. Innovation, in the design of the mobile device, connectivity, and service, which considers a knowledge and understanding of these categories of motivators, is vital in promoting an aesthetic appeal in a sea of *experiential distribution*.

---

* Picard, R., "Affective computing for HCI," in *International Conference on Human–Computer Interaction*. Munich, Germany: Lawrence Erlbaum Associates, Inc., pp. 829–833, 1999.
† Norman, D.A., *The Design of Everyday Things*. London, U.K.: MIT Press, 1998.
‡ Tian, L., "Aesthetic user experience and apparent space dimensions," Dept. of Industrial Design, Huazhong University of Science and Technology, Wuhan, P.R. China, 2010.
§ Eighmey, J. and McCord, L., "Adding value in the information age: Uses and gratifications of sites on the world wide web," *Journal of Business Research*, vol. 41, pp. 187–194, 1998.

The functional considerations involve capabilities, such as information search, shopping, travel, exploring, research, and education, which are specific tasks,* extrinsic in nature. The fulfillment of an intended task is a positive affirmation of the function provided and is an element of experience. This influences the adoption[†] of the technology that provides the function.

While the functional consideration provides extrinsic value, the hedonic motivator is intrinsic in that it fosters a multifaceted experience, crafted in emotion, memory, and consciousness, through entertainment, fun, and leisure. Hedonic usage scenarios are spontaneous, such as browsing or imagination, where the intention for use is independent of any specific task. It is a self-motivated type of usage, where the *handiness* of the mobile device provides a natural pathway for spontaneity in the virtual world. Studies[‡] have shown that the enabling of spontaneity in usage is a catalyst in the adoption of the mobile device, connectivity, and service.

The social element is deeply rooted in the human ethos and is a natural motivator in the propagation of mobile communications. Social interactions[§]—relationships, networking, conversations—are an extrinsic type of motivator, which affect experience indirectly through the quality of the mobile device interface, connectivity, and service.

The motivators—independently and interdependently—shape the *experiential distribution* and serve as insights into new horizons in the rendering of mobile device, connectivity, and service. The Second Life project reveals a plethora of perspectives and is an exemplification of these motivating considerations in mobile evolution—anytime, anyplace virtual presence.

While the *experiential distribution* is characterized by the sensory portal in concert with the quality, tenor, and texture of mobile connectivity and service, the mobile accessory in its role as an intimate possession is an integral portion of a holistic communication experience, and it is distinct in that it provides a localization of aesthetics in terms of form and function. The mobile connectivity and service provide the aesthetics of experience through the communicated content and context transactions.

The interdependence across usage behavior, cognition, and emotions shapes the *experiential distribution* in the horizons of mobile evolution. Here, aesthetics[¶]—symbols of beauty, elegance, and appeal—in a holistic sense is a compelling factor in user's attachment to the mobile device, connectivity, and service. Perceptions of beauty, elegance, and appeal have a strong influence on the portal—mobile device—to connectivity and service. The *handiness* of the mobile device naturally evokes an emotional attachment[**] to the portal, where the related aesthetics is a power catalyst, where the functional capability is assumed and is not a differentiator of perception.

---

* Babin, B.J., Darden, W.R., and Griffin, M., "Work and/or fun: Measuring hedonic and utilitarian shopping value," *Journal of Consumer Research*, vol. 20, no. 4, pp. 644–656, 1994.

† Sangwan, S., "Virtual community success: A uses and gratifications perspective," in *Proceedings of the 38th Hawaii International Conference on System Sciences*, Honolulu, HI, 2005.

‡ Sheehan, K.B., "Of surfing, searching, and newshounds: A typology of internet users' online sessions," *Journal of Advertising Research*, vol. 42, no. 5, pp. 62–71, 2002.

§ Stafford, T.F., Stafford, M.R., and Schkade, L.L., "Determining uses and gratifications for the internet," *Decision Sciences*, vol. 35, no. 2, pp. 259–288, 2004.

¶ Hassenzahl, M., "The interplay of beauty, goodness, and usability in interactive products," *Human–Computer Interaction*, vol. 19, no. 4, pp. 319–349, 2004.

** Rafaeli, A. and Vilnai-Yavetz, I., "Emotion as a connection of physical artifacts and organizations," *Organization Science*, vol. 15, no. 6, pp. 671–686, 2004.

5.4.3 Resonance

The experience of connectivity and service shapes the associated attractiveness, of these twin pillars of mobility, colored by usage behaviors, interests, choices, and lifestyles. The perception and the utility of the Internet take on a natural orientation through mobility, which is a reflection of an unchanging human heritage in a changing technology-mediated world of mobile communications. While connectivity in the mobility paradigm—an untethered time—space communication conduit—is implicitly aligned with the human social attribute, it is the service that offers boundless horizons of evolution in the realm of imagination. It is the elixir that captivates through the vivid creation and consumption of multimedia content, realized in a fabric of interconnection and interdependence. Service holds the keys to unlock opportunities in an unbounded space of innovation and evolution, beyond the plateaus and silos of commoditization. Differentiation and market expansion thrive, where users are enabled to spontaneously infer and inspire mobile evolution, incubated in an *experiential distribution*.

The *on-demand* nature of connectivity and service, coupled with the ease of *anytime, anyplace*, is a fundamental note in the adoption and proliferation of mobile connectivity and service that distinguishes *mobility* from the *fixed* modality for access to interaction and information. Among the dominant themes of interaction and information services are real-time entertainment—P2P/broadcast streaming or buffered audio/video—social networking, messaging, and conversations.

The possibilities in the mobile service arena are virtually infinite over mobile connectivity—embellished through the flexibility and the distributed nature of the LTE EPS architectural framework—a realization of the concepts explored in the LTE SAE framework vision—with its interworking capabilities, across an assortment of macro and local area access technologies. Operational support systems, business support systems, and information traffic management capabilities provide intelligent foundations to carve and customize a rich service experience, at the individual user level—an affirmation of one-size-does-not-all—while simultaneously promoting market expansion and revenue potential.

Mobile connectivity and service, inherently as distributed as the producers and consumers of information communication, offer a range of value creation and options—from a few prominent mainstream capabilities to unbounded niche possibilities. Voice communications are representative of a prominent service, while exotic mash-ups of content customized to suit individual choices that shape niche models of value. Information circumvents the traditional barriers and boundaries that encumber storage, distribution, and consumption of materials in the physical world. Its virtual nature allows incremental possibilities of customizing connectivity and service, in a distributed fashion, without adversely affecting cost.

The mobile Internet is rooted in decentralization, both from a technology perspective and from the usage paradigm view. Content sharing capabilities—leveraging decentralization—such as Kazaa pioneered by Niklas Zennstrom, who later conceived a globally ubiquitous distribution of conversations in the form of Skype, shed light on the enormous possibilities afforded by Internet technologies. Packets of information—voice, video, or data—traversed autonomously from any source to any destination, through a labyrinth of globally distributed routers and applications servers. This reflected a transformation of the telephony service paradigm, which like any human endeavor in recent history—monarchies, bureaucracies, and large corporations—reflected a movement away from a centralized command–control model. The shift continues to shed light on the possibilities in the evolution of the mobile ecosystem, in ways that demand a profound change in architectural style, design, implementation, value creation, business models, and usage. The changes usher an era of incremental and interdependent value creation and consumption.

The power of decentralization is as primal as the origins of life in the universe. As this model proliferates, changes, and evolves in the world of nature around us, it is central to the human ethos. The search for information, its creation, expansion, and consumption, elucidates and reflects this theme in our daily lives. The Google search framework plays this theme through decentralized contributions from users, who serve as cocreators in this ceaselessly evolving universal symphony. A self-sustaining and largely self-audited system of information grows abundantly out of this open and universal enterprise. The markers of information quality appear in the form of the frequency of visits associated with an information topic. In this model, the creator, combiner, collaborator, and cooperator dance spontaneously to direct and produce endless symphonies—rich in variety—evoking diverse and universal participation in information's harvest. These waves of change are compelling considerations that must be absorbed in the ideas that shape mobile connectivity and service.

In these corridors of thought, there is a resonance across the actors that influence and are influenced in a growing and changing harvest of possibilities propelled through the texture and tenor of information in a nomadic world. Figure 5.17 depicts a model of actors with roles that resonate with one another in a dance of ideas and experience that weave an evolving fabric of mobile connectivity and service.

The creator and the cooperator in Figure 5.17 are interchangeable in a user-centric world, where institutionalized centers of expertise and understanding of subject matter fade into a continuum of numerous localities—individuals. The information age reveals a landscape of unbounded learning potential through the power of distribution, where mobility is a natural catalyst in the propagation of information. Openness emerges as organic constituent in this field of evolving horizons, together with interoperability, and together they serve as the twin engines of dynamic change and evolution. It is a feedback system that moves experience and consciousness—in standardization, innovation, and implementation—toward a self-managed, self-correcting, and self-optimizing world.

The relevance, applicability, and compatibility with the nature of the information age, in its confluence with the human ethos, highlight significant clues in the foundations of architectural styles and concepts associated with mobile service and connectivity—foundations steeped in innovative perspectives. The LTE EPS embraces the notions of decentralization in its building

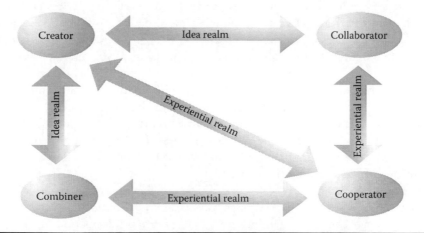

Figure 5.17 Resonance in mobile connectivity and service.

blocks of mobile connectivity, to allow a variety of implementations, together with interoperability across a variety of different connectivity technologies. The use of carefully selected choices of open interfaces and procedures for connectivity promote a symphony of possibilities that espouse a resonance across the creator, combiner, collaborator, and the cooperator. Exemplifications of the actors are as follows: creator, network access provider; combiner, network broker, across different network access providers; collaborator, value-added connectivity services (e.g., traffic and performance measuring entities); and cooperator, participative users, providing connectivity quality indications to an access network.

The guiding principles of mobile connectivity technologies, in the proliferation of information, mirror a localization of connectivity while providing a globally ubiquitous access to information. These principles reflect the four elements—creation, combination, collaboration, and cooperation—of a resonance across an unfolding landscape of connectivity and service. The LTE-Advanced connectivity system—radio and core network—embraces these elements through a distributed architectural style and radio spectrum aggregation models. Interoperability, implanted in an evolving suite of procedures and protocols pursued through global, collaborative, and cooperative standardization endeavors, espouses a resonance between mobile connectivity and service. The flexibility of bandwidth and quality of the mobile wireless information highways, in concert with the IP gateways in the LTE EPC, marks the advent of possibilities in multimedia services, created in the limitless realms of imagination. The architectural styles in the EPC embody the enabling of a distribution of logical functions that define and enforce policies, charging, provisioning, configuration, and security in the spirit of SON principles. The mobile broadband tenet of connectivity and service—anytime, anyplace—naturally requires SON capabilities for a viable, evolutionary, and sustainable paradigm for mobile service rendering and expansion.

The catalytic ingredients of distribution, autonomy, and coordination serve as the conduits of evolutionary mobile services—one where there is a natural confluence of technology and art. These insights highlight the significance of user expression and experience in a symphony of artful agility and adaptability to underscore the long-tail mobile service model of "selling less to more."

Data—serving as a generic template for assorted types of information—are the universal vehicle for transport over a mobile Internet. With the widening of the mobile data highways, the connectivity to service of machine interfacing and human interfacing devices is on an expansive inflection phase in the creation and consumption of information. A natural consequence of this connectivity growth is the need for compatible addressing techniques that match this evolutionary and virtually unbounded potential for information expansion. IPv6 embodies this fundamental demand for addressing M2M device, smartphones, tablets, laptops, etc., through the use of larger data sizes and link quality enhancements. The LTE EPS architecture and interfaces use IPv6 addressing to enable ubiquitous and expansive connectivity demands, while allowing incremental migrations through the use of network address translation (NAT), and private address pools for a preservation of the smaller addressing data sizes afforded by the legacy IPv4 protocol, during a migration phase, where connectivity segments—traversed by an information packet—may have different vintages of IP addressing (IPv4 or IPv6).

The Internet of Things includes M2M interactions, when a multitude of objects—devices and sensors—behave in a coordinated and collaborative fashion, where the information is augmented cooperatively through context containing space—time crafted information. These entities resonate with one another to create a value-added web of information that enables enormous improvement in the mobile service experience. The exponential growth and diversity of context in the IoT promotes the evolution of a biosphere of IoE (Internet of Everything), where the dance of trillions and more of sensors and actuators enrich individual and collective experience.

5.4.4 Imagination

"Ideas shape the course of history," John Maynard Keynes stated. As in the course of history, so it is in the evolution of mobile communications of profound and growing significance in the human ethos. Notions that profoundly shift from a focus on the symptoms to the source through the application of insight are quintessential in the sustainable and organic evolution of the mobile ecosystem—from ideas, standardization, and innovation to practical application. The enormous benefits that accrue from a shift in thought from symptoms to the source extend beyond the enabling nuggets of the mobile ecosystem across all facets of human endeavor.

The impact of experiential dynamics is at the core of change and evolution, in the next-generation system of mobile connectivity and service. It is this aspect that must be leveraged in the realms of imagination to shape the winds of innovation and leadership across the value chain—technology, business, and human factors. It is a stairway that subsumes competition and collaboration toward unbridled cooperation across the value chain. Connectivity, service, and human factors interact harmoniously through cooperative partnerships of complementary value to promote new opportunities and experiences. It is the road to an unbounded experiential realm where users benefit and contribute, while markets expand, incubated in inspirational leadership.

"For knowledge is limited to all we now know and understand, while imagination embraces the entire world, and all there ever will be to know and understand," stated Albert Einstein. It is the imagination that creates the architectural frameworks, procedures, and protocols, which enable a plethora of familiar and differentiated service experiences. The subsequent logical analysis, collaboration, cooperation, and implementation are the realization. The market of rendered services and the levels of user endorsement shape the competitive landscape, across the providers of connectivity and service. The dynamic nature of the unfolding mobile ecosystem demands ideas that encompass multidimensional elements across technology and leadership—ideas that both inspire and incubate in a sea of cooperative human endeavor. It also implies oneness—a bridge that spans the physical and the metaphysical. The mobile ecosystem serves as a dynamic global stage of communications, where the stage of evolution influences, while at the same time being influenced by the actors on this stage—human beings.

Transformation of paradigms in the mobile communications world—circuit-switched to packet-switched—is reflective of the transformation necessary in the leadership that serves to promote the evolution and implementation of technologies. One that permits and promotes excellence without the stifling pressure of legislation or regulation, toward allowing diversity, where one size does not fit all.

The universal truths allow the natural proliferation of choices, akin to the design trade-offs, which influence the selection of choices hinging on technology, business, and human factors. The existence of these choices—vast and different—has survived through the ages despite artificial attempts to measure, capture, quantify, or control them through a labyrinth of rules and regulations. Innovative designs must espouse graceful and flexible approaches that allow the selection of specific choices, without the need to eliminate other choices. Self-organizing principles that are embodied in the evolution of next-generation mobile communication systems are reflective of autonomous behaviors, such as self-healing, in a localized fashion, amidst a web of interconnected and interdependent networks and services. These principles are both viable and necessary ingredients of leadership in the information age.

Humans are a distribution of the universal expression—self-directing, self-optimizing, and self-sustaining. A departure from aberrant leadership is vital in the transformation away from the shadows of pedantic gains that may appeal to the unaware.

The information age empowered with mobility projects a distributed experience through the proliferation of virtual content. The availability of content transported over the wireless waves synthesizes experience across the chasms of time and space. Organic collaboration and cooperation are the instruments of innovation in the creation of unique and universal content, where the roles of the consumer and creator become fluid and interchangeable. The publishing and distribution of mobile content continues to be simplified through the enabling elements of connectivity, service, and search capabilities. The building blocks of these capabilities are subject to varying levels of commoditization impact. The segments that are easily replicated, such as in the case of basic connectivity to a mobile device, or are of a common nature such as service enablers are foundational and are prone to commoditization. These enabling elements are vital to create a fabric of universal interoperability and consistent behaviors in a mobile landscape.

The proliferation of both human- and machine-type mobile connectivity and service weaves a dynamic web of information. Untethered connectivity offered by technologies such as NFC serves as a bridge between the virtual and the physical worlds. These bridges of interactions span a multitude of mobile services, such as payments, content sharing, content discovery, advertising, coupon transactions, ticketing for transportation, and entertainment, to mention just a few. Enabling technologies such as NFC and other heterogeneous connectivity footprints, with varying degrees of space—time separation, serve as distributed and coordinated information lanes and highways across the unfolding horizons of mobile evolution.

The empowerment of the user is universally realized through a vast array of multifaceted collaborative tools*—e-mail, Wiki, syndication, Facebook, LinkedIn, Twitter, etc., to name a few, together with content sharing and distribution platforms (e.g., YouTube). The dimension of mobility adds the natural dynamic of movement central to the theme of human communications in the social fabric. The role of the user in this emerging social fabric augmented with mobility promotes an inversion of infrastructures toward widespread decentralization, inspiring further innovation to mine the enabling building blocks of connectivity and service for the creation of flexible mobile services that thrive in the horizons of the long tail—a paradigm that serves less to more.

The infrastructure inversion, concomitant with distributed architectural styles of connectivity and service, embellished with the enabling building-block tools that power the social communications fabric, neutralizes the traditional benefits of large institutions, modeled on the principles of Industrial Revolution era. These benefits continue to fade in the dawn of the information age, which thrives on decentralization and prolific user-level participation through collaborative and cooperative models in the creation and distribution of mobile services. Figure 5.18 presents a reflect, render, and realize model that is relevant in the exploration of possibilities to harness the

Figure 5.18 Insights to implementation—paths in mobile evolution.

---

* Allen, C., "Tracing the evolution of social software," *Life With Alacrity*, 2004.

potential insights embedded in the ethos of human communications that inspire imagination and innovation in the realm of mobile connectivity and service evolution.

Insights are invisible building blocks that channel thought in a field of imagination, driven by inspiration. They are vital for an effective realization in the world of form and function, across all aspects of human endeavor, and are significant in the information landscape fueled by the possibilities inherent in technology-mediated mobile evolution. It is an evolution that involves a transformation of a cast of actors—technology, business, and human—to realize the enormous potential of the mobile information age. The world of nature is replete with illustrations of transformation that typically occur through a phase of turbulence and chaos—much like collaborative thought and ideas filtered through study, discussion, and debate culminating in consensus—leading to a symphony of cooperation and coordination.

Codification works in relatively simple, well-structured scenarios—reminiscent of the few decades since the Industrial Revolution; it is only needed to function at the level of silos, where the details within the silo are valuable but are also prone to more commoditization and replication afforded by the progress of technology.

On the other hand—in the information age—interdependence is central to the human condition; silos are now commodities; the value has shifted to the examination and study of interdependence and its impact on our lives, in terms of innovation, evolution, value, etc. Insights derived through research, standardization, and innovation across intersections of disparate domains of subject matter provide the nascent and incremental foundations for potential inflection points of knowledge, understanding, and applicability. The progression is similar to natural paradigms, such as the phase transition of water from the liquid phase to gaseous (steam), which occur instantly after a sustained period of no perceptible change of state during the water heating process.

As Albert Einstein mused: "The significant problems we face cannot be solved at the same level of thinking we were at when we created them." This notion applies universally, particularly in the information age, where the tangible is a product of the intangible—the formless world of unconstrained and creative thought—one that transcends specific subject details to reveal a hidden landscape of infinite insight that has no boundaries or partitions.

The manner in which a caterpillar transforms into a butterfly is one such illustration. The nature of the *imaginal cells* that begin to develop within the caterpillar is entirely different and distinct from the native cells of the caterpillar. The immune system of the caterpillar attacks and attempts to destroy the unknown, unfamiliar cells—the *imaginal cells*. In this turbulence and chaos, the *imaginal cells* proliferate unabated in the presence of attacks from the immune system of the caterpillar. These new cells exhibit interdependence and although autonomous begin to behave in a coordinated and collaborative manner in the evolution and processing of information for a new reality—characterized by new behaviors and capabilities—enabling the transformation of the caterpillar into a butterfly, where the latter in its evolved state has no resemblance to its predecessor.

This metaphor illustrates potential for transformation in the mobile information age, which is innately customized for unique user-oriented experiences—unique for each individual rooted in the formless and malleable but unbounded realm of thought. Consequently, it has a natural propensity for transformation through innovation propelled by imagination, where interdependence incubated in technology-mediated mobile evolution is a treasure chest of insights that have an inexorable potential to enrich and augment human communications, experiences, and cultures on a local and global scale.

Think different and the same, in an incessant evolution of interdependence in moving seas of information.

Index